Symmetry in Chaotic Systems and Circuits

Symmetry in Chaotic Systems and Circuits

Editor

Christos Volos

MDPI • Basel • Beijing • Wuhan • Barcelona • Belgrade • Manchester • Tokyo • Cluj • Tianjin

Editor
Christos Volos
Aristotle University of Thessaloniki
Greece

Editorial Office
MDPI
St. Alban-Anlage 66
4052 Basel, Switzerland

This is a reprint of articles from the Special Issue published online in the open access journal *Symmetry* (ISSN 2073-8994) (available at: https://www.mdpi.com/journal/symmetry/special_issues/Symmetry_Nonlinear_Electrical_Circuits).

For citation purposes, cite each article independently as indicated on the article page online and as indicated below:

LastName, A.A.; LastName, B.B.; LastName, C.C. Article Title. *Journal Name* **Year**, *Volume Number*, Page Range.

ISBN 978-3-0365-5387-0 (Hbk)
ISBN 978-3-0365-5388-7 (PDF)

Contents

About the Editor

Christos Volos

Christos Volos received his Physics Diploma degree (1999), an M.Sc. degree in electronics (2002), and a Ph.D. degree (2008) in chaotic electronics from the Physics Department, Aristotle University of Thessaloniki, Greece. He also currently serves there as an Associate Professor, and he is a member of the Laboratory of Nonlinear Systems-Circuits & Complexity (LaNSCom).

His current research interests include chaos; nonlinear systems; mem-elements; the design of analog and mixed signal electronic circuits; neural networks, chaotic electronics, and their applications (secure communications, cryptography, robotics); and chaotic synchronization and control.

Dr. Volos has published 183 papers in international journals, 55 book chapters, 101 international conference papers, and 20 national conferences papers in addition to editing 10 books and 2 books as an author on the topic of nonlinear circuits and systems. Furthermore, he is an Editorial Board Member of 11 international journals and has been a Guest Editor of 27 Special Issues in international journals. Dr. Volos is also closely associated with several international journals, where he serves as a reviewer.

Editorial

Symmetry in Chaotic Systems and Circuits

Christos Volos

Laboratory of Nonlinear Systems, Circuits & Complexity (LaNSCom), Department of Physics, Aristotle University of Thessaloniki, 54124 Thessaloniki, Greece; volos@physics.auth.gr

Citation: Volos, C. Symmetry in Chaotic Systems and Circuits. *Symmetry* **2022**, *14*, 1612. https://doi.org/10.3390/sym14081612

Received: 30 June 2022
Accepted: 2 August 2022
Published: 5 August 2022

Publisher's Note: MDPI stays neutral with regard to jurisdictional claims in published maps and institutional affiliations.

Nowadays, Chaos theory consists one of the most fascinating fields in modern science, revolutionizing our understanding of order and pattern in Nature. On the other hand, Symmetry is a traditional and highly developed area of Mathematics, which seems to lie at the opposite end of the spectrum. However, in the last few years, scientists have found connections between these two areas, connections which can have profound consequences for our understanding of the complex behavior in many physical, chemical, biological, and mechanical systems.

Therefore, symmetry can play an important role in the field of nonlinear systems and especially in the field of designing nonlinear circuits that produce chaos. In more detail, from designing chaotic systems and circuits with symmetric nonlinear terms to the study of system's equilibria with symmetry, in the case of self-excited attractors, or symmetric line of equilibria, in the case of hidden attractors, the feature of symmetry can play significant role in this kind of systems.

The overall purpose of this Special Issue lies in presenting the latest scientific advances in nonlinear chaotic systems and circuits that introduce various kinds of symmetries. Applications of chaotic systems and circuits with symmetries, or the deliberate lack of symmetry, is also presenting in this Special Issue. The volume has 14 published papers, where the authors are from geographically distributed countries (Algeria, Bulgaria, China, Greece, Iran, Iraq, Malaysia, Mexico, Russia, Taiwan, Thailand, Turkey, Vietnam). This reflects the high impact of the proposed topic and the seniority in organization of this special issue.

In the first paper of this Special Issue entitled "Symmetry Evolution in Chaotic System", by C. Li et al. presents a comprehensive exploration of symmetry and conditional symmetry from the evolution of symmetry. Unlike other chaotic systems of conditional symmetry, in this work it is derived from the symmetric diffusionless Lorenz system. Transformation from symmetry and asymmetry to conditional symmetry is examined by constant planting and dimension growth, which proves that the offset boosting of some necessary variables is the key factor for reestablishing polarity balance in a dynamical system [1].

The paper "A Nonlinear Five-Term System: Symmetry, Chaos, and Prediction", by Vo Phu Thoai et al. presents a simple symmetrical system including only five nonlinear terms. By using various tools from nonlinear theory, such as phase portraits, bifurcation diagrams, Lyapunov exponents, and entropy, system's rich dynamical behavior is discovered. Interestingly, multi-stability is also observed when changing system's initial conditions. Chaotic behavior of such a system is predicted by applying a machine learning approach based on a neural network [2].

The paper "A Symmetric Controllable Hyperchaotic Hidden Attractor", by Xin Zhang et al. presents that by introducing a simple feedback, a hyperchaotic hidden attractor is found in the newly proposed Lorenz-like chaotic system. Some variables of the equilibria-free system can be controlled in amplitude and offset by an independent knob. Furthermore, a circuit experiment based on Multisim, which has been presented, is proved to be consistent with the theoretic analysis and numerical simulation [3].

Xinhe Zhu and Wei-Shih Du in the paper entitled "New Chaotic Systems with Two Closed Curve Equilibrium Passing the Same Point: Chaotic Behavior, Bifurcations, and

Synchronization", propose a chaotic system with infinitely many equilibrium points laying on two closed curves passing the same point. The proposed system belongs to a class of systems with hidden attractors. The dynamical properties of the new system were investigated by means of phase portraits, equilibrium points, Poincaré section, bifurcation diagram, Kaplan-Yorke dimension, and Maximal Lyapunov exponents. The anti-synchronization of systems is obtained using the active control. This study broadens the current knowledge of systems with infinite equilibria [4].

In the paper "Three-Saddle-Foci Chaotic Behavior of a Modified Jerk Circuit with Chua's Diode", by Pattrawut Chansangiam the chaotic behavior of a modified jerk circuit with Chua's diode is investigated. The Chua's diode considered in this work is a nonlinear resistor having a symmetric piecewise linear voltage-current characteristic. To describe the system, we apply fundamental laws of electrical circuit theory in order to formulate a mathematical model in terms of a third-order (jerk) nonlinear differential equation, or equivalently, a system of three first-order differential equations. The system's analysis shows that it has three collinear equilibrium points. Furthermore, numerical simulation illustrates that the system's oscillations are dense, have no period, are highly sensitive to initial conditions, and have a chaotic hidden attractor [5].

In the next paper "Two New Asymmetric Boolean Chaos Oscillators with No Dependence on Incommensurate Time-Delays and Their Circuit Implementation", Jesus M. Munoz-Pacheco et al. propose two new chaotic oscillators based on Autonomous Boolean Networks (ABN), preserving asymmetrical logic functions. That means that the ABNs require a combination of XOR-XNOR logic functions. The two ABNs do not have fixed points, and therefore, can evolve to Boolean chaos. Using the Lyapunov exponent's method the chaotic behavior of the proposed oscillators is proved to be insensitive to incommensurate time-delays paths. As a result, they can be implemented using distinct electronic circuits. More specifically, logic-gates-, GAL-, and FPGA-based implementations verify the theoretical findings. An integrated circuit using a CMOS 180nm fabrication technology is also presented to get a compact chaos oscillator with relatively high-frequency. Dynamical behaviors of those implementations are analyzed using time-series, time-lag embedded attractors, frequency spectra, Poincaré maps, and Lyapunov exponents [6].

The paper entitled "The Effect of a Non-Local Fractional Operator in an Asymmetrical Glucose-Insulin Regulatory System: Analysis, Synchronization and Electronic Implementation", by Jesus M. Munoz-Pacheco et al. analyzes the dynamics of a glucose-insulin regulatory system by applying a non-local fractional operator in order to represent the memory of the underlying system, and whose state-variables define the population densities of insulin, glucose, and β-cells, respectively. The authors have focused mainly on four parameters that are associated with different disorders (type 1 and type 2 diabetes mellitus, hypoglycemia, and hyperinsulinemia) to determine their observation ranges as a relation to the fractional-order. Like many preceding works in biosystems, the resulting analysis showed chaotic behaviors related to the fractional-order and system parameters. Subsequently, an active control scheme for forcing the chaotic regime (an illness) to follow a periodic oscillatory state (i.e., a disorder-free equilibrium) is proposed. Finally, the electronic realization of the fractional glucose-insulin regulatory model to prove the conceptual findings is also presented [7].

In the next paper "A New Hyperchaotic Map for a Secure Communication Scheme with an Experimental Realization", Nadia M. G. Al-Saidi et al. present a new 2D chaotic map, namely, the 2D Infinite-Collapse-Sine Model (2D-ICSM). By using various metrics, including Lyapunov exponents and bifurcation diagrams complex dynamics and robust hyperchaotic behavior of the 2D-ICSM is demonstrated. Furthermore, the cross-correlation coefficient, phase space diagram, and Sample Entropy algorithm prove that the 2D-ICSM has a high sensitivity to initial values and parameters, extreme complexity performance, and a much larger hyperchaotic range than existing maps. Finally, in order to empirically verify the efficiency and simplicity of the 2D-ICSM in practical applications, a symmetric secure communication system using the 2D-ICSM is presented [8].

In the next paper "A Two-Parameter Modified Logistic Map and Its Application to Random Bit Generation", L. Moysis et al. propose a modified logistic map based on the system previously proposed by Han in 2019. The constructed map exhibits interesting chaos-related phenomena like antimonotonicity, crisis, and coexisting attractors. In addition, the Lyapunov exponent of the map can achieve higher values, so the behavior of the proposed map is overall more complex compared to the original. The map is then successfully applied to the problem of random bit generation using techniques like the comparison between maps, XOR and bit reversal. The proposed algorithm passes all the NIST tests, shows good correlation characteristics, and has a high key space [9].

The paper entitled "A Simple Chaotic Flow with Hyperbolic Sinusoidal Function and Its Application to Voice Encryption", by S. Mobayen et al. studies a new chaotic system with hyperbolic sinusoidal function. The proposed chaotic system provides a new category of chaotic flows, which gives better perception of chaotic attractors. In more detail, in the proposed chaotic flow, according to the changes of parameters of the system, a self-excited attractor and two forms of hidden attractors are occurred. Dynamic behavior of the proposed chaotic flow is studied through eigenvalues, bifurcation diagrams, phase portraits, and spectrum of Lyapunov exponents. Moreover, the existence of double-scroll attractors in real word is considered via the Orcard-PSpice software through an electronic execution of the new chaotic flow and illustrative results between the numerical simulation and Orcard-PSpice outcomes are obtained. Furthermore, Random Number Generator (RNG) design based on the proposed system is also presented and a novel voice encryption algorithm is proposed [10].

In the next paper "Symmetric Key Encryption Based on Rotation-Translation Equation", Borislav Stoyanov and Gyurhan Nedzhibov propose an improved encryption algorithm based on numerical methods and rotation-translation equation. The new encryption-decryption algorithm is developed by using the concept of symmetric key instead of public key. The goal in this work is to improve an existing encryption algorithm by using a faster convergent iterative method, providing secure convergence of the corresponding numerical scheme, and improved security by a using rotation-translation formula [11].

G. Zhang et al., in the paper entitled "Image Encryption Algorithm Based on Tent Delay-Sine Cascade with Logistic Map", present a new chaotic map combined with delay and cascade, called Tent Delay-Sine Cascade with Logistic map (TDSCL). Compared with the original one-dimensional simple map, the proposed map has increased initial value sensitivity and internal randomness and a larger chaotic parameter interval. The chaotic sequence generated by TDSCL has pseudo-randomness and is suitable for image encryption. Based on this chaotic map, an image encryption algorithm with a symmetric structure, which can achieve confusion and diffusion at the same time, is proposed. Simulation results show that after encryption using the proposed algorithm, the entropy of the cipher is extremely close to the ideal value of eight, and the correlation coefficients between the pixels are lower than 0.01, thus the algorithm can resist statistical attacks. Moreover, the Number of Pixel Change Rate (NPCR) and the Unified Average Changing Intensity (UACI) of the proposed algorithm are very close to the ideal value, which indicates that it can efficiently resist chosen-plain text attack [12].

In the next paper "A Novel Method for Performance Improvement of Chaos-Based Substitution Boxes", Fırat Artuğer and Fatih Özkaynak examine the chaotic behavior in the field of information security. In this direction a novel method is proposed in order to improve the performance of chaos-based substitution box structures. Substitution box structures have a special role in block cipher algorithms, since they are the only nonlinear components in substitution permutation network architectures. However, the substitution box structures used in modern block encryption algorithms contain various vulnerabilities to side-channel attacks. Recent studies have shown that chaos-based designs can offer a variety of opportunities to prevent side-channel attacks. The problem of chaos-based designs is that substitution box performance criteria are worse than designs based on mathematical transformation. Therefore, in this work, a post processing algorithm is proposed to improve the performance of chaos-based designs. The analysis results show that the proposed method can improve the performance criteria. The importance of these results is that chaos-based designs may offer opportunities for other practical applications in addition to the prevention of side-channel attacks [13].

The last paper entitled "Dynamic Symmetry in Dozy-Chaos Mechanics", by Vladimir V. Egorov presents all kinds of dynamic symmetries in dozy-chaos (quantum-classical) mechanics by taking into account the chaotic dynamics of the joint electron-nuclear motion in the transient state of molecular "quantum" transitions. The reason for the emergence of chaotic dynamics is associated with a certain new property of electrons, consisting in the provocation of chaos (dozy chaos) in a transient state, which appears in them as a result of the binding of atoms by electrons into molecules and condensed matter and which provides the possibility of reorganizing a very heavy nuclear subsystem as a result of transitions of light electrons. Various dynamic symmetries appearing in theory are associated with the emergence of dynamic organization in electronic-vibrational transitions, in particular with the emergence of an electron-nuclear-reorganization resonance (the so-called Egorov resonance) and its antisymmetric (chaotic) "twin", with direct and reverse transitions, as well as with different values of the electron-phonon interaction in the initial and final states of the system. All these dynamic symmetries are investigated using the simplest example of quantum-classical mechanics, namely, the example of quantum-classical mechanics of elementary electron-charge transfers in condensed media [14].

The Guest Editor hopes you will delight in reading this Special Issue focused on cutting-edge research on symmetry in chaotic systems and circuits. We expect the collected works will motivate researchers to strive for further advances in this emerging area.

Finally, the Guest Editor would like to thank the authors of all the papers submitted to this special issue, as well as the anonymous reviewers, some of whom helped with multiple review assignments. Additionally, the Guest Editor would like to thank the journal's Editorial Board for being very encouraging and accommodative regarding this special issue.

Funding: This research received no external funding.

Conflicts of Interest: The author declares no conflict of interest.

References

1. Li, C.; Sun, J.; Lu, T.; Lei, T. Symmetry evolution in chaotic system. *Symmetry* **2020**, *12*, 574. [CrossRef]
2. Thoai, V.P.; Kahkeshi, M.S.; Huynh, V.V.; Ouannas, A.; Pham, V.T. A nonlinear five-term system: Symmetry, chaos, and prediction. *Symmetry* **2020**, *12*, 865. [CrossRef]
3. Zhang, X.; Li, C.; Lei, T.; Liu, Z.; Tao, C. A symmetric controllable hyperchaotic hidden attractor. *Symmetry* **2020**, *12*, 550. [CrossRef]
4. Zhu, X.; Du, W.S. New chaotic systems with two closed curve equilibrium passing the same point: Chaotic behavior, bifurcations, and synchronization. *Symmetry* **2020**, *11*, 951. [CrossRef]
5. Chansangiam, P. Three-Saddle-Foci Chaotic Behavior of a Modified Jerk Circuit with Chua's Diode. *Symmetry* **2020**, *12*, 1803. [CrossRef]
6. Munoz-Pacheco, J.M.; García-Chávez, T.; Gonzalez-Diaz, V.R.; de La Fuente-Cortes, G.; Gómez-Pavón, L.D. Two new asymmetric Boolean chaos oscillators with no dependence on incommensurate time-delays and their circuit implementation. *Symmetry* **2020**, *12*, 506. [CrossRef]
7. Munoz-Pacheco, J.M.; Posadas-Castillo, C.; Zambrano-Serrano, E. The effect of a non-local fractional operator in an asymmetrical glucose-insulin regulatory system: Analysis, synchronization and electronic implementation. *Symmetry* **2020**, *12*, 1395. [CrossRef]
8. Al-Saidi, N.M.; Younus, D.; Natiq, H.; Ariffin, M.R.; Asbullah, M.A.; Mahad, Z. A new hyperchaotic map for a secure communication scheme with an experimental realization. *Symmetry* **2020**, *12*, 1881. [CrossRef]
9. Moysis, L.; Tutueva, A.; Volos, C.; Butusov, D.; Munoz-Pacheco, J.M.; Nistazakis, H. A two-parameter modified logistic map and its application to random bit generation. *Symmetry* **2020**, *12*, 829. [CrossRef]
10. Mobayen, S.; Volos, C.; Çavuşoğlu, Ü.; SKaçar, S. A simple chaotic flow with hyperbolic sinusoidal function and its application to voice encryption. *Symmetry* **2020**, *12*, 2047. [CrossRef]
11. Stoyanov, B.; Nedzhibov, G. Symmetric key encryption based on rotation-translation equation. *Symmetry* **2020**, *12*, 73. [CrossRef]
12. Zhang, G.; Ding, W.; Li, L. Image encryption algorithm based on tent delay-sine cascade with logistic map. *Symmetry* **2020**, *12*, 355. [CrossRef]

13. Artuğer, F.; Özkaynak, F. A novel method for performance improvement of chaos-based substitution boxes. *Symmetry* **2020**, *12*, 571. [CrossRef]
14. Egorov, V.V. Dynamic symmetry in dozy-chaos mechanics. *Symmetry* **2020**, *12*, 1856. [CrossRef]

Article

Symmetry Evolution in Chaotic System

Chunbiao Li [1,2,3,*], **Jiayu Sun** [1,2], **Tianai Lu** [1,2] **and Tengfei Lei** [3]

[1] Jiangsu Collaborative Innovation Center of Atmospheric Environment and Equipment
    Technology (CICAEET), Nanjing University of Information Science and Technology, Nanjing 210044, China;
    jiayusun@nuist.edu.cn (J.S.); lutianai950404@nuist.edu.cn (T.L.)
[2] Jiangsu Key Laboratory of Meteorological Observation and Information Processing, Nanjing University of
    Information Science and Technology, Nanjing 210044, China
[3] Collaborative Innovation Center of Memristive Computing Application (CICMCA), Qilu Institute
    of Technology, Jinan 250200, China; leitengfei2017@qlit.edu.cn
*   Correspondence: chunbiaolee@nuist.edu.cn; Tel.: +86-139-1299-3098

Received: 29 February 2020; Accepted: 22 March 2020; Published: 5 April 2020

Abstract: A comprehensive exploration of symmetry and conditional symmetry is made from the evolution of symmetry. Unlike other chaotic systems of conditional symmetry, in this work it is derived from the symmetric diffusionless Lorenz system. Transformation from symmetry and asymmetry to conditional symmetry is examined by constant planting and dimension growth, which proves that the offset boosting of some necessary variables is the key factor for reestablishing polarity balance in a dynamical system.

Keywords: symmetry; asymmetry; offset boosting; chaotic system

---

1. Introduction

The system structure is a fundamental topological constraint to the dynamical evolution, which determines how the attractor stretches in phase space. Symmetric systems give birth to attractors with a symmetrical face [1–5]. When symmetry is broken, the attractor splits into a symmetric pair of attractors [6–8] or is preserved by doubling coexisting attractors [9]. Asymmetric systems seem to give a single asymmetric attractor in most cases, although sometimes it hatches coexisting asymmetric attractors [10–14] under a set of combined parameters. However, many asymmetric systems have coexisting attractors of conditional symmetry with the new polarity balance from the offset boosting.

Furthermore, symmetric structure does not reject conditional symmetry. In this paper, the symmetry evolution in chaotic systems is analyzed, as shown in Figure 1. From the start of the variable polarity reversal, if a dynamical system can establish its own polarity balance from itself, the system is symmetric, or else losing the polarity balance indicates the asymmetric structure. If a system recovers its polarity balance from a step with offset boosting, the derived system is of conditional symmetry. From this observation, we can conclude that a system, whether it is symmetric or asymmetric, can be transformed to be of conditional symmetry. In Section 2, the early proposed chaotic systems of conditional symmetry are collected. In Section 3, conditional symmetry is coined in a symmetric system. In Section 4, the collapse of polarity balance is thoroughly explored in two directions, one of which is from the constant planting, and the other of which is from the dimension growth. Conditional symmetry is therefore in the primary road where the offset-boosting-induced polarity balance is well preserved. The conclusion is given in the last section.

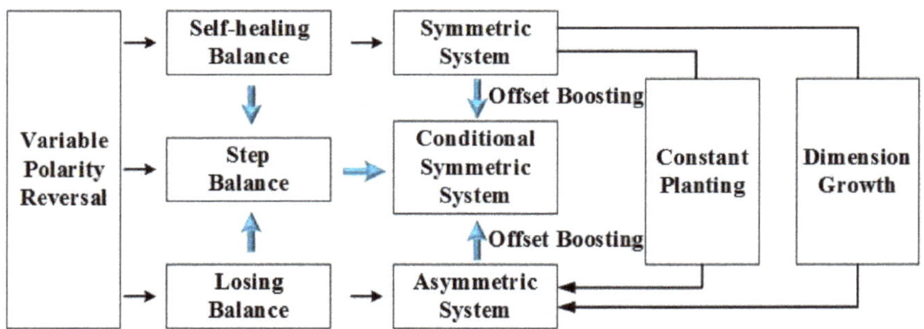

Figure 1. Relationship among symmetry, asymmetry and conditional symmetry.

2. Conditional Symmetry from Asymmetry

As we know, for a dynamical system $\dot{X} = F(X) = (f_1(X), f_2(X), \ldots, f_N(X))^T$, $(X = (x_1, x_2, \ldots, x_N)^T)$, if there exists a variable substitution $u_{i_1} = -x_{i_1}, u_{i_2} = -x_{i_2}, \cdots, u_{i_k} = -x_{i_k}, u_i = x_i$, (here $1 \leq i_1, \cdots, i_k \leq N, i_1, \cdots, i_k$ are not identical, $i \in \{1, 2, \ldots, N\} \backslash \{i_1, \cdots, i_k\}$) satisfying $\dot{U} = F(U)$ $(U = (u_1, u_2, \ldots, u_N))$, then the system $\dot{X} = F(X)$ $(X = (x_1, x_2, \ldots, x_N))$ is symmetric. Conditional symmetry is a new terminology to describe the polarity balance from offset boosting [15–18]. For a differential dynamical system, $\dot{X} = F(X) = (f_1(X), f_2(X), \ldots, f_N(X))^T$, $(X = (x_1, x_2, \ldots, x_N)^T)$, the substitution $u_{i_0} = x_{i_0} + c$ $(i_0 \in \{1, 2, \ldots, N\}$ (c is an arbitrary constant) brings the offset boosting in the variable x_{i_0}, where the new constant c will change the average value of the variable x_{i_0}. For a dynamical system, if there exists a variable substitution, $u_{i_0} = x_{i_0} + c_0$, $u_i = x_i$ (here c_0 is a non-zero constant, then $i_0 \in \{1, 2, \ldots, N\}$, and $i \in \{1, 2, \ldots, N\} \backslash \{i_0\}$), which makes the deduced system $\dot{U} = F^*(U) = \left(f_1^*(U), f_2^*(U), \ldots, f_N^*(U) \right)$ $(U = (u_1, u_2, \ldots, u_N))$ asymmetric, but when $f_{j_0}^*(U)$ $(1 \leq j_0 \leq N, j_0 \neq i_0)$ is revised, the system becomes symmetric, and then system $\dot{X} = F(X)$ $(X = (x_1, x_2, \ldots, x_N))$ is conditionally symmetric. Some early proposed chaotic systems of conditional symmetry [19,20] are listed in Table 1. All the coexisting attractors of conditional symmetry are shown in Figure 2. As we can see, all these systems are asymmetric ones but give twin attractors.

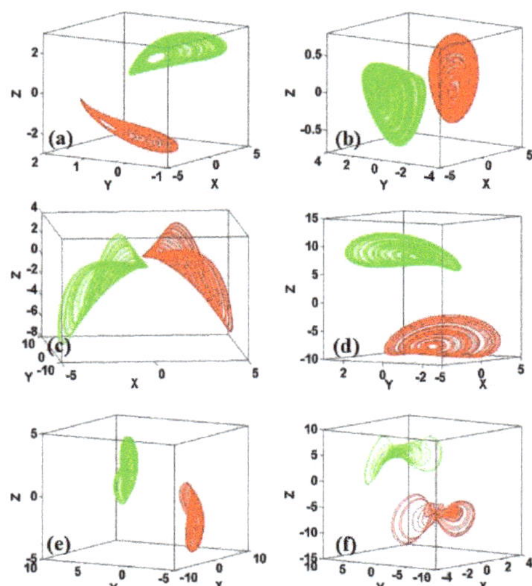

Figure 2. Coexisting twin attractors in chaotic systems in Table 1: (**a**) CS1, (**b**) CS2, (**c**) CS3, (**d**) CS4, (**e**) CS5, (**f**) CS6.

Table 1. Early explored typical chaotic systems of conditional symmetry.

Cases	System Equations	Parameters	Initial Condition	Lyapunov Exponents
CS1	$\dot{x} = y^2 - az^2,$ $\dot{y} = -z^2 - by + c,$ $\dot{z} = yz + F(x),$ $F(x) = \|x\| - 3$	$a = 0.4,$ $b = 1.75,$ $c = 3$	$(3, -1.5, -2)$ $(3, -1.5, 1)$	0.1191, 0, -1.2500
CS2	$\dot{x} = y^2 - a,$ $\dot{y} = bz,$ $\dot{z} = -y - z + F(x),$ $F(x) = \|x\| - 3$	$a = 1.22,$ $b = 8.48$	$(3, 1, 0.5)$ $(-3, 1, 0.5)$	0.2335, 0, -1.2335
CS3	$\dot{x} = F(y),$ $\dot{y} = z,$ $\dot{z} = -x^2 - az + b(F(y))^2 + 1,$ $F(y) = \|y\| - 4$	$a = 2.6,$ $b = 2$	$(0.5, 4, -1)$ $(0.5, -4, -1)$	0.0463, 0, -2.6463
CS4	$\dot{x} = y,$ $\dot{y} = F(z),$ $\dot{z} = x^2 - ay^2 + bxy + xF(z),$ $F(z) = \|z\| - 8$	$a = 1.24,$ $b = 1$	$(4, 0.8, -2)$ $(-4, 0.8, 2)$	0.0645, 0, -1.2582
CS5	$\dot{x} = 1 - G(y)z,$ $\dot{y} = az^2 - G(y)z,$ $\dot{z} = F(x),$ $F(x) = \|x\| - 3$ $G(y) = \|y\| - 5$	$a = 0.22$	$(-1, 1, -1)$ $(2, 6, -1)$	0.0729, 0, -1.6732
CS6	$\dot{x} = F(y),$ $\dot{y} = xG(z),$ $\dot{z} = -axF(y) - bxG(z) - x^2 + (F(y))^2,$ $F(y) = \|y\| - 5$ $G(z) = \|z\| - 5$	$a = 3,$ $b = 1.2$	$(0, -6\ -6)$ $(0, 6, 6)$	0.0506, 0, -0.2904

3. Constructing Conditional Symmetry from Symmetry

Interestingly, a symmetric structure also gives the chance for hosting an offset-boosting-assisted polarity balance and leading to conditional symmetry. Taking the diffusionless Lorenz system [21,22], for example,

$$\begin{cases} \dot{x} = y - x + n, \\ \dot{y} = -xz + m, \\ \dot{z} = xy - R. \end{cases} \tag{1}$$

where the parameters m and n are introduced for later discussion. When $m = n = 0$, $R = 1$, the system has a chaotic attractor with Lyapunov exponents (0.2101, 0, −1.2101) and a corresponding Kaplan–Yorke dimension $D_{KY} = 2.1736$ under initial conditions (−1, 0, −1). In this work, for obtaining representative Lyapunov exponents rather than absolute ones [23–25], all the finite-time Lyapunov exponents (LEs) are computed for the time interval $[0, 10^7]$ for the initial points on the attractor based on the Wolf algorithm. It is a simple matter to determine the Kaplan–Yorke dimension from the spectrum of Lyapunov exponents by $k + (LE_1 + \ldots + LE_k)/|LE_{k+1}|$ (here $LE_1 + \ldots + LE_k \geq 0$, and $LE_1 + \ldots + LE_{k+1} \leq 0$). System (1) is of rotational symmetry since the system is invariant under the transformation $(x, y, z) \rightarrow (-x, -y, z)$ when $m = n = 0$, corresponding to a 180° rotation about the z-axis. In this case, system (1) has a symmetric oscillation or a symmetric pairs of twin attractors under different initial condition (IC), as shown in Figure 3.

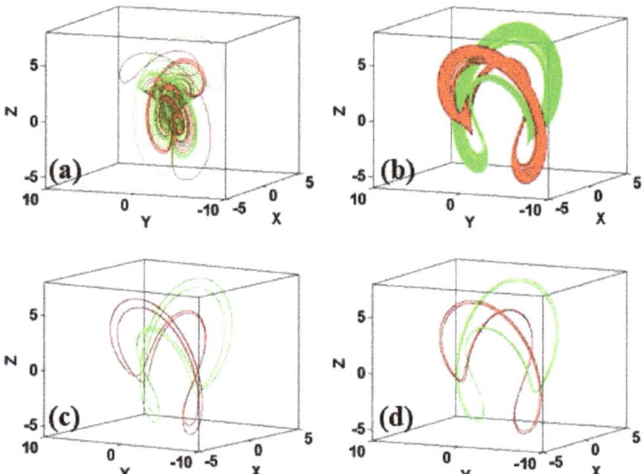

Figure 3. Symmetric attractor or symmetric pairs of attractors of system (1) with $m = n = 0$, IC = (1, 1, 1) is red and IC = (1, −1, 1) is green: (**a**) R =1, (**b**) R =4.9, (**c**) R = 5.2, (**d**) R = 5.4.

Taking a further function introducing,

$$\begin{cases} \dot{x} = F(y) - x + n, \\ \dot{y} = -xG(z) + m, \\ \dot{z} = xF(y) - R. \end{cases} \tag{2}$$

where $F(y) = |y| - 6$, $G(z) = |z| - 8$, $m = n = 0$, $R = 1$, system (2) gives birth to twin coexisting attractors of conditional symmetry, as shown in Figure 4. Compared with the rotational symmetry with system (1), system (2) is of conditional reflection symmetry since it is invariant under the transformation $(x, y, z) \rightarrow (-x, y + c_1, z + c_2)$ (c_1, c_2 stand for calling a polarity reverse from the absolute value function). We can compare these twin attractors; each one is symmetrically different from the above cases.

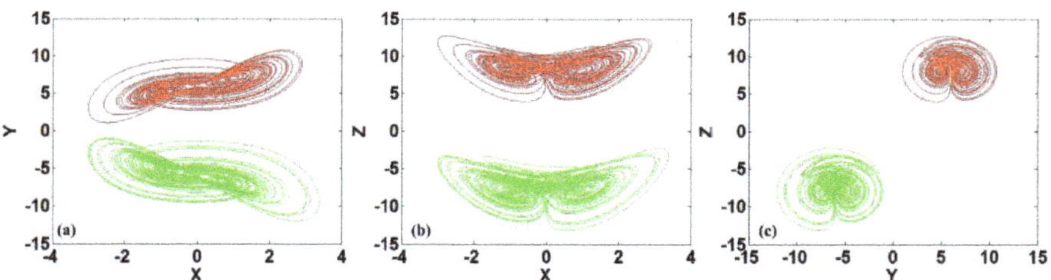

Figure 4. Coexisting twin attractors of system (2) with $F(y) = |y| - 6$, $G(z) = |z| - 8$, $m = n = 0$, $R = 1$, IC = (1, 7, 9) is red, and IC = (−1, −6, −7) is green.

4. Recovering Conditional Symmetry from Destroyed Symmetry

4.1. Symmetry Destroyed by the Constant Planting

For observing the effect to conditional symmetry owing to the symmetric structure, two additional constants are introduced in the diffusionless Lorenz system. The constant term, like a polarity fire extinguisher, revises the polarity balance. As shown in Figures 5 and 6, when m and n vary, system (1) switches between symmetric attractors and asymmetric ones for the compound structure with Lorenz attractor. Note that any constant m or n removes the polarity balance, which identifies that system (1) loses symmetry when $m \neq 0$, or $n \neq 0$. However, for system (2), the situation is different. If $m = 0$, $n \neq 0$, system (2) does not keep conditional symmetry. However, if $n = 0$, $m \neq 0$, system (2) maintains conditional symmetry, giving two coexisting bifurcations, as shown in Figure 7. Unlike the attractors shown in Figures 3 and 4, now all the coexisting attractors of conditional symmetry reside in the asymmetric structure, as shown in Figure 8. Two typical pairs of chaotic signals are shown in Figure 9, where the signals lose symmetry but stand steadily in the form of conditional symmetry.

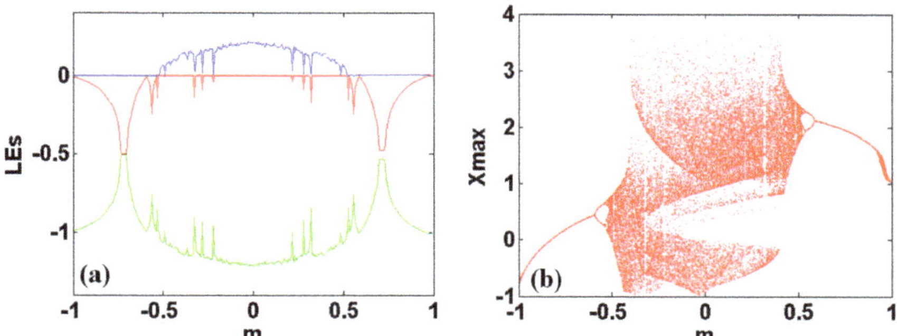

Figure 5. Dynamical evolvement in system (1) with $n = 0$, $R = 1$ and initial conditions (1, 1, 1): (**a**) Lyapunov exponents (LEs), and (**b**) bifurcation diagram.

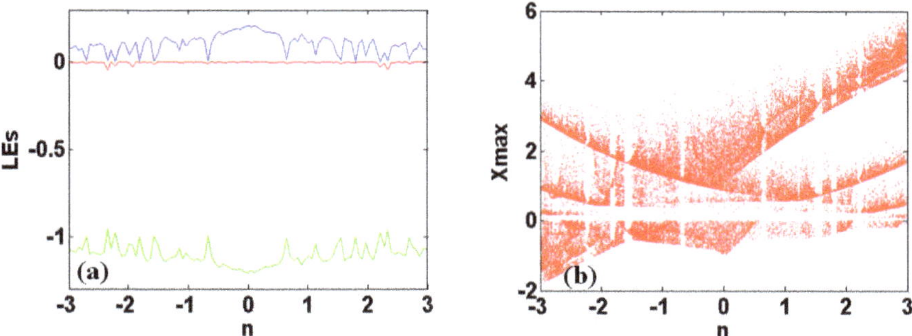

Figure 6. Dynamical evolvement in system (1) with $m = 0$, $R = 1$ and initial conditions (1, 1, 1): (**a**) Lyapunov exponents, and (**b**) bifurcation diagram.

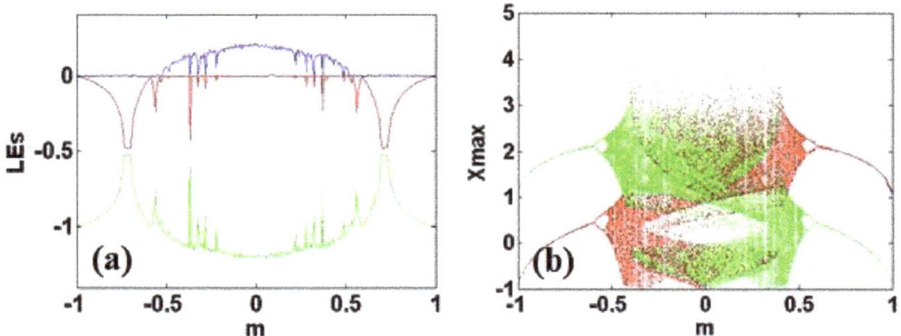

Figure 7. Dynamical evolvement in system (2) with $n = 0$, $R = 1$: (**a**) Lyapunov exponents, and (**b**) bifurcation diagram.

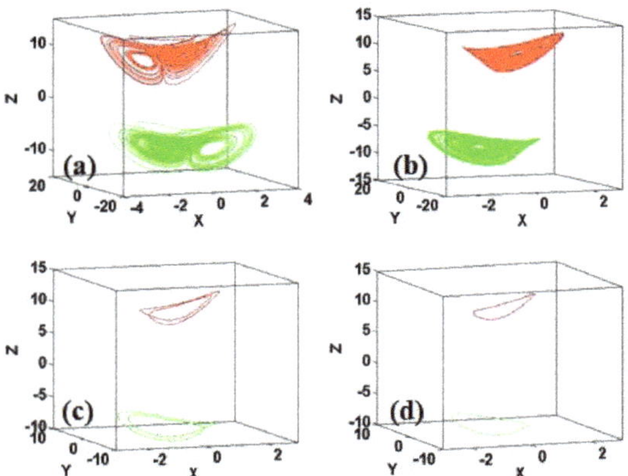

Figure 8. Conditional symmetric pairs of attractors in system (2) with $n = 0$, $R = 1$, IC $= (1, 7, 9)$ is red and $(-1, -6, -7)$ is green: (**a**) $m = 0.25$, (**b**) $m = 0.45$, (**c**) $m = 0.55$, (**d**) $m = 0.7$.

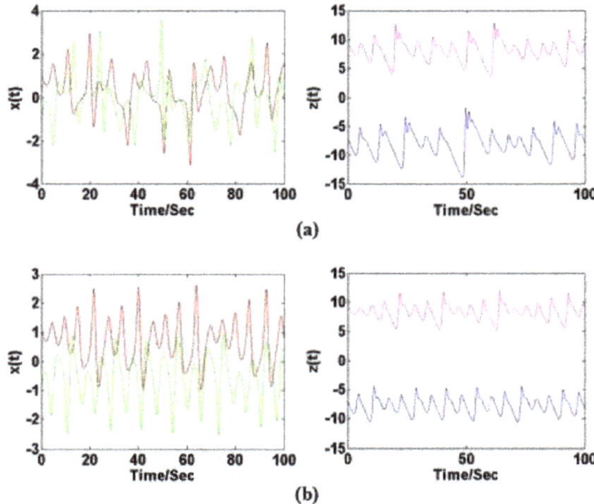

Figure 9. Conditional symmetric pairs of signals in system (2) with $n = 0$, $R = 1$, IC $= (1, 7, 9)$ is red and $(-1, -6, -7)$ is green: (**a**) $m = 0.25$, (**b**) $m = 0.45$.

4.2. Symmetry Evolution Induced by the Dimension Growth

The influence of dimension growth to polarity balance is complicated, some of which may preserve or destroy the polarity balance of the original system. Taking the following system, for example,

$$\begin{cases} \dot{x} = y - x, \\ \dot{y} = -xz, \\ \dot{z} = xy - R + axu, \\ \dot{u} = bx. \end{cases} \tag{3}$$

In this case, system (3) is still symmetric, since it is invariant under the transformation $(x, y, z, u) \rightarrow (-x, -y, z, -u)$. Now system (3) has a symmetric chaotic attractor with Lyapunov exponents $(0.2609, 0, -0.0079, -1.2530)$ and corresponding Kaplan–Yorke dimension $D_{KY} = 3.2019$, is shown in Figure 10.

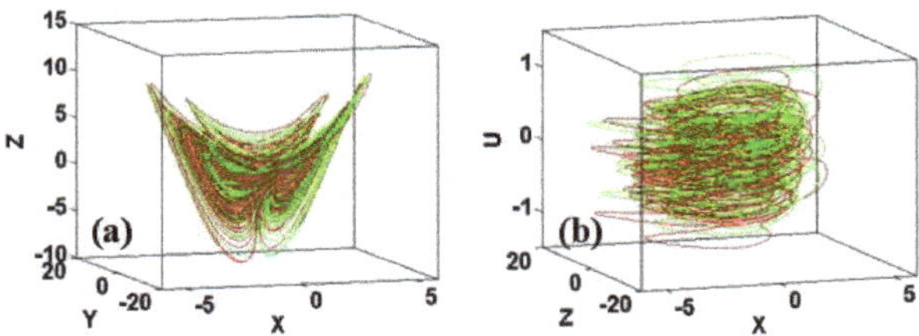

Figure 10. Symmetric attractor of system (3) with $a = 0.5$, $b = 0.1$, $R = 3$, IC = (1, 1, 1, 2) is red, IC = (−1, −1, 1, −2) is green: (**a**) *x-y-z* space, (**b**) *x-z-u* space.

System (3) is also a seed system for hosting conditional symmetry,

$$\begin{cases} \dot{x} = F(y) - x, \\ \dot{y} = -xG(z), \\ \dot{z} = xF(y) - R + axu, \\ \dot{u} = bx. \end{cases} \tag{4}$$

where $F(y) = |y| - 15$, $G(z) = |z| - 15$, $a = 0.5$, $b = 0.1$, $R = 3$; system (4) gives birth to twin coexisting attractors of conditional symmetry, as shown in Figure 11. System (4) is of conditional rotational symmetry since it is invariant under the transformation $(x, y, z, u) \rightarrow (-x, y+c_1, z+c_2, -u)$ (c_1, c_2 stand for calling a polarity reverse from the absolute value function).

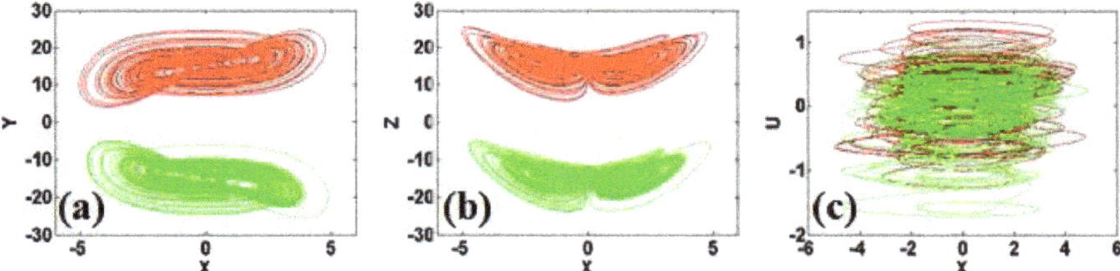

Figure 11. Coexisting conditional symmetric attractors in system (4) with $F(y) = |y| - 15$, $G(z) = |z| - 15$, $a = 0.5$, $b = 0.1$, $R = 3$, IC = (1, 16, 16, 2) is red, IC = (−1, −14, −14, −2) is green.

The dimension growth sometimes changes the polarity balance of the original system.

$$\begin{cases} \dot{x} = y - x - axu, \\ \dot{y} = -xz, \\ \dot{z} = xy - R, \\ \dot{u} = bx. \end{cases} \tag{5}$$

System (5) becomes asymmetric since it is changed under the polarity transformation. When $a = 0.1$, $b = 0.1$, $R = 3$, system (5) has chaotic attractor with Lyapunov exponents (0.0432, 0, −0.1083, −2.8978) and corresponding Kaplan–Yorke dimension $D_{KY} = 2.3989$ under initial conditions (1, 1, 1, 2). Interestingly, this time the variable u is positive, and therefore the absolute value symbol of u can be introduced for hatching coexisting attractors, as shown in Figure 12.

$$\begin{cases} \dot{x} = y - x - ax|u|, \\ \dot{y} = -xz, \\ \dot{z} = xy - R, \\ \dot{u} = bx. \end{cases} \tag{6}$$

where $a = 0.1$, $b = 0.1$, $R = 3$; system (6) has a symmetric pair of coexisting chaotic attractors. Interestingly, here these coexisting attractors are unlike the cases shown in reference [9]. In the fourth dimension of system (6), the polarity balance is recovered by the out variable x rather than by an extra imported signum function.

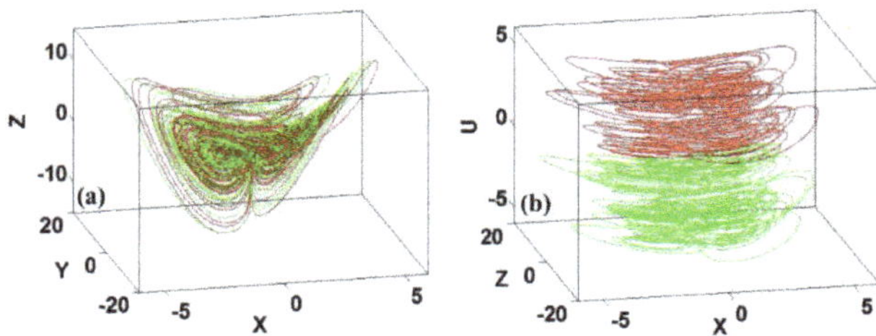

Figure 12. Symmetric attractor of system (6) with $a = 0.5$, $b = 0.1$, $R = 3$, IC = (1, 1, 1, 2) is red, IC = (-1, -1, 1, -2) is green.

Furthermore, based on the above case, the dimension growth also leaves the possibility for hosting conditional-symmetry-like coexisting attractors. Taking a further function introducing to system (6).

$$\begin{cases} \dot{x} = F(y) - x - ax|H(u)|, \\ \dot{y} = -xG(z), \\ \dot{z} = xF(y) - R, \\ \dot{u} = bx. \end{cases} \tag{7}$$

where $F(y) = |y| - 15$, $G(z) = |z| - 15$, $H(u) = |u| - 10$, $a = b = 0.1$, $R = 3$; system (7) gives birth to twin coexisting attractors, which have the features of conditional symmetry, as shown in Figure 13. However, system (7) is not of conditional rotational symmetry since it seems not invariant under the transformation $(x, y, z, u) \rightarrow (-x, y + c_1, z + c_2, u + c_3)$ (c_1, c_2, c_3 stand for calling a polarity reverse from the absolute value function). The mechanism of the coexistence of attractors hides in the same balance ability from the structure (6).

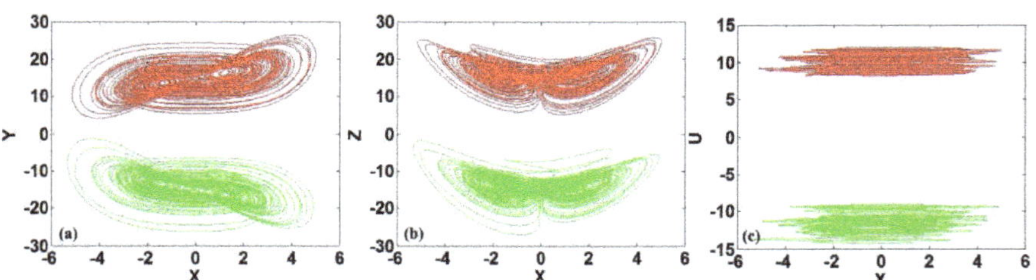

Figure 13. Coexisting attractors in systems (7) with $F(y) = |y| - 15$, $G(z) = |z| - 15$, $H(u) = |u| - 10$, $a = b = 0.1$, $R = 3$, IC = (1, 16, 16, 11) is red, IC = (-1, -14, -14, -10) is green.

5. Conclusions

Conditional symmetry is a more flexible symmetry, which can be derived from both symmetry and asymmetry. In fact, in the physical world symmetric structure is prone to be destroyed by a newly introduced constant or by the dimension growth. However, asymmetric systems have enough space for conditional symmetry if the offset-boosting assisted polarity balance is established. Conditional symmetric systems are more promising than symmetric ones, which have reliable twin attractors rather than a broken butterfly. In those chaos-based communications, conditional symmetry or symmetry usually indicates that the corresponding system has double monopolar chaotic signals, which meets the needs of engineering application to a large extent.

Author Contributions: Conceptualization. C.L.; Data curation, C.L. and J.S.; formal analysis, T.L. (Tianai Lu); funding acquisition, C.L.; investigation, C.L., J.S., T.L. (Tianai Lu) and T.L. (Tengfei Lei); methodology, C.L.; project administration, C.L.; resources, C.L.; software, T.L. (Tianai Lu); supervision, C.L.; validation, C.L. and T.L. (Tengfei Lei); visualization, J.S.; writing—original draft, C.L.; writing—review and editing, C.L. and T.L. (Tengfei Lei). All authors have read and agreed to the published version of the manuscript.

Funding: This work was supported financially by the National Natural Science Foundation of China (Grant No.: 61871230), the Natural Science Foundation of Jiangsu Province (Grant No.: BK20181410), and a project funded by the Priority Academic Program Development of Jiangsu Higher Education Institutions.

Conflicts of Interest: The authors declare no conflict of interest.

References

1. Lai, Q.; Chen, S. Generating multiple chaotic attractors from Sprott B system. *Int. J. Bifurc. Chaos* **2016**, *26*, 1650177. [CrossRef]
2. Bao, B.C.; Bao, H.; Wang, N.; Chen, M.; Xu, Q. Hidden extreme multistability in memristive hyperchaotic system. *Chaos Solit Fractals* **2017**, *94*, 102–111. [CrossRef]
3. Zhang, X.; Wang, C.H. Multiscroll Hyperchaotic System with Hidden Attractors and Its Circuit Implementation. *Int. J. Bifurc. Chaos* **2019**, *29*, 1950117. [CrossRef]
4. Deng, Q.L.; Wang, C.H. Multi-scroll hidden attractors with two stable equilibrium points. *Chaos* **2019**, *29*, 093112. [CrossRef] [PubMed]
5. Zhao, X.; Liu, J.; Liu, H.J.; Zhang, F.F. Dynamic Analysis of a One-parameter Chaotic System in Complex Field. *IEEE Access* **2020**, *8*, 28774–28781. [CrossRef]
6. Sprott, J.C. *Elegant Chaos: Algebraically Simple Chaotic Flows*; World Scientific: Singapore, 2010; pp. 1–40.
7. Sprott, J.C. Simplest chaotic flows with involutional symmetries. *Int. J. Bifurc. Chaos* **2014**, *24*, 1450009. [CrossRef]
8. Zhang, X.; Wang, C.H.; Yao, W.; Lin, H.R. Chaotic system with bondorbital attractors. *Nonlinear Dyn.* **2019**, *97*, 2159–2174. [CrossRef]
9. Li, C.; Lu, T.; Chen, G.; Xing, H. Doubling the coexisting attractors. *Chaos* **2019**, *29*, 051102. [CrossRef]
10. Barrio, R.; Blesa, F.; Serrano, S. Qualitative analysis of the Rössler equations: Bifurcations of limit cycles and chaotic attractors. *Physica D* **2009**, *238*, 1087–1100. [CrossRef]
11. Sprott, J.C.; Li, C. Asymmetric bistability in the Rössler system. *Acta Phys. Pol. B* **2017**, *48*, 97–107. [CrossRef]
12. Sprott, J.C.; Wang, X.; Chen, G. Coexistence of point, periodic and strange attractors. *Int. J. Bifurc. Chaos* **2013**, *23*, 1350093. [CrossRef]
13. Jafari, S.; Ahmadi, A.; Panahi, S.; Rajagopal, K. Extreme multistability: When imperfection changes quality. *Chaos Solitons Fractals* **2018**, *108*, 182–186. [CrossRef]
14. Karthikeyan, R.; Jafari, S.; Karthikeyan, A.; Srinivasan, A.; Ayele, B. Hyperchaotic Memcapacitor Oscillator with Infinite Equilibria and Coexisting Attractors. *Circuits Syst. Signal. Process.* **2018**, *37*, 1–23.
15. Li, C.; Sprott, J.C. Variable-boostable chaotic flows. *Opt. Int. J. Light Electron. Opt.* **2016**, *127*, 10389–10398. [CrossRef]
16. Gu, Z.; Li, C.; Iu, H.H.C.; Min, F.; Zhao, Y.B. Constructing hyperchaotic attractors of conditional symmetry. *Eur. Phys. J. B* **2019**, *92*, 221. [CrossRef]
17. Lu, T.; Li, C.; Jafari, S.; Min, F. Controlling Coexisting Attractors of Conditional Symmetry. *Int. J. Bifurc. Chaos* **2019**, *29*, 1950207. [CrossRef]
18. Zhang, X. Constructing a chaotic system with any number of attractors. *Int. J. Bifurc. Chaos* **2017**, *27*, 1750118. [CrossRef]
19. Li, C.; Sprott, J.C.; Xing, H. Constructing chaotic systems with conditional symmetry. *Nonlinear Dyn.* **2017**, *87*, 1351–1358. [CrossRef]
20. Li, C.; Sprott, J.C.; Liu, Y.; Gu, Z.; Zhang, J. Offset Boosting for Breeding Conditional Symmetry. *Int. J. Bifurc. Chaos* **2018**, *28*, 1850163. [CrossRef]
21. Schrier, G.V.D.; Maas, L.R.M. The diffusionless Lorenz equations; Shil'nikov bifurcations and reduction to an explicit map. *Physica D* **2000**, *141*, 19–36. [CrossRef]

22. Li, C.; Sprott, J.C.; Thio, W. Linearization of the Lorenz System. *Phys. Lett. A* **2015**, *379*, 888–893. [CrossRef]
23. Leonov, G.A.; Kuznetsov, N.V.; Mokaev, T.N. Homoclinic orbits, and self-excited and hidden attractors in a Lorenz-like system describing convective fluid motion. *Eur. Phys. J. Spec. Top.* **2015**, *224*, 1421–1458. [CrossRef]
24. Kuznetsov, N.V.; Leonov, G.A.; Mokaev, T.N.; Prasad, A.; Shrimali, M.D. Finite-time Lyapunov dimension and hidden attractor of the Rabinovich system. *Nonlinear Dyn.* **2018**, *92*, 267–285. [CrossRef]
25. Kuznetsov, N.V.; Mokaev, T.N. Numerical analysis of dynamical systems: Unstable periodic orbits, hidden transient chaotic sets, hidden attractors, and finite-time Lyapunov dimension. *J. Phys. Conf. Ser.* **2019**, *1205*, 012034. [CrossRef]

Article

A Nonlinear Five-Term System: Symmetry, Chaos, and Prediction

Vo Phu Thoai [1], Maryam Shahriari Kahkeshi [2], Van Van Huynh [3], Adel Ouannas [4] and Viet-Thanh Pham [5,*]

[1] Faculty of Electrical and Electronics Engineering, Ton Duc Thang University, Ho Chi Minh City, Vietnam; vophuthoai@tdtu.edu.vn

[2] Faculty of Engineering, Shahrekord University, Shahrekord 64165478, Iran; m.shahriyarikahkeshi@alumni.iut.ac.ir

[3] Modeling Evolutionary Algorithms Simulation and Artificial Intelligence, Faculty of Electrical and Electronics Engineering, Ton Duc Thang University, Ho Chi Minh City, Vietnam; huynhvanvan@tdtu.edu.vn

[4] Laboratory of Mathematics, Informatics and Systems (LAMIS), University of Laarbi Tebessi, Tebessa 12002, Algeria; ouannas.adel@univ-tebessa.dz

[5] Nonlinear Systems and Applications, Faculty of Electrical and Electronics Engineering, Ton Duc Thang University, Ho Chi Minh City, Vietnam; phamvietthanh@tdtu.edu.vn

* Correspondence: phamvietthanh@tdtu.edu.vn

Received: 19 April 2020; Accepted: 22 May 2020; Published: 25 May 2020

Abstract: Chaotic systems have attracted considerable attention and been applied in various applications. Investigating simple systems and counterexamples with chaotic behaviors is still an important topic. The purpose of this work was to study a simple symmetrical system including only five nonlinear terms. We discovered the system's rich behavior such as chaos through phase portraits, bifurcation diagrams, Lyapunov exponents, and entropy. Interestingly, multi-stability was observed when changing system's initial conditions. Chaos of such a system was predicted by applying a machine learning approach based on a neural network.

Keywords: chaos; symmetry; entropy; prediction

---

1. Introduction

The study of chaos in nonlinear systems has attracted significant attention in recent research [1–4]. Interestingly, new chaotic systems have continued to be proposed. A memristor-based chaotic circuit showing multi-stability was constructed by Song et al. [1]. Azar and Serrano [2] designed port Hamiltonian systems with chaos while Askar and Al-khedhairi mentioned a chaos of duopoly game for a player's actions [5]. Chaos appeared in fractional systems, fractional-order maps, and discrete-time systems [6–8]. Chen et al. discovered entropy for indicating early-warning signals of zero-eigenvalue chaotic systems [9]. Disturbance observer control was applied to synchronize a chaotic system having one constant term and no equilibrium [10]. The special characteristics of chaos provide useful applications such as cryptography, transmission, security, and fractional chaotic memory [11–13]. Image encryption was developed by using a chaos of Farhan's system [11] while a combination of compressed sensing and chaos in encryption scheme was introduced in [14]. Parallel mode of chaotic cryptography provided a transmission efficiency and resisted dangerous attacks [15]. Xie et al. developed image restoration for chaos-based transmission, which is effective to reduce devices' consumption [12]. S-boxes were constructed with special systems' chaos [16,17]. Ouannas et al. investigated MIMO communications using chaos synchronization [18]. By applying constant phase elements, Petrzela implemented chaotic memory [13].

Increased interest in symmetry in chaotic system has been reported in recent works [19]. Zhu and Du presented a chaotic system with a symmetrical curve equilibrium [20]. Chaotic oscillators were built with asymmetrical logic functions, illustrating the feasibility for integrated circuits [21]. It is noted that a symmetrical hyperchaotic attractor was able to control [19]. Especially, Li et al. examined comprehensively the evolution of symmetry [22]. From the viewpoint of information security, the vital roles of symmetry are verified by the improvement of substitution box structures [23], image encryption [24], and symmetric key encryption [25]. Discovering symmetrical chaotic systems is still an open topic.

When studying chaotic systems, discovering simple systems and counterexamples with chaotic behaviors is a vital research topic [26]. Common three-dimensional chaotic systems often have more than six terms and five-term chaotic systems are the most elegant ones [26]. Moreover, chaotic systems without linear terms have rarely been reported [27]. The purpose of our work was to investigate a novel five-term chaotic system without a linear term. Table 1 is provided for comparison with the results of recent studies.

Table 1. Numbers of linear and nonlinear terms in some three-dimensional chaotic systems.

System	Linear Term	Nonlinear Term	Total Term
[17]	6	4	10
[1]	6	2	8
[20]	1	7	8
[27]	0	8	8
[16]	3	4	7
[28]	2	5	7
[11]	2	4	6
[9]	2	3	5
This work	0	5	5

This work focuses on the aim to study a symmetrical system with chaos. Its simple form and rich dynamics are presented in Section 2. Section 3 presents an entropy measurement of the system while chaos prediction using neural network is reported in Section 4. The last section concludes our work.

2. System without Linearity

We investigate a system with no linear terms:

$$\begin{aligned}
\dot{x} &= ayz, \\
\dot{y} &= 1 - z^2, \\
\dot{z} &= bx^3 + yz.
\end{aligned} \tag{1}$$

In system (1), a and b are positive parameters ($a, b > 0$). Interestingly, five terms of the system (1) are nonlinear ones. Only few systems without linear terms have been studied [27]. Simple chaotic systems/circuits have attracted considerable attention because of their elegance [26]. From the viewpoint of terms, the simplest chaotic systems are five-term ones [26]. Therefore, we would like to consider system (1), which includes five nonlinear terms. In addition, the system can be implemented physically by using common electronic elements such as resistors, capacitors, operational amplifiers, and analog multipliers. A practical implementation of system (1) is illustrated in Figure 1. In the design, the circuit of system (1) includes five resistors, three capacitors, three operational amplifiers, and four analog multipliers. However, corresponding equations of the system do not describe certain events.

Considering coordinate transformation (2)

$$(x, y, z) \rightarrow (-x, y, -z), \tag{2}$$

system (1) is invariant. Therefore, system (1) is symmetric. It is worth noting that symmetry in nonlinear systems has attracted interest in recent years [19–21].

By solving

$$\begin{cases} ayz = 0, \\ 1 - z^2 = 0, \\ bx^3 + yz = 0, \end{cases} \tag{3}$$

we get two equilibria of system (1)

$$E_1(0,0,1), \tag{4}$$

$$E_2(0,0,-1). \tag{5}$$

The Jacobian matrix of system (1) is given by

$$J = \begin{bmatrix} 0 & az & ay \\ 0 & 0 & -2z \\ 3bx^2 & z & y \end{bmatrix}. \tag{6}$$

Because of symmetry, by considering the Jacobian matrix at the equilibrium E_1, we get the characteristic equation

$$\lambda^3 + 2\lambda = 0. \tag{7}$$

and two eigenvalues

$$\lambda_1 = 0, \tag{8}$$

$$\lambda_{2,3} = \pm j\sqrt{2}. \tag{9}$$

Therefore, this calculation shows that the system (1) is at a *critical case* for E_1 and E_2.

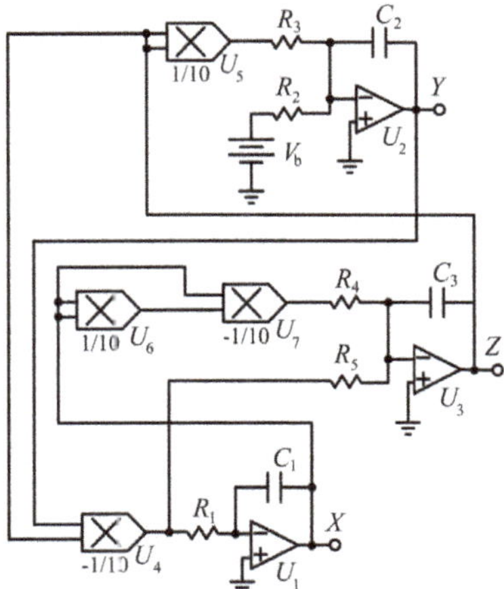

Figure 1. Illustration of a circuit, which is designed to realize system (1). Voltages at the outputs of three operational amplifiers X, Y, Z correspond to three state variables x, y, z of system (1).

System (1) displays rich dynamics when varying a. The bifurcation diagram in Figure 2 shows windows of chaos, which are also verified by maximum Lyapunov exponents (see Figure 3). Chaos can be found in ranges, for example $[1, 1.21]$, and $[1.903, 2.275]$. Illustration of chaos is presented in Figure 4 for $a = 1$, and $b = 0.05$. The maximum Lyapunov exponent equals to 0.02714. Chaotic dynamics is similar to the observed chaotic one of the Lorenz system [29]. The Lorenz system describes the atmospheric convection and includes seven terms (with five linear terms). Our system has five terms (without linear terms).The waveforms of the variables x and z in Figure 5 display slow–fast dynamics. Slow–fast dynamics are important to get the autowaves [30].

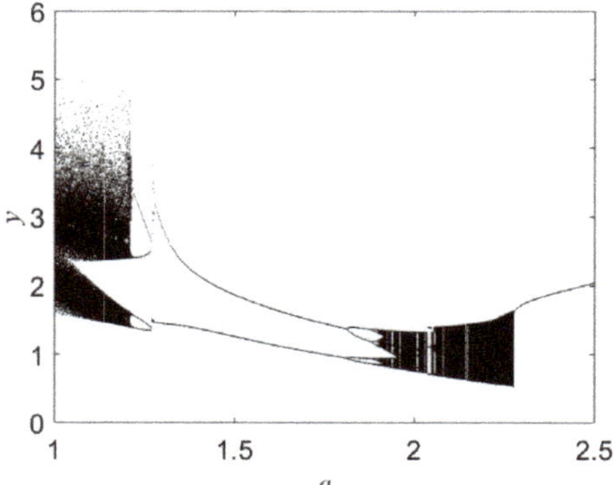

Figure 2. Bifurcation diagram. We change parameter a while keeping $b = 0.05$, and initial conditions $(0.5, 1, 0.5)$.

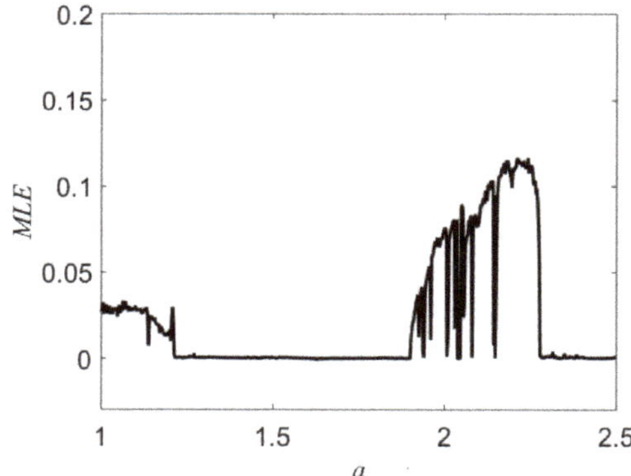

Figure 3. Maximum Lyapunov exponents. We change a while keeping $b = 0.05$, and initial conditions $(0.5, 1, 0.5)$.

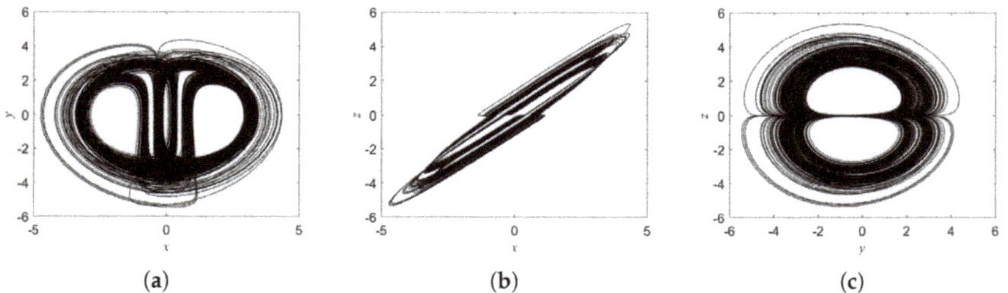

(a)   (b)   (c)

Figure 4. Attractors observed in three planes illustrating chaos in system (1) for $a = 1$: (**a**) $x - y$, (**b**) $x - z$, and (**c**) $y - z$.

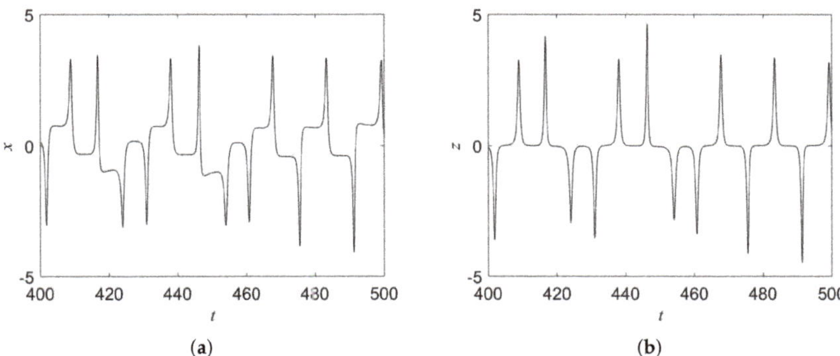

Figure 5. (**a**) Waveform of x, and (**b**) waveform of z observed in system (1) for $a = 1$, and $b = 0.05$.

The symmetrical property of system (1) leads to the appearance of multistabily, which has been investigated in Figure 6. We plot simultaneously two bifurcation diagrams for initial conditions $(\pm 0.5, 1, \pm 0.5)$. Different coexisting attractors are reported in Figure 7.

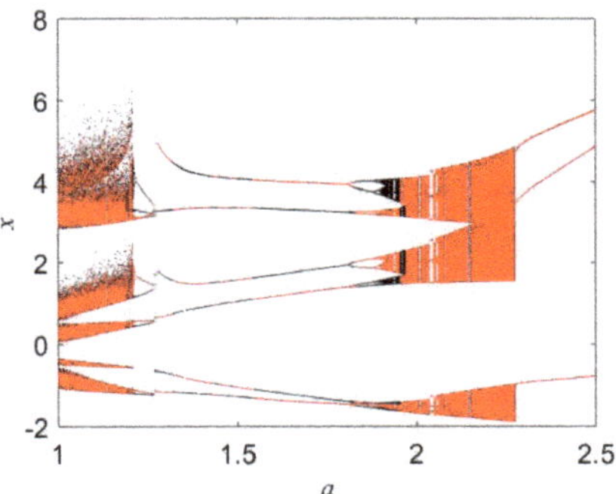

Figure 6. Bifurcation diagrams for initial conditions $(0.5, 1, 0.5)$ (black) and $(-0.5, 1, -0.5)$ (red) while keeping $b = 0.05$.

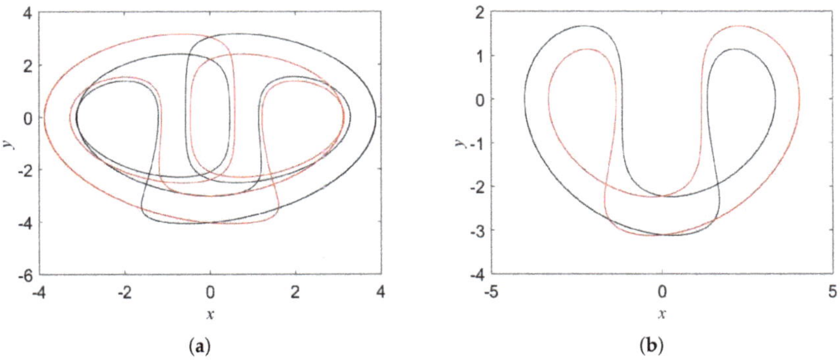

Figure 7. *Cont.*

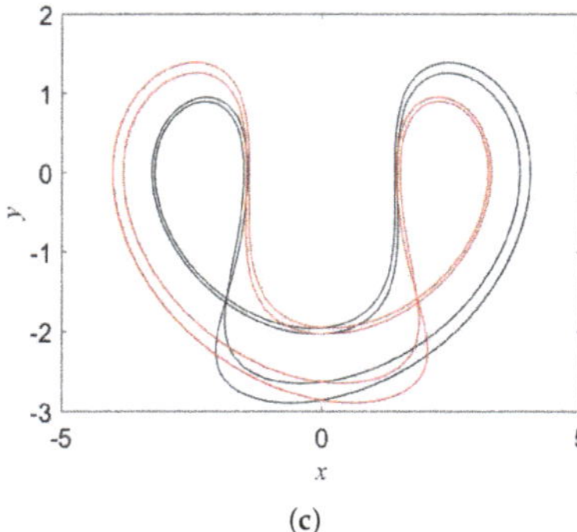 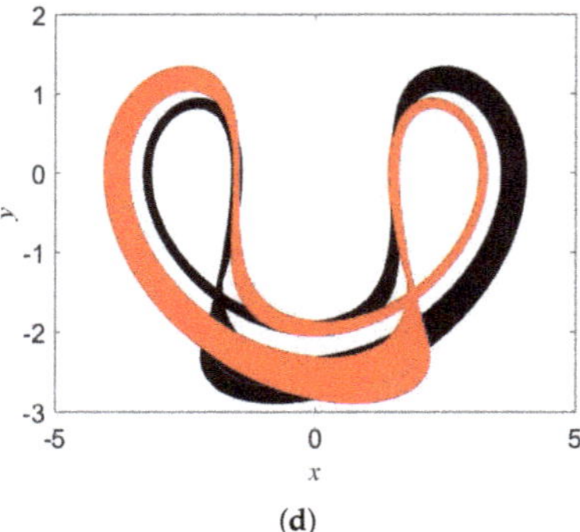

(c)   (d)

Figure 7. Coexisting attractors in system (1) for $(x(0), y(0), z(0)) = (0.5, 1, 0.5)$ (black) and $(x(0), y(0), z(0) = (-0.5, 1, -0.5)$ (red): (**a**) $a = 1.23$, (**b**) $b = 1.6$, (**c**) $b = 1.845$, and (**d**) $a = 1.93$.

3. System's Entropy

Entropy is an important tool not only in information theory but also in nonlinear works [31]. Entropy represents quantifies of information in a particular information system. It is useful for researches to describe nonlinear system's complexity with entropy. Interestingly, entropy measurement has been witnessed for chaotic systems for last years [28,32]. Memristor-based chaotic oscillator has been developed by Liu et al. to achieve a high spectral entropy [3]. Chen et al. has indicated the usage of entropy as early warning indexes of chaotic signals [9].

We calculate entropy of system (1) to consider its complexity. The approximate entropy (ApEn) is measured for x variable. ApEn highlights advantages such as small samples demand, simple computation, and noise reduction [33].

Approximate entropy calculation [33] is presented briefly as follows. Firstly, we take a set of data $x(1)$, $x(2)$, ..., $x(n)$ from system (1). Vectors $X(j)$ for $j = 1, \ldots, n - m + 1$ are constructed by $X(j) = (x(j), \ldots, x(j + m - 1))$ with a given m. The distance between vectors $X(i)$ and $X(j)$ is given by $d(X(i), X(j))$. As a result, we get the relative frequency of $X(i)$ being similar to $X(j)$:

$$C_i^m(r) = \frac{K}{n - m + 1},\tag{10}$$

where K is the number of j satisfying $d(X(i), X(j)) \leq r$ for a given $X(i)$.

We obtain the approximate entropy

$$ApEn = \phi^m(r) - \phi^{m+1}(r),\tag{11}$$

in which

$$\phi^m(r) = \frac{1}{n - m - 1} \sum_{i=1}^{n-m+1} \log C_i^m(r).\tag{12}$$

Figure 8 depicts the approximate entropy [33] for parameter a. ApEn measures regularity and unpredictability of x. Significantly small values of ApEn indicate regular signals. As shown in Figure 8, system's complex behavior can be witnessed for two ranges of a ([1, 1.21], and [1.903, 2.275]). Table 2 reports three examples of calculated ApEn values for a. The values of ApEn in the case 1 (0.2162), and the case 3 (0.3526) verify the complexity of system (1). Tiny value of ApEn ($7.418 \times 10^{-7} \approx 0$) confirms the periodical behavior of the system in the case 2.

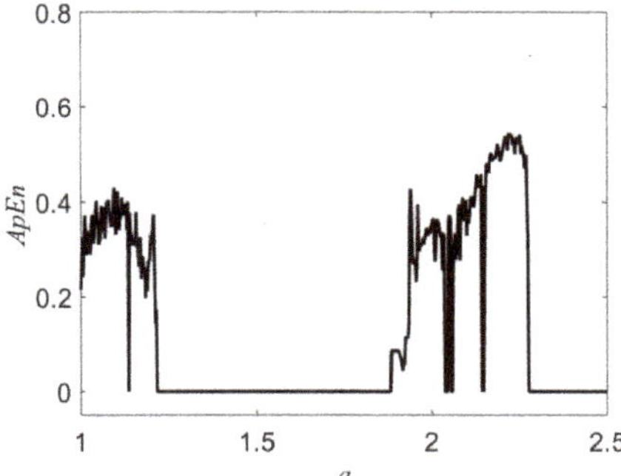

Figure 8. Approximate entropy (ApEn) of system (1) calculated for $a \in [1, 2.5]$.

Table 2. Examples of calculated ApEn values.

Cases	a	ApEn
1	1	0.2162
2	1.5	7.418×10^{-7}
3	2	0.3526

4. Chaos Prediction

Artificial neuron network is constructed by connecting many neurons [34,35]. Numerous applications of neural networks in practice have been found in computer vision, pattern recognition, natural language processing, and robotics [36,37]. Ability of neuron network to represent nonlinear system has been investigated and attracted considerable interest [38–40]. Predicting chaotic system is challenge due to its sensitivity with initial conditions.

In this section, we build a simple feed-forward neural network (see Figure 9) to predict signals of system (1). As illustrated in Figure 9, the artificial neuron network includes four layers: input layer, two hidden layers, and output layer. It is considered as a deep neural network because there are multiple layers before the output layer [41]. The input layer includes three neurons. Each hidden layer is composed of ten neurons while there are only three neurons in the output layer. The numbers of hidden neurons and hidden layers are selected by considering the specific dynamics of the system such as multistability, and slow-fast dynamics. The computational roles of hidden layers are similar in order to model the dynamical system from its time series. It is worth noting that the hardest task of machine learning, choosing the suitable balance between model complexity and simplicity, must be considered seriously. The effective and robust architecture of the neural network as well as the optimization of network's parameters guarantee the good performance of the network. In this work, we construct a simple feed-forward neural network. Compared with advanced networks, for example convolutional neural network, recurrent neural network, liquid state machine, and echo state network, the proposed architecture is effective and robust when being applied to system (1).

A dataset is generated by running system (1) with different initial conditions. The proposed neural network is trained with the data set (x, y, z) by applying the Levenberg–Marquardt algorithm. It is noted that there are different training algorithms such as the Levenberg–Marquardt algorithm, Bayesian Regularization algorithm, Scaled Conjugate Gradient algorithm, and the Fletcher–Powell Conjugate Gradient [37,42]. In this work, we use the Levenberg–Marquardt algorithm because of its good convergence and robustness. The obtained performance is 2.3051×10^{-9}. After the training process, we achieve a network matching with the data set. Outputs of the network present expected signals. Figure 10 illustrates the prediction results (X, Y, Z) compared with actual data (x, y, z). The agreement of the prediction results with the actual data shows the capability of the network for predicting chaos of system (1) in short term. In comparison to other works [43–45], the neural network is simple and displays good performance.

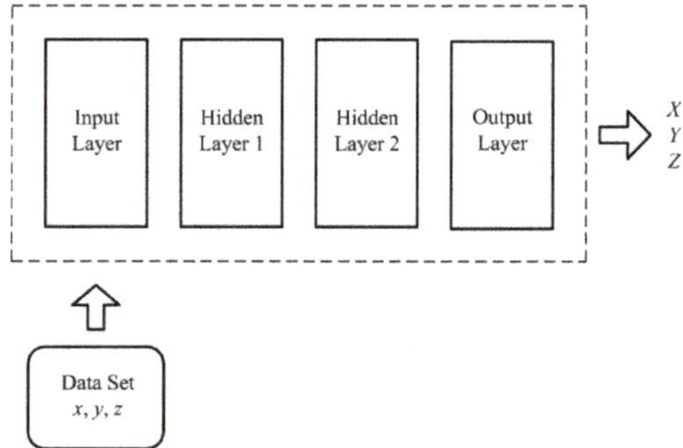

Figure 9. Neuron network includes four layers. Data set is provided by system (1) and is used for training.

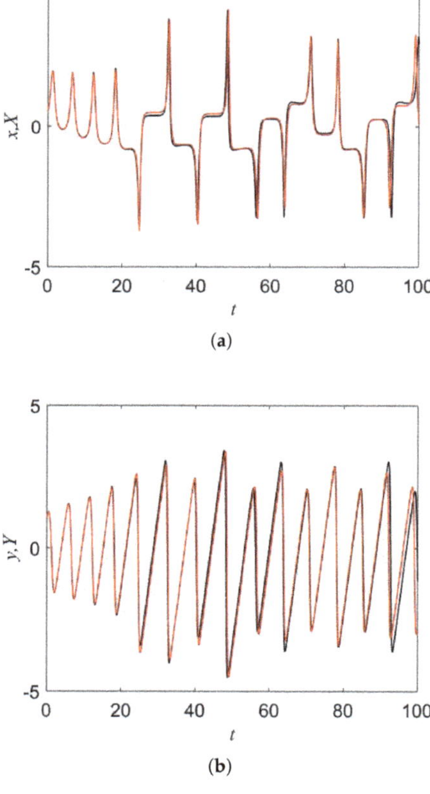

(a)

(b)

Figure 10. *Cont.*

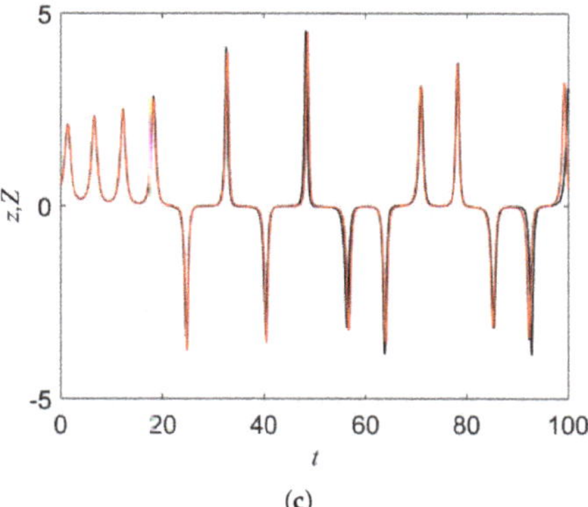

(c)

Figure 10. Signals x, y, z of system (1) (black color) and desired signals X, Y, Z at the output of the neural network (red color): (**a**) x and X, (**b**) y and Y, (**c**) z and Z.

5. Conclusions

Our work introduces a symmetry nonlinear system with remarkable dynamics. There are only five nonlinear terms in the system, which generates chaos. By considering the initial conditions, we find coexisting attractors in the system verifying its multistability feature. Entropy measurement also indicates the system's complexity. We believe that our work contributes to the known list of chaotic systems with algebraic simplicity. It is possible for us to apply a modified version of such a system to describe turbulent flows [46,47]. We implemented a neural-based approach to predict a system's chaos in short-term. Long-term prediction of such chaotic signals should be considered. In addition, prediction results will be applied to control chaos in our future investigation. In addition, realization of the system for practical chaos-based applications will be studied in our future works.

Author Contributions: Data curation, V.V.H.; Formal analysis, V.P.T.; Funding acquisition, M.S.K.; Investigation, V.P.T. and V.V.H.; Methodology, A.O. and V.-T.P.; Project administration, M.S.K.; Resources, V.-T.P.; Software, M.S.K.; Supervision, A.O.; Visualization, V.V.H.; Writing—original draft, V.P.T. and A.O.; Writing—review & editing, V.-T.P. All authors have read and agreed to the published version of the manuscript.

Conflicts of Interest: The authors declare no conflict of interest.

References

1. Song, Y.; Yuan, F.; Li, Y. Coexisting attractors and multistability in a simple memristive Wien-Bridge chaotic circuit. *Entropy* **2019**, *21*, 678. [CrossRef]
2. Azar, A.T.; Serrano, F.E. Stabilization of port Hamiltonian chaotic systems with hidden attractors by adaptive terminal sliding mode control. *Entropy* **2020**, *22*, 122. [CrossRef]
3. Liu, L.; Du, C.; Liang, L.; Zhang, X. A high Spectral Entropy (SE) memristive hidden chaotic system with multi-type quasi-periodic and its circuit. *Entropy* **2019**, *21*, 1026. [CrossRef]
4. Danca, M.F. Puu system of fractional order and its chaos suppression. *Symmetry* **2020**, *12*, 340. [CrossRef]
5. Askar, S.S.; Al-khedhairi, A. Dynamic effects arise due to consumers' preferences depending on past choices. *Entropy* **2020**, *22*, 173. [CrossRef]
6. Ouannas, A.; Khennaoui, A.A.; Bendoukha, S.; Vo, T.P.; Pham, V.T.; Huynh, V.V. The fractional form of the Tinkerbell map is chaotic. *Appl. Sci.* **2018**, *8*, 2640. [CrossRef]
7. Huynh, V.V.; Ouannas, A.; Wang, X.; Pham, V.T.; Nguyen, X.Q.; Alsaadi, F.E. Chaotic map with no fixed points: Entropy, implementation and control. *Entropy* **2019**, *21*, 279. [CrossRef]
8. Ouannas, A.; Khennaoui, A.A.; Momani, S.; Grassi, G.; Pham, V.T. Chaos and control of a three-dimensional fractional order discrete-time system with no equilibrium and its synchronization. *AIP Adv.* **2020**, *10*, 045310. [CrossRef]

9. Chen, L.; Nazarimehr, F.; Jafari, S.; Tlelo-Cuautle, E.; Hussain, I. Investigation of early warning indexes in a three-dimensional chaotic system with zero eigenvalues. *Entropy* **2020**, *22*, 341. [CrossRef]

10. Wang, S.; Yousefpour, A.; Yusuf, A.; Jahanshahi, H.; Alcaraz, R.; He, S.; Munoz-Pacheco, J.M. Synchronization of a non-equilibrium four-dimensional chaotic system using a disturbance-observer-based adaptive terminal sliding mode control method. *Entropy* **2020**, *22*, 271. [CrossRef]

11. Farhan, A.K.; Al-Saidi, N.M.; Maolood, A.T.; Nazarimehr, F.; Hussain, I. Entropy analysis and image encryption application based on a new chaotic system crossing a cylinder. *Entropy* **2019**, *21*, 958. [CrossRef]

12. Xie, Y.; Yu, J.; Chen, X.; Ding, Q.; Wang, E. Low-element image restoration based on an out-of-order elimination algorithm. *Entropy* **2020**, *22*, 1192. [CrossRef]

13. Petrzela, J. Fractional-order chaotic memory with wideband constant phase elements. *Entropy* **2020**, *22*, 422. [CrossRef]

14. Xie, Y.; Yu, J.; Guo, S.; Ding, Q.; Wang, E. Image encryption scheme with compressed sensing based on new three-dimensional chaotic system. *Entropy* **2019**, *21*, 819. [CrossRef]

15. Yu, J.; Guo, S.; Song, X.; Xie, Y.; Wang, E. Image parallel encryption technology based on sequence generator and chaotic measurement matrix. *Entropy* **2020**, *22*, 76. [CrossRef]

16. Wang, X.; Akgul, A.; Cavusoglu, U.; Pham, V.T.; Hoang, D.V.; Nguyen, X.Q. A chaotic system with infinite equilibria and its S-Box constructing application. *Appl. Sci.* **2018**, *8*, 2132. [CrossRef]

17. Wang, X.; Cavusoglu, U.; Kacar, S.; Akgul, A.; Pham, V.T.; Jafari, S.; Alsaadi, F.E.; Nguyen, X.Q. S-Box based image encryption application using a chaotic system without equilibrium. *Appl. Sci.* **2019**, *9*, 781. [CrossRef]

18. Ouannas, A.; Debbouche, N.; Wang, X.; Pham, V.T.; Zehrour, O. Secure Multiple-Input Multiple-Output communications based on F-M synchronization of fractional-order chaotic systems with non-identical dimensions and orders. *Appl. Sci.* **2018**, *8*, 1746. [CrossRef]

19. Zhang, X.; Li, C.; Lei, T.; Liu, Z.; Tao, C. A symmetric controllable hyperchaotic hidden attractor. *Symmetry* **2020**, *12*, 550. [CrossRef]

20. Zhu, X.; Du, W.S. New chaotic systems with two closed curve equilibrium passing the same point: Chaotic behavior, bifurcations, and synchronization. *Symmetry* **2019**, *11*, 951. [CrossRef]

21. Munoz-Pacheco, J.M.; García-Chávez, T.; Gonzalez-Diaz, V.R.; de La Fuente-Cortes, G.; del Carmen Gómez-Pavón, L. Two new asymmetric Boolean chaos oscillators with no dependence on incommensurate time-delays and their circuit implementation. *Symmetry* **2020**, *12*, 506. [CrossRef]

22. Li, C.; Sun, J.; Lu, T.; Lei, T. Symmetry evolution in chaotic system. *Symmetry* **2020**, *12*, 574. [CrossRef]

23. Artuğer, F.; Özkaynak, F. A novel method for performance improvement of chaos-based substitution Boxes. *Symmetry* **2020**, *12*, 571. [CrossRef]

24. Zhang, G.; Ding, W.; Li, L. Image encryption algorithm based on tent delay-sine cascade with logistic map. *Symmetry* **2020**, *12*, 355. [CrossRef]

25. Stoyanov, B.; Nedzhibov, G. Symmetric key encryption based on rotation-translation equation. *Symmetry* **2020**, *12*, 73. [CrossRef]

26. Sprott, J.C. *Elegant Chaos Algebraically Simple Chaotic Flows*; World Scientific: Singapore, 2010.

27. Mobayen, S.; Kingni, S.T.; Pham, V.T.; Nazarimehr, F.; Jafari, S. Analysis, synchronisation and circuit design of a new highly nonlinear chaotic system. *Int. J. Syst. Sci.* **2018**, *49*, 617–630. [CrossRef]

28. Xu, G.; Shekofteh, Y.; Akgul, A.; Li, C.; Panahi, S. A new chaotic system with a self-excited attractor: Entropy measurement, signal encryption, and parameter estimation. *Entropy* **2018**, *20*, 86. [CrossRef]

29. Lorenz, E.N. Deterministic nonperiodic flow. *J. Atmos. Sci.* **1963**, *20*, 130–141. [CrossRef]

30. Arena, P.; Caponetto, R.; Fortuna, L.; Manganaro, G. Cellular neural networks to explore complexity. *Soft Comput.* **1997**, *1*, 120–136. [CrossRef]

31. Volos, C.K.; Jafari, S.; Kengne, J.; Munoz-Pacheco, J.M.; Rajagopal, K. Nonlinear dynamics and entropy of complex systems with hidden and self-excited attractors. *Entropy* **2019**, *21*, 370. [CrossRef]

32. Liu, L.; Du, C.; Zhang, X.; Li, J.; Shi, S. Dynamics and entropy analysis for a new 4-D hyperchaotic system with coexisting hidden attractors. *Entropy* **2019**, *21*, 287. [CrossRef]

33. Pincus, S.M. Approximate entropy as a measure of system complexity. *Proc. Natl. Acad. Sci. USA* **1991**, *88*, 2297–2301. [CrossRef] [PubMed]

34. Lytton, W.W. *From Computer to Brain: Foundations Of Computational Neuroscience*; Springer: Berlin/ Heidelberg, Germany, 2002.

35. Haykin, S.O. *Neural Networks and Learning Machines*; Pearson: London, UK, 2008.
36. Fausett, L.V. *Fundamentals of Neural Networks: Architectures, Algorithms And Applications*; Pearson: London, UK, 1993.
37. Aggarwal, C.C. *Neural Networks and Deep Learning: A Textbook*; Springer: Berlin/Heidelberg, Germany, 2018.
38. Jaeger, H.; Haas, H. Harnessing nonlinearity: predicting chaotic systems and saving energy in wireless communication. *Science* **2004**, *304*, 78–80. [CrossRef] [PubMed]
39. Du, C.; Cai, F.; Zidan, M.A.; Ma, W.; Lee, S.H.; Lu, W.D. Reservoir computing using dynamic memristors for temporal information processing. *Nat. Commun.* **2017**, *8*, 2204. [CrossRef] [PubMed]
40. Moon, J.; Ma, W.; Shin, J.H.; Cai, F.; Du, C.; Lee, S.H.; Lu, W.D. Temporal data classification and forecasting using a memristor-based reservoir computing system. *Nat. Electron.* **2019**, *2*, 480–487. [CrossRef]
41. Goodfellow, I.; Bengio, Y.; Courville, A. *Deep Learning*; MIT Press: Cambridge, MA, USA, 2016.
42. Du, K.L.; Swamy, M.N.S. *Neural Networks and Statistical Learning*; Springer: Berlin/Heidelberg, Germany, 2019.
43. Woolley, J.W.; Agarwal, P.K.; Baker, J. Modeling and prediction of chaotic systems with artificial neural networks. *Int. J. Numer. Methods Fluids* **2010**, *63*, 989–1004. [CrossRef]
44. Lamamra, K.; Vaidyanathan, S.; Azar, A.T.; Salah, C.B. Chaotic System Modelling Using a Neural Network with Optimized Structure. In *Fractional Order Control and Synchronization of Chaotic Systems*; Azar, A.T., Vaidyanathan, S., Ouannas, A., Eds.; Springer International Publishing: Cham, Switzerland, 2017; pp. 833–856.
45. Lellep, M.; Prexl, J.; Linkmann, M.; Eckhardt, B. Using machine learning to predict extreme events in the Hénon map. *Chaos* **2019**, *30*, 013113. [CrossRef]
46. McDonough, J.M. Three-dimensional poor man's Navier-Stokes equation: A discrete dynamical system exhibiting $k^{-5/3}$ inertial subrange energy scaling. *Phys. Rev. E* **2009**, *79*, 065302. [CrossRef]
47. Alberti, T.; Consolini, G.; Carbone, V. A discrete dynamical system: The poor man's magnetohydrodynamic (PMMHD) equations. *Chaos* **2019**, *29*, 103107. [CrossRef]

Article

A Symmetric Controllable Hyperchaotic Hidden Attractor

Xin Zhang [1,2], **Chunbiao Li** [1,2,3,*], **Tengfei Lei** [3], **Zuohua Liu** [4] **and Changyuan Tao** [4]

[1] Jiangsu Collaborative Innovation Center of Atmospheric Environment and Equipment Technology (CICAEET), Nanjing University of Information Science & Technology, Nanjing 210044, China; xinzhang96@nuist.edu.cn

[2] Jiangsu Key Laboratory of Meteorological Observation and Information Processing, Nanjing University of Information Science & Technology, Nanjing 210044, China

[3] Collaborative Innovation Center of Memristive Computing Application (CICMCA), Qilu Institute of Technology, Jinan 250200, China; leitengfei2017@qlit.edu.cn

[4] State Key Laboratory of Coal Mine Disaster Dynamics and Control, Chongqing University, Chongqing 400044, China; liuzuohua@cqu.edu.cn (Z.L.); taocy@cqu.edu.cn (C.T.)

[*] Correspondence: goontry@126.com or chunbiaolee@nuist.edu.cn; Tel.: +86-13912993098

Received: 5 March 2020; Accepted: 20 March 2020; Published: 4 April 2020

Abstract: By introducing a simple feedback, a hyperchaotic hidden attractor is found in the newly proposed Lorenz-like chaotic system. Some variables of the equilibria-free system can be controlled in amplitude and offset by an independent knob. A circuit experiment based on Multisim is consistent with the theoretic analysis and numerical simulation.

Keywords: hidden attractor; amplitude control; offset boosting

1. Introduction

As we all know, chaos is ubiquitous in nature and human society, and has great potential in engineering applications. However, there exists great challenge in conditioning broadband chaotic signals, and appropriate amplitude and polarity are the key specifications for chaos generation and transmission [1–3], and therefore, recently great efforts have been made on the research of amplitude control and offset boosting. Normally, the amplitude of system variable requires further adjusting a couple of parameters. In many cases, a unipolar signal is more suitable for transmitting in inter-coupled integrated circuits. Such a challenge exists in the conversion from the bipolar signal to unipolar signal. An independent non-bifurcation parameter to rescale the signal without revising the Lyapunov exponents is important for chaos application. Suitable signal control saves the modulator in chaos-based applications [4,5], including amplitude control [6,7] and offset boosting [8,9].

In addition, hidden attractors exist in chaos, but one cannot find them from the neighborhood of any equilibrium point. Thus, it is of great value in theoretical and physical significance and engineering application to study the realization method of hidden attractors. The Chua system, Lorenz-like systems, and the chaotic systems with stable equilibria [10–15], line equilibria [16–18], or no equilibria [19–24] give us impressive points. Hyperchaos with higher complexity is beneficial to secure communication, so some research extends to hyperchaos. A hyperchaotic system with a hidden attractor was proposed by Wang et al. [25]; Chlouverakis and Sprott [26] claimed the simplest hyperchaotic snap system in algebra; and Yuan et al. [27] showed a memristive multi-scroll hyperchaotic system. Other many hyperchaotic systems have come out of the Lorenz-like system [28–31]. Some other hyperchaotic ones have been proposed, including a memristive hyperchaotic system [32,33], fractional order hyperchaotic system [34,35] or hyperchaotic multi-wing system [36,37]. To the best of our knowledge, there is no relevant research on a hyperchaotic hidden attractor with geometric control. Based on a three-dimensional Lorenz-like system, Wang et al. [38] put forward a hyperchaotic system for producing multi-wing attractors; while in this work, the proposed system has four features as follows:

I)   There exists a parameter to control amplitude and frequency of signals in a small range.
II) Amplitude of x and y can be controlled simultaneously.
III) There is an offset boosting controller.
IV) A special parameter can realize both amplitude and offset control of one system variable.

As shown in Figure 1, the proposed hyperchaotic system has multiple independent geometric controllers including controllers for rescaling amplitude, frequency and offset. Some of the reported 4-D hyperchaotic Lorenz-like systems are listed in Table 1. In the paper, the system controllers are signal controllers and multistability observers as well. In Section 2, the mathematical model of the hyperchaotic system is given. In Section 3, complex dynamic behavior is analyzed. The process of amplitude control and offset boosting is discussed in Section 4. In Section 5, multistability is investigated. In Section 6, the analog circuit is given. Finally, we give the conclusions and discussion.

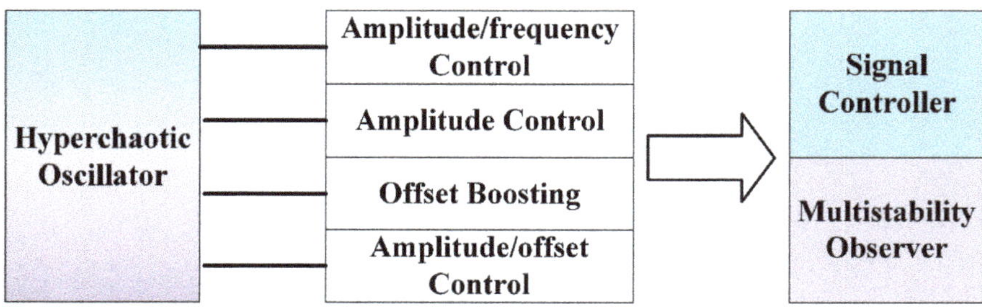

Figure 1. Hyperchaotic attractor with multiple independent controllers.

Table 1. Comparison of the Lorenz-like hyperchaotic systems.

Reference	Number of Terms	Number of Equilibrium	Amplitude/Frequency Control	Amplitude/Offset Control
[15]	9	one	no	not mentioned
[28]	10	line equilibrium	no	not mentioned
[30]	10	one	no	not mentioned
[31]	9	one	no	not mentioned
[38]	9	no	no	not mentioned
this work	9	no	yes	yes

2. Model Description

A 3-D Lorenz-like chaotic system is proposed by Cang et al [39], which is,

$$\begin{cases} \dot{x} = -ay - xz, \\ \dot{y} = -x + xz, \\ \dot{z} = -d - xy. \end{cases} \tag{1}$$

System (1) has a simple rotational symmetric structure with six terms. Based on system (1), a new hyperchaotic system is proposed as,

$$\begin{cases} \dot{x} = -ay - xz - u, \\ \dot{y} = -cx + xz, \\ \dot{z} = -b - mxy, \\ \dot{u} = kx - y. \end{cases} \tag{2}$$

where x, y, z, u are system variables, and a, b, c, k are bifurcation parameters of system (2). When $a = 5$, $b = 4, c = 1, k = 0.5$ and $m = 1$, system (2) has a hyperchaotic attractor with Lyapunov exponents (0.3606, 0.1222, 0, −1.4827) and a Kaplan-Yorke dimension of $D_{KY} = 3.3256$ under initial conditions (1, −1, −1, 1), as shown in Figure 2.

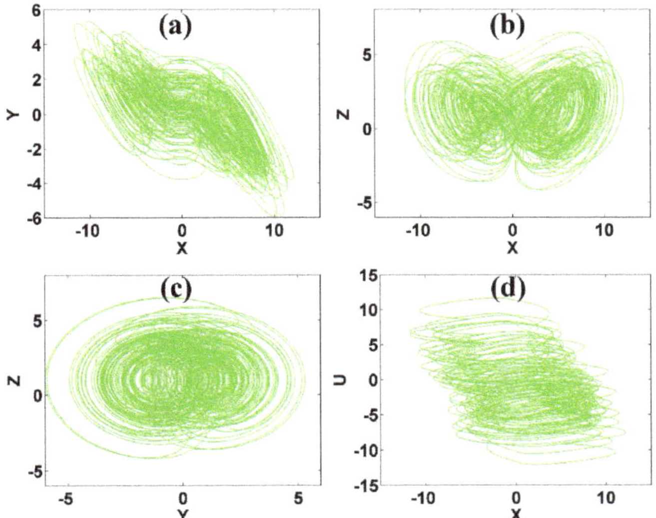

Figure 2. Hyperchaotic attractor of system (2) with $a = 5, b = 4, c = 1, k = 0.5, m = 1$ and initial conditions [1, −1, −1, 1]: (**a**) x-y plane, (**b**) x-z plane, (**c**) y-z plane, (**d**) x-u plane.

The hyper-volume contraction is

$$\nabla V = \frac{\partial \dot{x}}{\partial x} + \frac{\partial \dot{y}}{\partial y} + \frac{\partial \dot{z}}{\partial z} + \frac{\partial \dot{u}}{\partial u} = -z \tag{3}$$

When $a = 5, b = 4, c = 1$ and $k = 0.5$, the dissipative curve of Equation (3) is as shown in Figure 3. The negative average of ∇V proves that system (2) is dissipative.

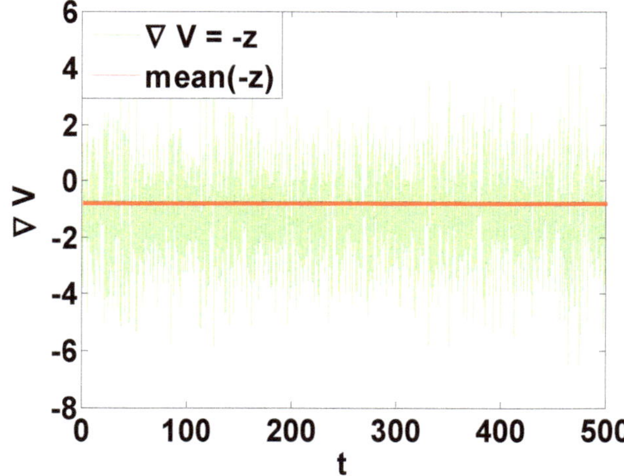

Figure 3. Dissipative curve of system (2).

3. Basic Dynamic Analysis

3.1. Anylis of Equilibria

For system (2), the equilibria can be solved by the following equation:

$$\begin{cases} -ay - xz - u = 0 \\ -cx + xz = 0 \\ -b - mxy = 0 \\ kx - y = 0 \end{cases} \tag{4}$$

The fourth equation indicates that $y = kx$, but the third equation means that $b = -mxy$, then $b = -mkx^2$, which means that there is no real solution, correspondingly the hyperchaotic attractor of system (2) is hidden.

3.2. Bifurcation Analysis

For system (2), the parameters modify the dynamics effectively. To make the demonstration simpler, we ignore the multistability caused by the special structure of symmetry. When $b = 4$, $c = 1$, $k = 0.5$, $m = 1$ under initial conditions $(1, -1, -1, 1)$, Lyapunov exponent spectra and bifurcation diagram when a varies in $[-10, 23.4]$ are shown in Figure 4, where a typical transition from period to chaos shows up and finally system (1) results in the state of hyperchaos. Typical phase trajectories are shown in Figure 5. Quasi-periodicity was not found in the examination interval of system (2). When $a = 5$, $c = 1$, $k = 0.5$, $m = 1$ and initial conditions are $(1, -1, -1, 1)$, when b varies in $[0, 15]$, system (2) heads to hyperchaos from chaos. Lyapunov exponent spectra and bifurcation diagrams are shown in Figure 6, which shows a robust hyperchaos. Both cases have almost linearly scaled Lyapunov exponents in specific regions indicating the function of frequency rescaling.

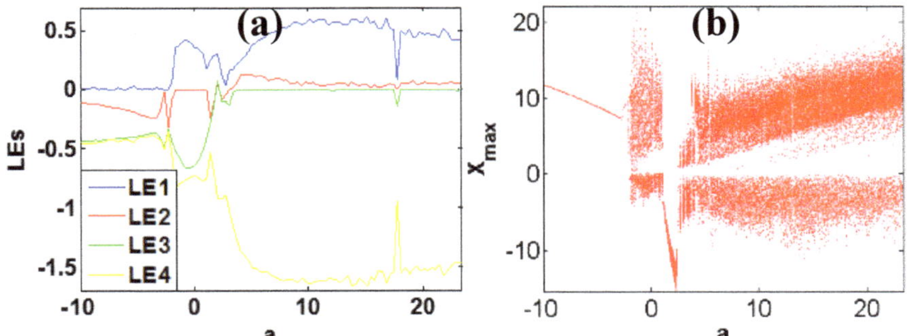

Figure 4. Dynamical behavior of system (2) with $b = 4$, $c = 1$, $k = 0.5$, $m = 1$ under initial conditions [1, -1, -1, 1]: (**a**) Lyapunov exponents, (**b**) bifurcation diagram.

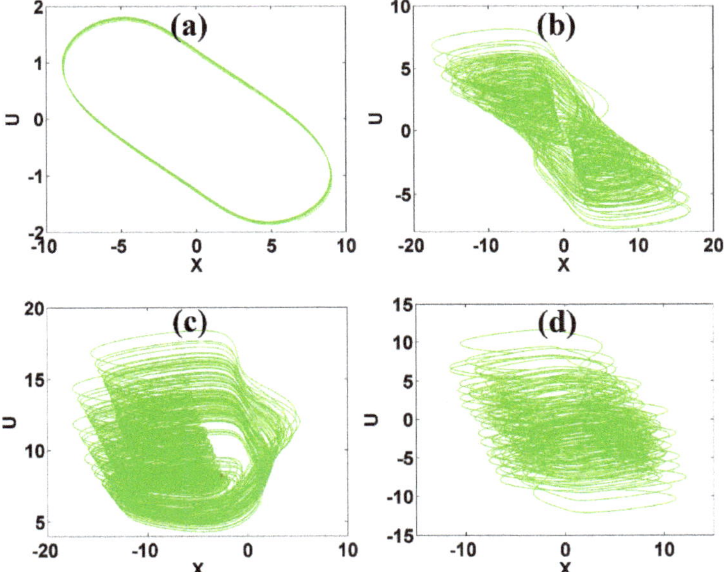

Figure 5. Typical phase trajectories of system (2) with $b = 4$, $c = 1$, $k = 0.5$, $m = 1$ under initial condition [1, -1, -1, 1] in the plane x-u: (**a**) $a = -5$ (period), (**b**) $a = -0.6$ (chaos), (**c**) $a = 3$ (chaos), (**d**) $a = 5$ (hyperchaos).

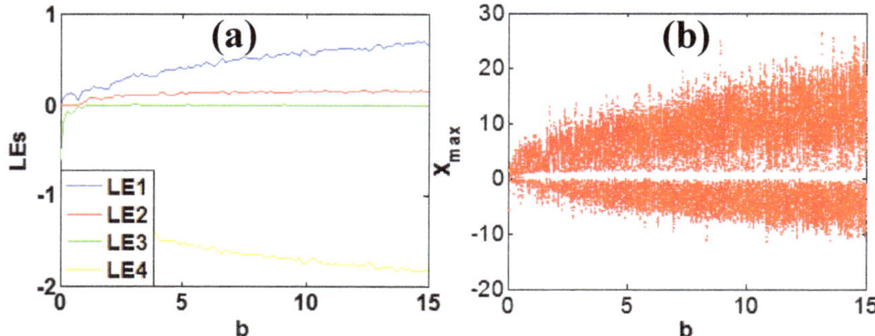

Figure 6. Dynamical behavior of system (2) with $a = 5, c = 1, k = 0.5, m = 1$ under initial condition $[1, -1, -1, 1]$: (**a**) Lyapunov exponents, (**b**) bifurcation diagram.

Comparing Figures 4 and 6, we can see that the parameter a or b visits chaos quickly but modifies the solution in its own way. The parameter a almost has positive correlation with amplitude in a limited range. Meanwhile parameter b has positive correlation with amplitude and frequency, which is distinct and different from other systems. Typical phase trajectories and waveforms are shown in Figure 7.

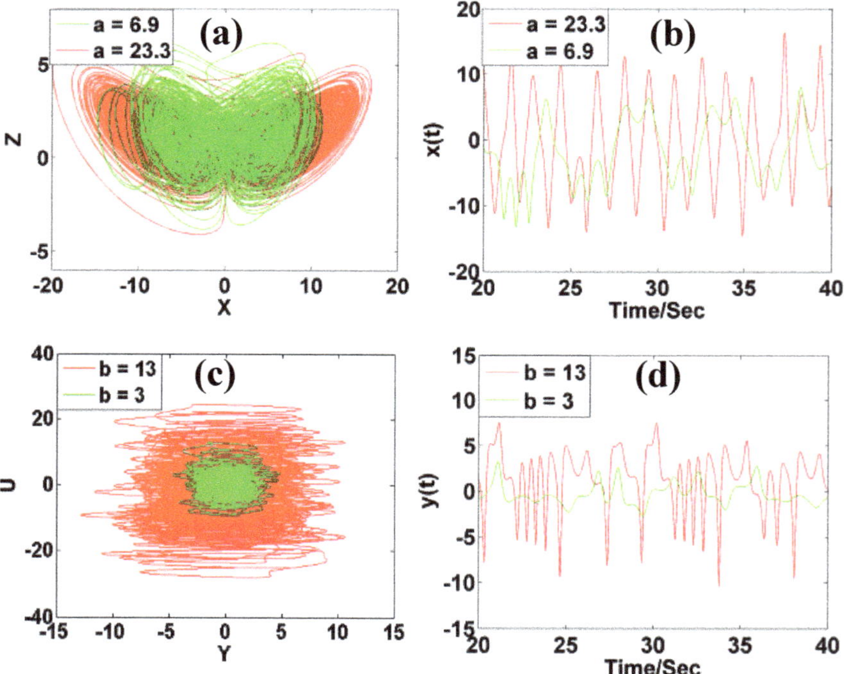

Figure 7. Chaotic oscillations of system (2) with $c = 1, k = 0.5, m = 1$ under initial condition $[1, -1, -1, 1]$: (**a**) phase trajectory in x-z ($b = 4$), (**b**) signal $x(t)$, (**c**) phase trajectory in y-u plane ($a = 5$), (**d**) signal $y(t)$.

Fix the parameters $a = 5, b = 4, k = 0.5, m = 1$, when parameter c varies in $[0, 1.7]$; the Lyapunov exponent spectra and bifurcation diagram are shown in Figure 8a,b. When c varies in $[0, 1.4]$, system (2) exhibits hyperchaos, while when c varies in $[1.4, 1.7]$, system (2) presents chaos. When $a = 5, b = 4$, $c = 1$ and $m = 1$, the parameter k varies in $[0.15, 7.8]$; the Lyapunov exponent spectra and bifurcation diagram are shown in Figure 8c,d. When k varies in [0.15, 1.82], system (2) keeps chaos, and when c varies in [1.82, 7.8], system (2) exhibits hyperchaos. Comparing the Lyapunov exponents controlled by parameters c and k, system (2) has relatively robust hyperchaos under the parameters c.

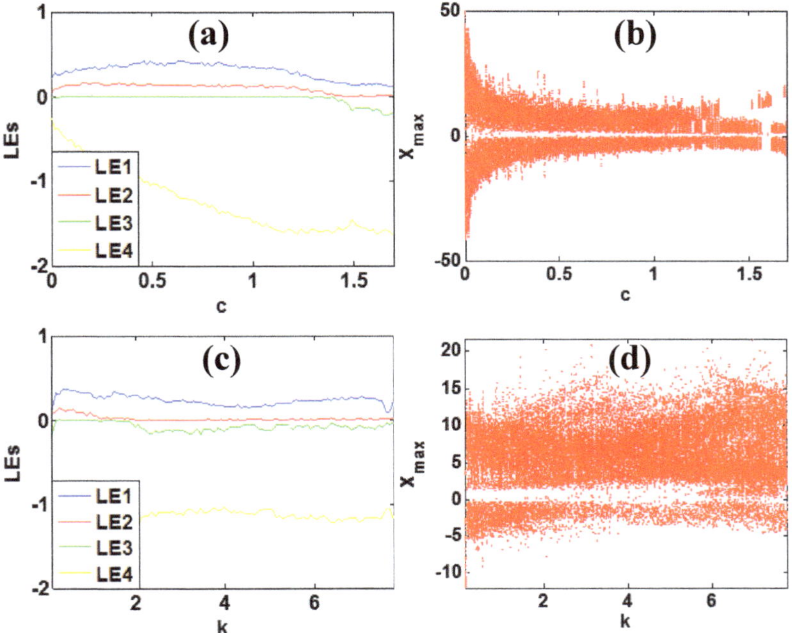

Figure 8. Dynamical behavior of system (2) with $a = 5$, $b = 4$, $m = 1$ under initial conditions [1, −1, −1, 1]: (**a,b**): Lyapunov exponents and bifurcation diagram of c when $k = 0.5$, (**c,d**): Lyapunov exponents and bifurcation diagram of k when $c = 1$.

3.3. Amplitude Control

Besides the above two control knobs, the parameter m in the third dimension in system (2) is a single non-bifurcation knob for amplitude control. To understand this rescaling mechanism, we turn back to the initial system (2). Here, we take the transformation: $x \rightarrow hx, y \rightarrow hy, z \rightarrow z,\ u \rightarrow hu(h > 0)$, which only leaves an additional coefficient in the third dimension:

$$\begin{cases} \dot{x} &= -ay - xz - u, \\ \dot{y} &= -cx + xz, \\ \dot{z} &= -b - mh^2 xy, \\ \dot{u} &= kx - y. \end{cases} \tag{5}$$

indicating that the amplitude of variables x, y and u can be controlled by the parameter m, with the signal z unchanged. It also has no effect on the frequency of the hyperchaotic chaotic signals.

The output signals are controlled by the non-bifurcation parameter m in system (2). As shown in Figure 9, the amplitude of the signals x, y and u are rescaled by the non-bifurcation parameter m. When $m = 0.25$, the amplitudes of the x, y and u signals are very large. The amplitudes of the x, y and u signals decrease with an inverse proportion to the parameter m without changing the amplitude of z. Figure 10 shows the phase trajectories on the planes of x-u and y-z when the control parameter m varies.

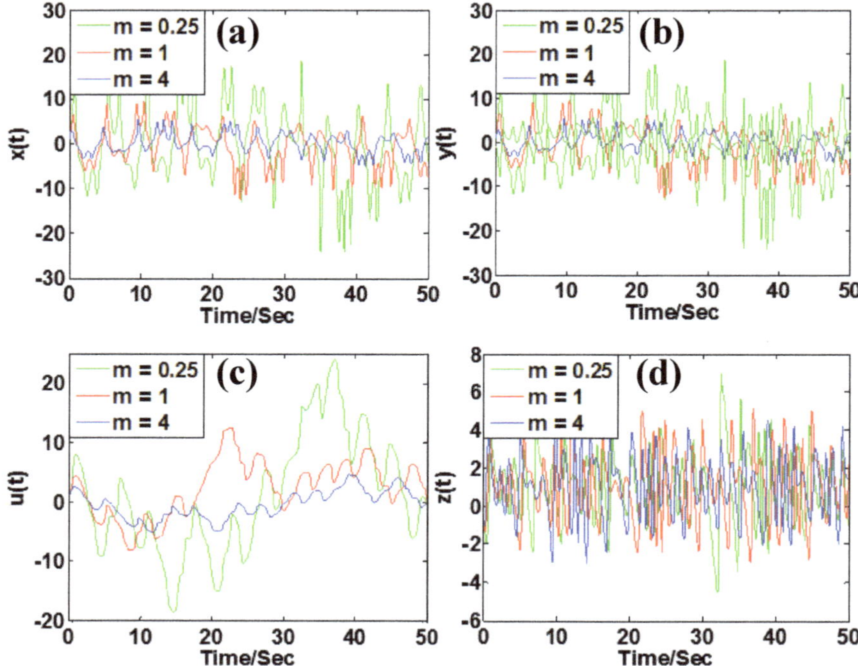

Figure 9. Rescaled variables in system (2) with $a = 5$, $b = 4$, $c = 1$, $k = 0.5$ under initial condition [1, −1, −1, 1]: (**a**) signal $x(t)$, (**b**) signal $y(t)$, (**c**) signal $u(t)$, (**d**) signal $z(t)$.

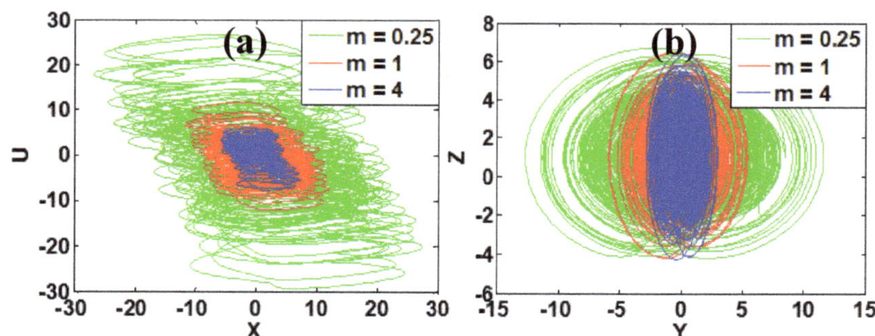

Figure 10. Phase trajectories of system (2) with $a = 5$, $b = 4$, $c = 1$, $k = 0.5$ under initial condition [1, −1, −1, 1]: (**a**) x-u, (**b**) y-z.

As we can see in Figure 11a, when the parameter m varies in [0, 5], the average of the absolute values of state variables x, y and u significantly decreases with an inverse proportion to m, while the average of signal z basically has no change. The corresponding Lyapunov exponent spectrum of parameter m varies in [0, 5] are shown in Figure 11b. It can be further proved that the parameter m of system (2) does not change the frequency of the signals.

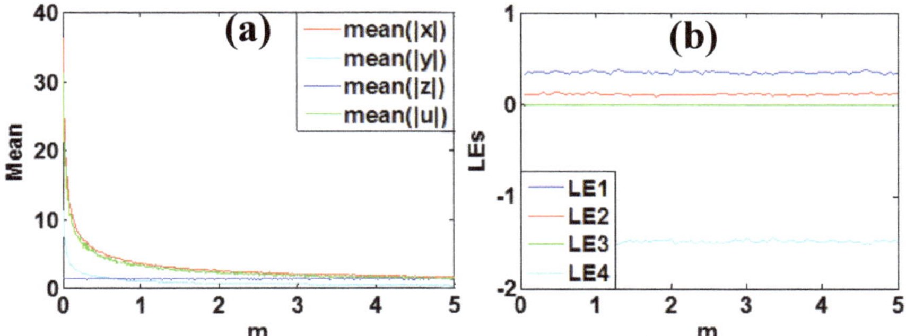

Figure 11. Dynamical evolution of system (2) with $a = 5, b = 4, c = 1, k = 0.5$ and initial condition $[1, -1, -1, 1]$: (**a**) average values of the absolute value of chaotic signals, (**b**) invariable Lyapunov exponents.

3.4. Offset Boosting

Since the derivative of a constant is zero, when a constant is added to a variable in a dynamical system, the system exhibits the same dynamics. To understand this, we turn back to the initial system (2). Here, we take the transformation: $u \to u - n$, which does not change the system equation but only leaves an additional constant in the first equation:

$$\begin{cases} \dot{x} = -ay - xz - u + n, \\ \dot{y} = -cx + xz, \\ \dot{z} = -b - mxy, \\ \dot{u} = kx - y. \end{cases} \tag{6}$$

When changing the variable u with $u - n$ (n is a constant), system (2) gives the same dynamics. Therefore, if the variable u does not show in the other equations in system (2), the introduced constant will give a boosting control of the variable u. The chaotic signal $u(t)$ can be revised from unipolar to bipolar or vice versa.

When $a = 5, b = 4, c = 1, k = 0.5$ and $m = 1$, the signal u is boosted from a bipolar to a unipolar one, which is indicated by the red and blue attractors in Figure 12a. The waveform of chaotic signal $u(t)$ is shown in Figure 12b. The change of parameter n causes the up and down translation of the signal $u(t)$. Some monostable systems have relatively large areas of basins of attraction; therefore, the initial conditions do not need to modify according to the variable which makes the offset control simpler, as shown in Figure 13.

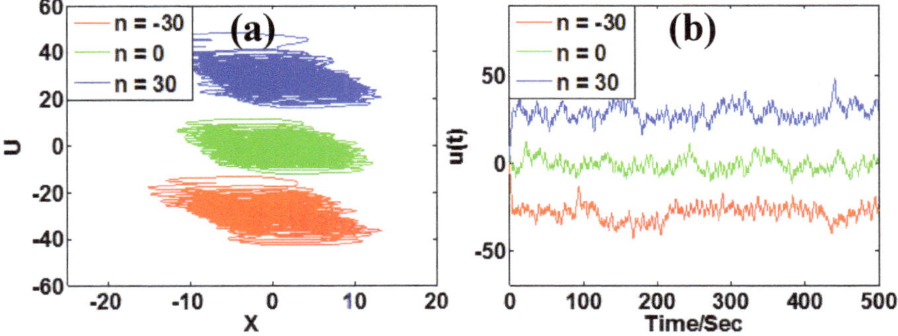

Figure 12. Typical chaotic oscillation of system (6) with $a = 5, b = 4, c = 1, k = 0.5, m = 1$ under initial condition $[1, -1, -1, 1]$: (**a**) phase trajectory in the plane of x-u, (**b**) waveform $u(t)$.

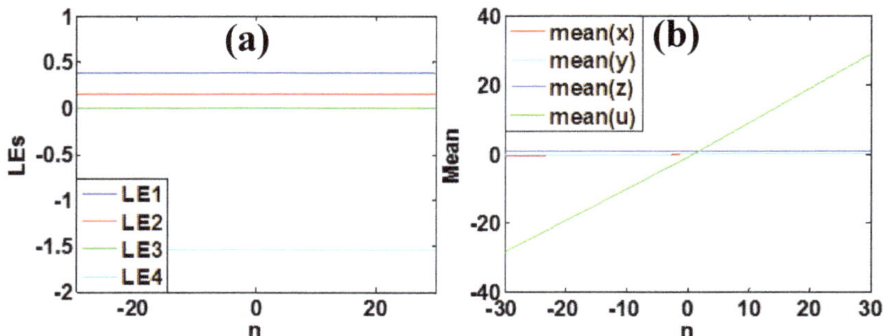

Figure 13. Dynamical evolution of system (6) with $a = 5, b = 4, c = 1, k = 0.5, m = 1$ under initial conditions [1, −1, −1, 1]: (**a**) Lyapunov exponent spectra of n, (**b**) average values of the hyperchaotic signal.

Here the offset of the variable u is boosted along the u-axis according to the constant n. When n is positive, u is moved in the positive direction, and negative n causes the opposite direction. When the boosting controller n is changed from −30 to 30, system (6) has the same Lyapunov exponents, which is shown in Figure 13. The average value of variable u changes linearly with the increase of parameter n, while others remain unchanged.

3.5. Mixed Geometric Control

More striking, parameter c almost has a positive correlation with the offset of signal z, almost without changing other signals, and also has a negative correlation with the amplitude of variable x and positive correlation with the amplitude of variable y. Figure 14 shows the typical phase trajectories and waveforms. Figure 15 shows the corresponding Lyapunov exponent spectra and average value of the x, y and z signals. Therefore, all in all, there are five parameters, a, b, c, m and n, rescaling the system variables, some of which are restricted in a specific region, as shown in Table 2.

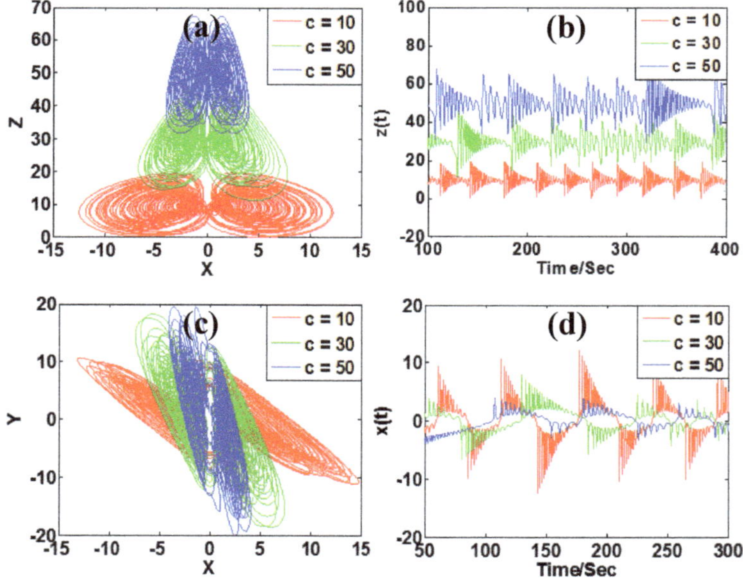

Figure 14. Typical chaotic oscillation of system (2) with $a = 5, b = 4, k = 0.5, m = 1$ under initial conditions [1, −1, −1, 1]: (**a**) phase trajectory in x-z, (**b**) signal $z(t)$, (**c**) phase trajectory in x-y, (**d**) signal $x(t)$.

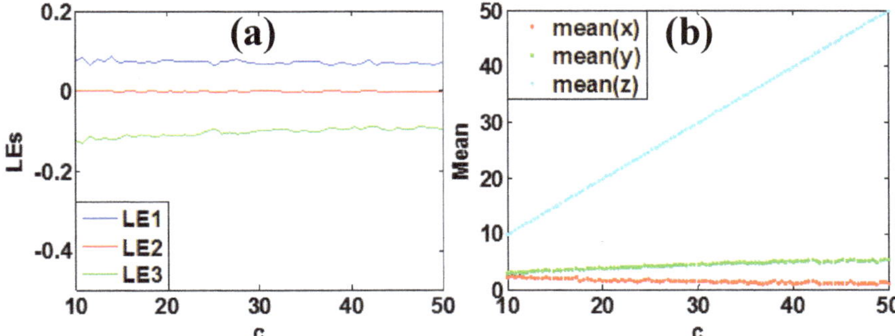

Figure 15. Dynamical evolution of system (2) with $a = 5$, $b = 4$, $k = 0.5$, $m = 1$ under initial conditions [1, −1, −1, 1]: (**a**) Lyapunov exponent spectra of c, (**b**) average values of the signals x, y and z.

Table 2. Five independent parameters in system (2) for geometric control.

Parameters	Execution Interval	Amplitude Control	Frequency Control	Offset Control
a	[6.9, 23.3]	positive control with x	positive	no
b	[3, 13]	positive control with x, y, z, u	positive	no
c	[10, 50]	positive control with y negative control with x	no	z
m	[0.1, 5]	Positive control with x, y, u	no	no
n	[−30, 30]	no	no	u

4. Bistability Analysis

In all the above analysis, we did not consider the multistability in each issue to simplify the demonstration. In fact, for the special structure of symmetry, coexisting attractors exist in their own basins of attraction in phase space. Specifically, for symmetrical systems, when the symmetry is broken, a pair of symmetrical coexisting attractors usually show up.

System (2) is a rotational symmetric system, which can be proved by the invariance of transformation $x \rightarrow -x, y \rightarrow -y, z \rightarrow z, u \rightarrow -u$. Symmetric systems are prone to show multistability due to the effect of broken symmetry. In general, predicting multistability seems not easy in theory. A common method to identify multistability is using the basins of attraction based on the ergodic initial conditions. Alternative methods can resort to non-bifurcation manipulation, in which a linear transformation is performed on the basin of attraction to generate a dynamical dispersion for a fixed initial condition, which can reveal different coexisting symmetrical pairs by generating different average values [40].

When offset boosting is introduced from the variables x and u,

$$\begin{cases} \dot{x} = -ay - (x+d)z - (u-d), \\ \dot{y} = (z-c)(x+d), \\ \dot{z} = -b - m(x+d)y, \\ \dot{u} = k(x+d) - y. \end{cases} \quad (7)$$

and the average values of variables x and u will change according to the offset control parameter d. The coexisting attractors are drawn into different areas of the basin since the basins of attraction of the coexisting symmetric pair of attractors also change according to the offset parameter, as shown in Figure 16. In Figure 16a, the averages of variables x and u are revised by the offset parameters, while the average values of variable y remains the same. From Figure 16b, the invariance of Lyapunov exponents indicates the same structure of the symmetric pair of coexisting attractors. The typical phase trajectories of the symmetrical attractors of the system (2) under a pair of opposite initial conditions are shown in Figure 17.

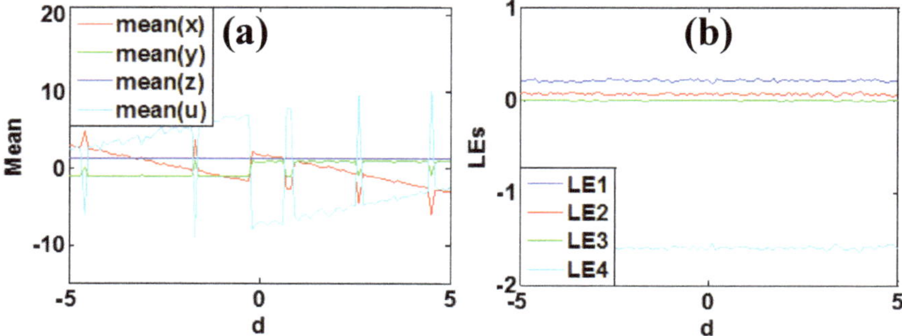

Figure 16. Dynamical behaviors of system (7) with $a = 5$, $b = 4$, $k = 0.5$, $m = 1$ and initial condition [1, −1, −1, 1], when parameter d varies in [−5, 5]: (**a**) average values of the signals, (**b**) Lyapunov exponents.

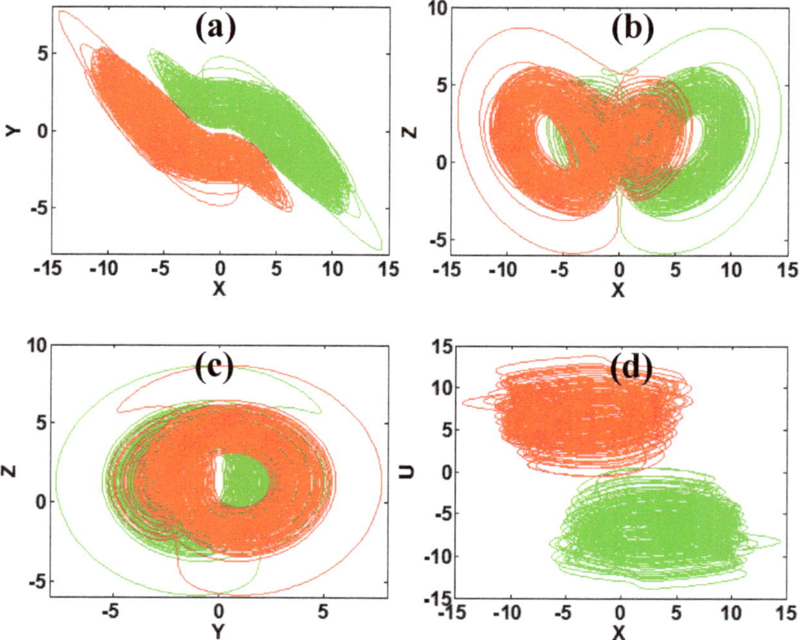

Figure 17. Coexisting symmetrical chaotic attractors of system (2) with $a = 5$, $b = 4$, $c = 1.3$, $k = 0.5$, $m = 1$ with initial conditions IC1 = (1, −1, −1, 1) (green); IC2 = (−1, 1, −1, −1) (red).

To further verify the multistability in system (2), the basin of attraction is shown in Figure 18, which has the predicted symmetry and complex fractal structure. To show the types of attractors more clearly and comprehensively, the similar chaotic attractors are presented using an identical color in the basin of attraction. It can be clearly seen that there are two areas in different colors in the basin. The corresponding Lyapunov exponents of the two attractors are (0.2137, 0.0623, 0, -1.5761), and the Kaplan-Yorke dimension is 3.1751.

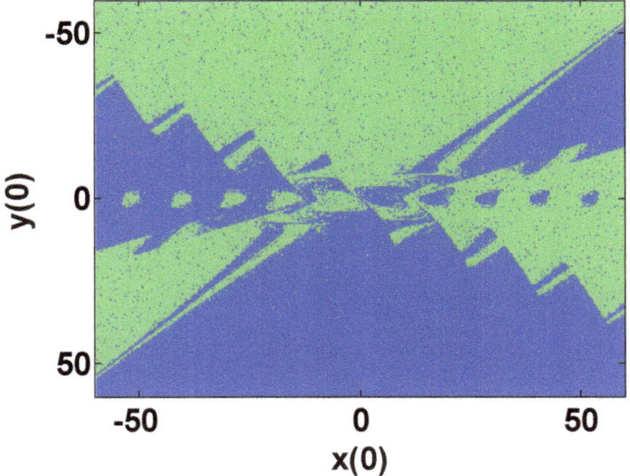

Figure 18. Basins of attraction of system (2) with $a = 5$, $b = 4$, $c = 1.3$, $k = 0.5$, $m = 1$ in plane of $z(0) = -1$ and $u(0) = 0$.

Both chaotic and hyperchaotic attractors show sensitivity to the initial condition, and furthermore, multistability and hyperchaos make the sensitivity more complicated. From two initial conditions in the same basin of attraction, even a slight difference results in great divergence in system (2), which is shown in Figure 19a. While from the two initial conditions in different basins of attraction, the slight difference leads to two separate phase trajectories, as shown in Figure 19b.

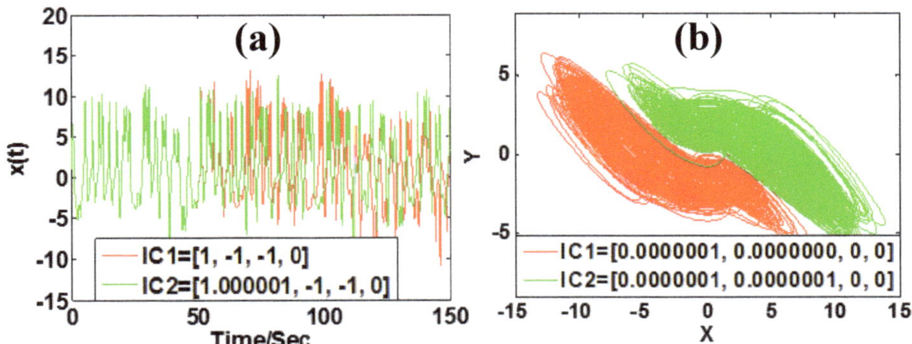

Figure 19. Dynamical behavior of system (2) with $a = 5$, $b = 4$, $k = 0.5$, $m = 1$ under different initial condition (**a**) $c = 1$; (**b**) $c = 1.3$.

5. Circuit Implementation

The analog circuit of system (2) is designed as shown in Figure 20 with the circuit equation:

$$\begin{cases} \dot{x} = -\dfrac{1}{R_1 C_1} y - \dfrac{1}{R_2 C_1} xz - \dfrac{1}{R_3 C_1} u \\ \dot{y} = -\dfrac{1}{R_4 C_2} x + \dfrac{1}{R_5 C_2} xz \\ \dot{z} = -\dfrac{1}{R_6 C_3} + \dfrac{1}{R_7 C_3} xy \\ \dot{u} = \dfrac{1}{R_8 C_4} x - \dfrac{1}{R_9 C_4} y \end{cases} \tag{8}$$

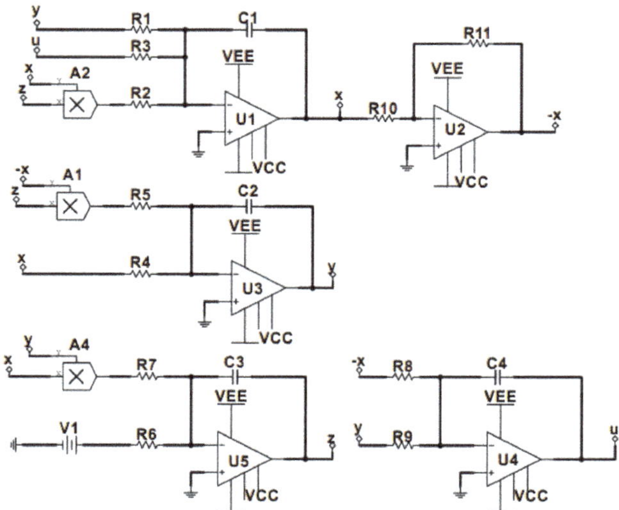

Figure 20. Circuit schematic of system (8).

Totally, the hyperchaotic circuit consists of four channels, which contain the integration, addition, subtraction, and nonlinear operations. The circuit is powered by 18V. The variables x, y, z and u in system (2) are the state voltages of the capacitors in four channels. The corresponding circuit element parameters can be selected as $C_1 = C_2 = C_3 = C_4 = 10nF$, $R_2 = R_5 = R_7 = 4k\Omega$, $R_3 = R_4 = R_9 = 40k\Omega$, $R_1 = 8k\Omega$, $R_6 = 100k\Omega$, $R_8 = 80k\Omega$, $R_{10} = R_{11} = 10k\Omega$. Here, a common time scale of 1000 is applied for better demonstration in the oscilloscope. The phase trajectories in circuit (8) under amplitude control are shown in Figure 21. Circuit experiment proves that the parameter m rescales the amplitude of x, y and u. Symmetric attractors are shown in Figure 22.

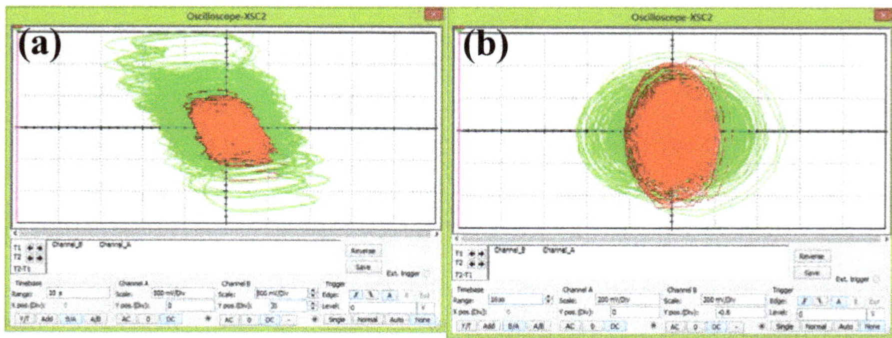

Figure 21. Circuit simulation of system (8) with $a = 5$, $b = 4$, $c = 1.3$, $k = 0.5$, $m = 1$ (green), $m = 4$ (red) under initial condition $[1, -1, -1, 1]$: (**a**) x-u plane, (**b**) y-z plane.

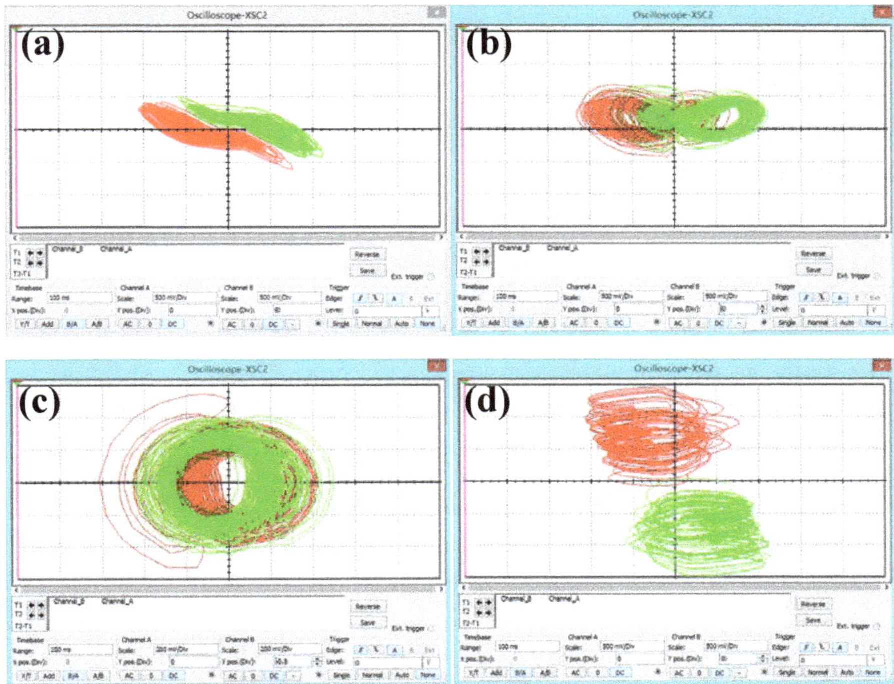

Figure 22. Circuit simulation of symmetric attractors in system (8) with $a = 5$, $b = 4$, $c = 1.3$, $k = 0.5$, $m = 1$ under initial conditions IC1= (1, −1, −1, 1)(green), IC2= (−1, 1, −1, −1)(red): (**a**) *x-y* plane, (**b**) *x-z* plane, (**c**) *y-z* plane, (**d**) *x-u* plane.

6. Discussion and Conclusions

A hidden hyperchaotic attractor is found, which has the property of amplitude control and offset boosting. The proposed system shares a symmetric structure, where one can find an independent knob for amplitude control. An extra introduced dimension leaves an opportunity for attractor shifting in phase space by an independent controller. Broken symmetry induced bistability is also well addressed in this work. All the coexisting symmetric attractors governed by the basin of attraction can be rescaled by the non-bifurcation parameter. Numerical and circuit simulation agree with each other proving the properties found in the hyperchaotic system.

Author Contributions: Conceptualization, C.L.; Data curation, X.Z.; formal analysis, X.Z.; funding acquisition, C.L., Z.L. and C.T.; investigation, C.L., X.Z. and T.L.; methodology, C.L.; project administration, C.L.; resources, C.L.; software, X.Z.; supervision, C.L.; validation, C.L., X.Z., T.L., Z.L. and C.T.; visualization, X.Z.; writing—original draft, X.Z.; writing—review and editing, C.L., Z.L. and C.T. All authors have read and agreed to the published version of the manuscript.

Funding: This work was supported financially by the National Natural Science Foundation of China (Grant No. 61871230, 51974045), the Natural Science Foundation of Jiangsu Province (Grant No.: BK20181410), and a project funded by the Priority Academic Program Development of Jiangsu Higher Education Institutions.

Conflicts of Interest: The authors declare no conflicts of interest.

References

1.   Li, C.; Sprott, J.C. Amplitude control approach for chaotic signals. *Nonlinear Dyn.* **2013**, *73*, 1335–1341. [CrossRef]

2. Li, C.; Sprott, J.C. Finding coexisting attractors using amplitude control. *Nonlinear Dyn.* **2014**, *78*, 2059–2064. [CrossRef]

3. Chen, H.; Bayani, A.; Akgul, A.; Jafari, M.-A.; Pham, V.-T.; Wang, X.; Jafari, S. A flexible chaotic system with adjustable amplitude, largest Lyapunov exponent, and local Kaplan–Yorke dimension and its usage in engineering applications. *Nonlinear Dyn.* **2018**, *92*, 1791–1800. [CrossRef]

4. Wang, C.; Liu, X.; Xia, H. Multi-piecewise quadratic nonlinearity memristor and its 2N-scroll and 2N + 1-scroll chaotic attractors system. *Chaos Interdiscip. J. Nonlinear Sci.* **2017**, *27*, 033114. [CrossRef] [PubMed]

5. Hu, W.; Akgul, A.; Li, C.; Zheng, T.; Li, P. A switchable chaotic oscillator with two amplitude-frequency controllers. *J. Circuits Syst. Comput.* **2017**, *26*, 1750158. [CrossRef]

6. Li, C.; Sprott, J.C.; Akgul, A.; Herbert, H.C.; Iu, H.H.C.; Zhao, Y. A new chaotic oscillator with free control. *Chaos* **2017**, *27*, 083101. [CrossRef] [PubMed]

7. Li, C.; Sprott, J.C.; Yuan, Z.; Li, H. Constructing chaotic systems with total amplitude control. *Int. J. Bifurc. Chaos* **2015**, *25*, 1530025. [CrossRef]

8. Li, C.; Sprott, J.C.; Xing, H. Constructing chaotic systems with conditional symmetry. *Nonlinear Dyn.* **2017**, *87*, 1351–1358. [CrossRef]

9. Li, C.; Sprott, J.C. Variable-boostable chaotic flows. *Optik—Int. J. Light Electron Opt.* **2016**, *127*, 10389–10398. [CrossRef]

10. Leonov, G.A.; Vagaitsev, V.I.; Kuznetsov, N.V. Localization of hidden Chua's attractors. *Phys. Lett. A* **2011**, *375*, 2230. [CrossRef]

11. Leonov, G.A.; Vagaitsev, V.I.; Kuznetsov, N.V. Hidden attractor in smooth Chua systems. *Phys. D* **2012**, *241*, 1482. [CrossRef]

12. Rocha, R.; Ruthiramoorthy, J.; Kathamuthu, T. Memristive oscillator based on Chua's circuit: stability analysis and hidden dynamics. *Nonlinear Dyn.* **2017**, *88*, 2577–2587. [CrossRef]

13. Bao, B.; Xu, Q.; Bao, H.; Chen, M. Extreme multistability in a memristive circuit. *Electron. Lett.* **2016**, *52*, 1008–1010. [CrossRef]

14. Lai, Q.; Nestor, T.; Kengne, J.; Zhao, X. Coexisting attractors and circuit implementation of a new 4D chaotic system with two equilibria. *Chaos Solitons Fractals* **2018**, *107*, 92–102. [CrossRef]

15. Wang, G.Y.; Zheng, Y.; Liu, J.B. A hyperchaotic Lorenz attractor and its circuit implementation. *Acta Phys. Sin.* **2007**, *56*, 3113–3120.

16. Jafari, S.; Sprott, J.C. Simple chaotic flows with a line equilibrium. *Chaos Solitons Fractals* **2013**, *57*, 79–84. [CrossRef]

17. Bao, H.; Wang, N.; Bao, B.C.; Chen, M.; Jin, P.P.; Wang, G.Y. Initial condition dependent dynamics and transient period in memristor-based hypogenetic jerk system with four line equilibria. *Commun. Nonlinear Sci.* **2018**, *57*, 264–275. [CrossRef]

18. Jafari, S.; Sprott, J.C.; Molaie, M. A simple chaotic flow with a plane of equilibria. *Int. J. Bifurc. Chaos* **2016**, *26*, 1650098. [CrossRef]

19. Jafari, S.; Sprott, J.C. Elementary quadratic chaotic flows with no equilibria. *Phys. Lett. Sect. A Gen. Atomic Solid State Phys.* **2013**, *377*, 699–702. [CrossRef]

20. Bao, B.C.; Bao, H.; Wang, N.; Chen, M.; Xu, Q. Hidden extreme multistability in memristive hyperchaotic system. *Chaos Solitons Fractals* **2017**, *94*, 102–111. [CrossRef]

21. Munmuangsaen, B.; Srisuchinwong, B. A hidden chaotic attractor in the classical lorenz system. *Chaos Solitons Fractals* **2018**, *107*, 61–66. [CrossRef]

22. Lai, Q.; Chen, S.M. Research on a new 3d autonomous chaotic system with coexisting attractors. *Optik—Int. J. Light Electron Opt.* **2016**, *127*, 3000–3004. [CrossRef]

23. Wang, C.; Wei, Z.; Yu, P.; Zhang, W.; Yao, M. Study of hidden attractors, multiple limit cycles from hopf bifurcation and boundedness of motion in the generalized hyperchaotic rabinovich system. *Nonlinear Dyn.* **2015**, *82*, 131–141.

24. Zhou, L.; Wang, C.; Zhou, L. A novel no-equilibrium hyperchaotic multi-wing system via introducing memristor. *Int. J. Circ. Theor. App.* **2018**, *46*, 84–98. [CrossRef]

25. Wang, Z.; Cang, S.; Ochola, E.O.; Sun, Y. A hyperchaotic system without equilibrium. *Nonlinear Dyn.* **2012**, *69*, 531–537. [CrossRef]

26. Chlouverakis, K.E.; Sprott, J.C. Chaotic hyperjerk systems. *Chaos Solitons Fractals* **2006**, *28*, 739–746. [CrossRef]

27. Yuan, F.; Wang, G.; Wang, X. Extreme multistability in a memristor- based multi-scroll hyperchaotic system. *Chaos* **2016**, *26*, 073107. [CrossRef]
28. Ruan, J.Y.; Sun, K.H.; Mou, J. Memristor-based Lorenz hyper-chaotic system and its circuit implementation. *Acta Phys. Sin.* **2016**, *65*, 190502.
29. Lai, Q.; Guan, Z.; Wu, Y.; Liu, F.; Zhang, D. Generation of multi-wing chaotic attractors from a lorenz-like system. *Int. J. Bifurc. Chaos* **2013**, *23*, 1650177. [CrossRef]
30. Si, G.; Cao, H.; Zhang, Y. A new four dimensional hyperchaotic Lorenz system and its adaptive control. *Chin. Phys. B* **2011**, *20*, 010509. [CrossRef]
31. Wang, H.; Cai, G.; Miao, S.; Tian, L. Nonlinear feedback control of a novel hyperchaotic system and its circuit implementation. *Chin. Phys. B* **2010**, *19*, 030509.
32. Zhou, L.; Wang, C.; Zhou, L.L. Generating Four-Wing Hyperchaotic Attractor and Two-Wing, Three-Wing, and Four-Wing Chaotic Attractors in 4D Memristive System. *Int. J. Bifurc. Chaos* **2017**, *27*, 1750027. [CrossRef]
33. Pham, V.-T.; Volos, C.; Gambuzza, L.V A memristive hyperchaotic system without equilibrium. *Sci. World J.* **2014**, *2014*, 368986. [CrossRef] [PubMed]
34. Xiao, J.; Ma, Z.Z.; Yang, Y.H. Dual synchronization of fractional-order chaotic systems via a linear controller. *Sci. World J.* **2013**, *2013*, 159194. [CrossRef] [PubMed]
35. Zhou, P.; Bai, R. One adaptive synchronization approach for fractional-order chaotic system with fractional-order. *Sci. World J.* **2014**, *2*, 490364. [CrossRef] [PubMed]
36. Zhang, X.; Wang, C.H. Multiscroll hyperchaotic system with hidden attractors and its circuit implementation. *Int. J. Bifurc. Chaos* **2019**, *29*, 1950117. [CrossRef]
37. Zhang, S.; Zeng, Y.C.; Li, Z.J.; Wang, M.J.; Xiong, L. Generating one to four-wing hidden attractors in a novel 4D no-equilibrium chaotic system with extreme multistability. *Chaos* **2018**, *28*, 013113. [CrossRef]
38. Wang, Z.L.; Ma, J.; Cang, S.J.; Wang, Z.H.; Chen, Z.Q. Simplified hyper-chaotic systems generating multi-wing non-equilibrium attractor. *Optik* **2016**, *127*, 2424–2431. [CrossRef]
39. Cang, S.; Wang, Z.; Chen, Z.; Jia, H. Analytical and numerical investigation of a new lorenz-like chaotic attractor with compound structures. *Nonlinear Dyn.* **2014**, *75*, 745–760. [CrossRef]
40. Li, C.; Wang, X.; Chen, G. Diagnosing multistability by offset boosting. *Nonlinear Dyn.* **2017**, *90*, 1334–1341. [CrossRef]

Article

New Chaotic Systems with Two Closed Curve Equilibrium Passing the Same Point: Chaotic Behavior, Bifurcations, and Synchronization

Xinhe Zhu [1] and Wei-Shih Du [2,*]

[1] School of Mathematical Sciences, Tianjin Polytechnic University, Tianjin 300387, China
[2] Department of Mathematics, National Kaohsiung Normal University, Kaohsiung 82444, Taiwan
* Correspondence: wsdu@mail.nknu.edu.tw

Received: 2 June 2019; Accepted: 16 July 2019; Published: 26 July 2019

Abstract: In this work, we introduce a chaotic system with infinitely many equilibrium points laying on two closed curves passing the same point. The proposed system belongs to a class of systems with hidden attractors. The dynamical properties of the new system were investigated by means of phase portraits, equilibrium points, Poincaré section, bifurcation diagram, Kaplan–Yorke dimension, and Maximal Lyapunov exponents. The anti-synchronization of systems was obtained using the active control. This study broadens the current knowledge of systems with infinite equilibria.

Keywords: chaos; bifurcation; closed curve equilibrium; synchronization

---

1. Introduction

Chaotic systems have been widely studied and used in various practical fields by mathematicians, physicists, scientists, and engineers in the past four decades; see [1–4] and the references therein. Many chaotic systems with different shapes of attractors have been reported, such as chaotic systems with butterfly attractors (see, e.g., [5]) and systems with multiscroll chaotic attractors (see, e.g., [6]). Recent developments include some different types of chaotic systems with no equilibrium points (see, e.g., [7]), with a single stable equilibrium (see, e.g., [8]), with a line of equilibrium points (see, e.g., [9]), with a circular equilibrium (see, e.g., [10]), with circle and square equilibrium (see, e.g., [11]), with rounded square loop equilibrium (see, e.g., [12]), and with different closed curve equilibrium (see, e.g., [13]). Furthermore, it has also been applied in many different areas including information processing (see, e.g., [14]) and chaotic masking communication (see, e.g., [15]).

According to a new classification of chaotic dynamics [16], there are two kinds of attractors: self-excited attractors and hidden attractors. Recall that an attractor is referred to as being *self-excited* if its basin of attraction intersects any arbitrarily small open neighborhood of an equilibrium, otherwise it is called a *hidden attractor*. The basin of attraction for a hidden attractor is not connected with any unstable fixed point. For example, hidden attractors are observed in the systems without fixed points, with no unstable fixed points, or with one stable fixed point. A system with infinitely many equilibrium points can be classified as one system with hidden attractors, for the reason that we do not know which part of the equilibria may be used to localize the hidden attractors, which should be treated in detail (see, e.g., [17]). Recent important investigations and developments in the study of chaotic dynamical systems with practical problems and challenges have been asked to satisfy at least one of the following criteria as Sprott mentioned in [18]: *(S1) The system should credibly model some important unsolved problem in nature and shed light on that problem; (S2) the system should exhibit some behavior previously unobserved; (S3) the system should be simpler than all other known examples exhibiting the observed behavior.* An important ongoing research topic is dedicated to discovering and developing new and novel chaotic systems with different shapes of closed curve equilibrium.

The main goal of this work is to present a new system with infinitely many equilibrium points arranged on two closed curves passing the same point, which extends the general knowledge about such systems. Our new chaotic system (see Section 2 below for details) is meaningful for satisfying two of the three conditions, (S1), (S2), and (S3), as well as there being a certain novelty value in this work. In Section 2, some dynamical properties of the proposed system, which have been studied using a bifurcation diagram, phase portrait, Poincaré section, maximal Lyapunov exponents, and Kaplan–Yorke dimension, are presented. The ability of anti-synchronization of the new system is also discussed in Section 3.

2. A New Family with Two Closed Curve Equilibrium

In this work, motivated by the known dynamic systems mentioned above, we study the following general model given by

$$\begin{aligned}
\dot{u} &= w, \\
\dot{v} &= -wf(u, v, w), \\
\dot{w} &= g(u, v),
\end{aligned} \tag{1}$$

where u, v and w are three state variables, $f(u, v, w)$ and $g(u, v)$ are two nonlinear functions. The equilibrium points in model (1) can be obtained by calculating

$$\begin{aligned}
w &= 0, \\
-wf(u, v, w) &= 0, \\
g(u, v) &= 0.
\end{aligned} \tag{2}$$

It is obtained that the equilibrium points locate on a curve described by $g(u, v) = 0$ in the plane $w = 0$. In fact, by selecting appropriate functions f and g, some known systems, both chaotic and with different closed curve equilibrium, can be constructed.

(*Example A*)

Take $f(u, v, w) = \alpha v + \beta v^2 + uw$ and $g(u, v) = u^2 + v^2 - 1$, then model (1) will deduce the following system

$$\begin{aligned}
\dot{u} &= w, \\
\dot{v} &= -w(\alpha v + \beta v^2 + uw), \\
\dot{w} &= u^2 + v^2 - 1,
\end{aligned} \tag{3}$$

which was introduced and studied by Gotthans, Sprott, and Petrzela [11] in 2016. The chaotic systems (3) has circle equilibrium (see Figure 1).

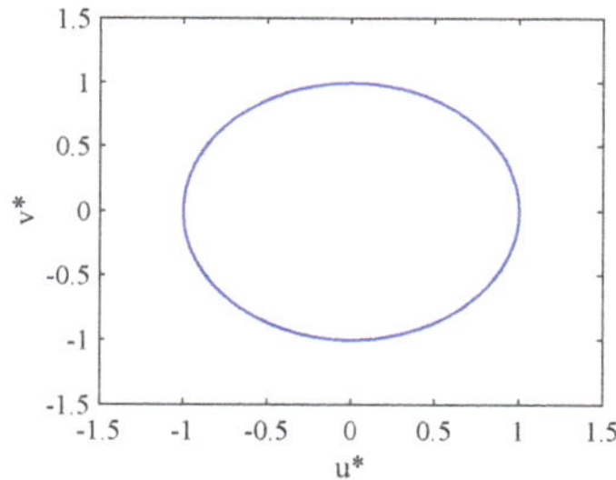

Figure 1. The circle-shape of equilibrium points of system (3) in the plane $w = 0$.

(*Example B*)

Let $f(u, v, w) = \alpha v + \beta v^2 + uw$ and $g(u, v) = u^2 - |uv| + v^2 - 1$. Then, model (1) deduces the following system:

$$
\begin{aligned}
\dot{u} &= w, \\
\dot{v} &= -w(\alpha v + \beta v^2 + uw), \\
\dot{w} &= u^2 - |uv| + v^2 - 1,
\end{aligned}
\tag{4}
$$

which was established by Wang, Pham, and Volos [19] in 2017. The chaotic system (4) has cloud-shaped curve equilibrium, as shown in Figure 2.

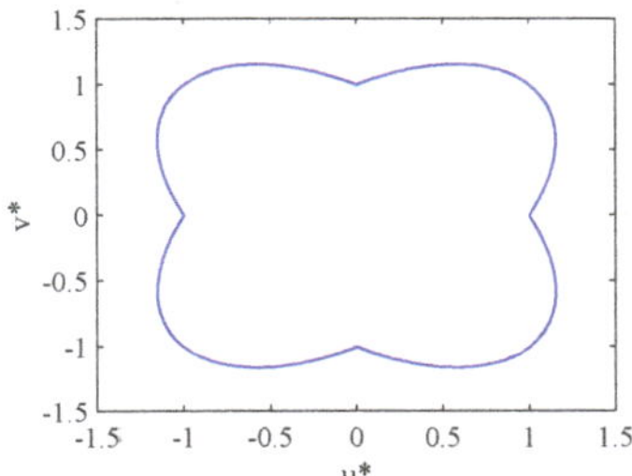

Figure 2. The cloud-shape of equilibrium points of system (4) in the plane $w = 0$.

(*Example C*)

Very recently, Zhu and Du [13] discovered and studied a new family of systems with different equilibrium (as shown in Figure 3) described by

$$
\begin{aligned}
\dot{u} &= w, \\
\dot{v} &= -w(\alpha v + \beta v^2 + uw), \\
\dot{w} &= |u|^k + |v|^k - 1,
\end{aligned}
\tag{5}
$$

where $k \in \mathbb{N}$. In fact, the chaotic system (5) can be obtained by putting $f(u, v, w) = \alpha v + \beta v^2 + uw$ and $g(u, v) = |u|^k + |v|^k - 1$ into model (1). In [13], Zhu and Du analyzed the dynamical properties of their proposed systems using the methods of equilibrium points, eigenvalues, phase portraits, maximal Lyapunov exponents, and Kaplan–Yorke dimension; see [13] for more details.

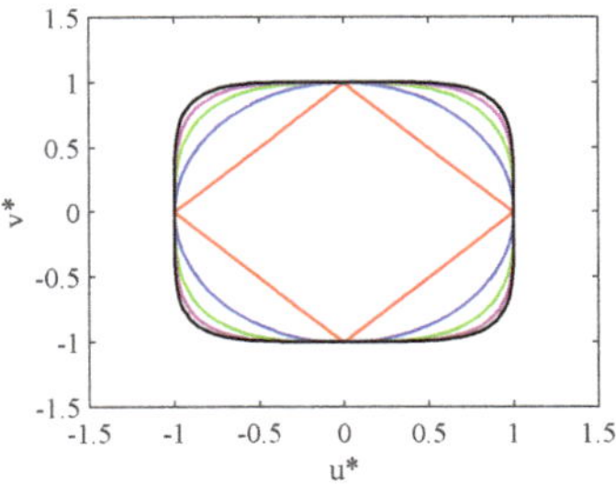

Figure 3. Different shapes of equilibrium points of system (5), $k = 1, 2, 3, 4, 5$, from the interior to the outside, respectively, in the plane $w = 0$.

The results established in [11,13,19] are very important for indicating the existence of chaotic systems with different shapes of equilibrium points (see Table 1). Note that the first two equations of the systems proposed in [13,19] are the same as in [11]. The difference is the third equation. When we choose a different third equation, we can get different systems to display new features, such as different shapes of equilibrium point and other dynamic properties.

Table 1. Chaotic systems with infinitely many equilibrium points.

System	Equilibrium	Closed Curve Equilibrium	Paper
Gotthans, Sprott, and Petrzela	$u^2 + v^2 - 1 = 0$	Circle	[11]
Zhu and Du	$|u|^k + |v|^k - 1 = 0$	Circle, Square, etc	[13]
Wang, Pham, and Volos	$u^2 - |uv| + v^2 - 1 = 0$	Cloud-shaped	[19]
New system (see below)	$u^2 - |u| + |v| + v^2 = 0$	Eye-shaped	This work

To the best of our knowledge, there is no paper devoted to the study of chaotic dynamical systems with eye-shaped curve equilibrium. Therefore, this study is an important ongoing research topic. In this paper, motivated and inspired by this, two functions, $f(u,v,w)$ and $g(u,v)$, are chosen in the following forms

$$
\begin{aligned}
f(u,v,w) &= \alpha v + \beta v^2 + uw, \\
g(u,v) &= u^2 - |u| + |v| + v^2,
\end{aligned}
\tag{6}
$$

where α and β are two positive parameters. Substituting (6) into system (1), our new system is described as

$$
\begin{aligned}
\dot{u} &= w, \\
\dot{v} &= -w(\alpha v + \beta v^2 + uw), \\
\dot{w} &= u^2 - |u| + |v| + v^2.
\end{aligned}
\tag{7}
$$

It is verified that system (7) has infinitely many equilibrium points (u^*, v^*). These equilibrium points are located on the curve in the coordinate plane described by

$$
(u^*)^2 - |u^*| + |v^*| + (v^*)^2 = 0.
\tag{8}
$$

It means that the new system (7) has eye-shaped curve equilibrium as shown in Figure 4. Note that the eye-shaped curve is different from some other shapes reported, such as line, square, circle, or cloud-shaped [11,19], and is symmetric about the u-axis, v-axis, and origin. Furthermore, system (7) has hidden attractors [17]. Above all, investigating system (7) will strengthen our understanding of hidden attractors.

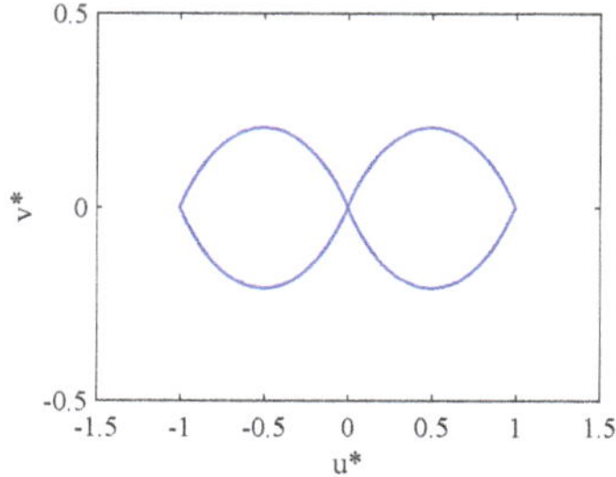

Figure 4. The eye-shape of equilibrium points of system (7) in the plane $w = 0$.

For $\alpha = 5, \beta = 30$, and initial conditions $(0.06, 0.01, 0.01)$, the new system (7) has chaotic attractors (see Figures 5 and 6). For the simulation, we used the Wolf et al. method to calculate the Lyapunov exponents [20], the time of computation was 1000, and we obtained the Lyapunov exponents $(0.0424, 0, -0.2484)$. The method of Wolf et al. is rooted conceptually in a previously developed technique that could only be applied to analytically defined model systems to monitor the long-term growth rate of small volume elements in an attractor. In addition, the corresponding Kaplan–Yorke dimension of system (7) is 2.1707. Poincaré return maps are often used to transform complicated behavior of a dynamic system in phase space to discrete maps in a lower dimensional space to reveal the complicated behaviors. Poincaré return maps corresponding with phase portraits in Figure 6 are presented in Figure 7; there are some dense points in the Poincaré section, and it can be determined that the motion is a chaotic state. These results reveal that the system is chaotic.

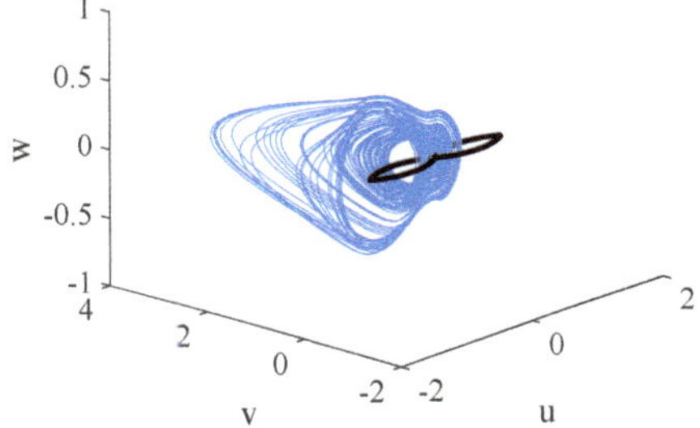

Figure 5. 3D view of the chaotic attractor and eye-shape of equilibrium points located in the plane $w = 0$ of system (7) for $\alpha = 5, \beta = 30$.

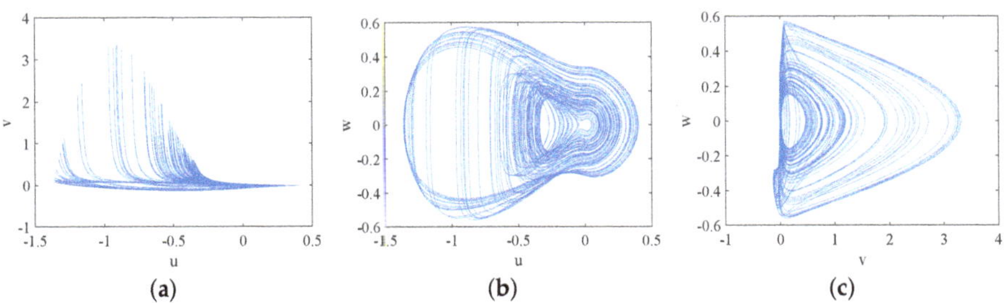

Figure 6. The projection of the trajectory of system (7) in (**a**) u-v plane, (**b**) u-w plane, (**c**) v-w plane for $\alpha = 5, \beta = 30$.

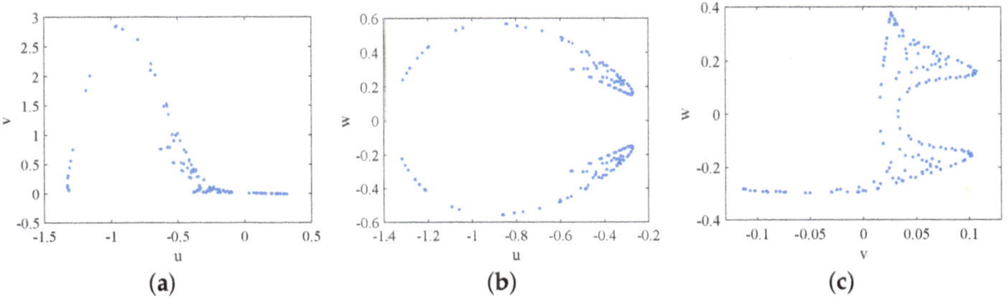

Figure 7. The Poincaré section of system (7) for (**a**) $z = 0.2$, (**b**) $y = 0.2$, (**c**) $x = -0.2$ for $\alpha = 5, \beta = 30$.

Gradually changing the value of the parameter β or α, the bifurcation plot of the system can be discovered in Figure 8. Figures 9 and 10 reveal the diagram of Maximal Lyapunov Exponents and the diagram of Kaplan–Yorke dimension of system (7) for $\alpha = 5, \beta \in [28, 48]$, respectively.

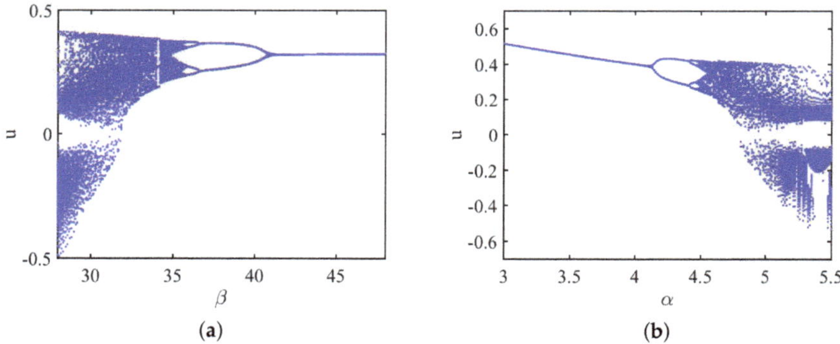

(a)

(b)

Figure 8. Bifurcation plot of system (7) for (**a**) $\alpha = 5, \beta \in [28, 48]$ and (**b**) $\beta = 30, \alpha \in [3, 5.5]$.

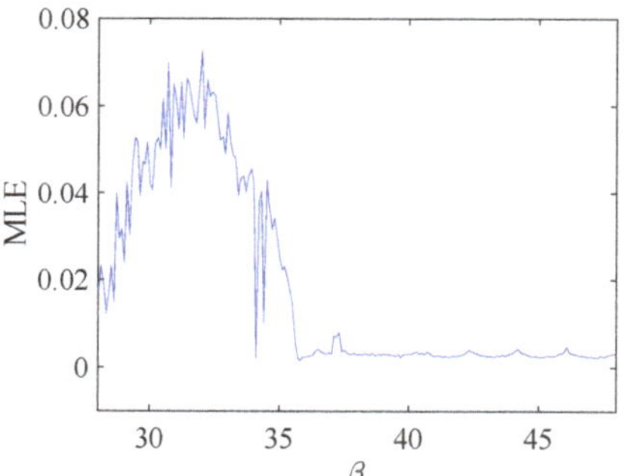

Figure 9. Maximal Lyapunov Exponents spectrum of system (7) for $\alpha = 5, \beta \in [28, 48]$.

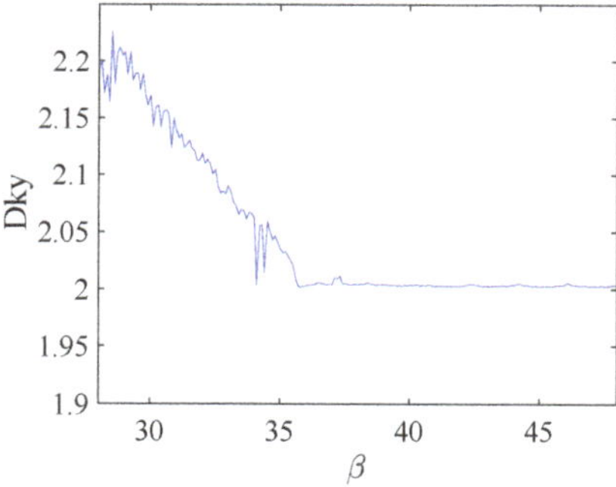

Figure 10. Kaplan–Yorke dimension of system (7) for $\alpha = 5, \beta \in [28, 48]$.

The new system with eye-shaped equilibrium has periodic behavior in the range $36 \leq \beta \leq 48$. For instance, the system can display period-1 behavior for $\alpha = 5, \beta = 45$, period-2 behavior for $\alpha = 5, \beta = 38$, and period-4 behavior for $\alpha = 5, \beta = 36$ (see Figure 11a–c, respectively).

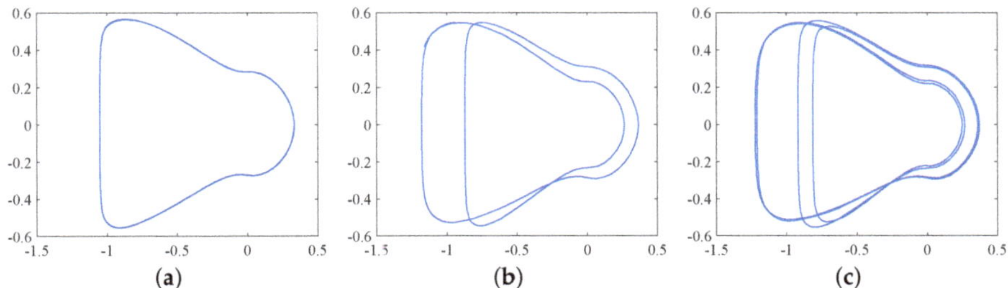

Figure 11. Periodic behavior of system (7) in the u-w plane: (**a**) period-1 ($\beta = 45$), (**b**) period-2 ($\beta = 38$), (**c**) period-4 ($\beta = 36$).

3. Anti-Synchronization of New Systems

Synchronization of chaos is a phenomenon that may occur when two, or more, dissipative chaotic systems are coupled. Since the pioneering work of Pecora and Carroll related to synchronization in chaotic systems [21], some methods of chaotic synchronization have been presented related to complete, generalized, lag, and imperfect phase synchronization [22]. Many papers on applications of chaos synchronization for cryptographic [23], kinetics [24], physiology [25], neural networks [26], and economics [27] have appeared.

In the following, we consider the anti-synchronization of the systems with eye-shaped equilibrium related to the driver-response system. The driver system with eye-shaped closed curve equilibrium is as follows:

$$\dot{u} = w,$$
$$\dot{v} = -w(\alpha v + \beta v^2 + uw),$$
$$\dot{w} = u^2 - |u| + |v| + v^2,$$

where u, v, and w are are three state variables, and the value of $\alpha = 5, \beta = 30$.

The response system is described as

$$\dot{u}_1 = w_1 + h_1,$$
$$\dot{v}_1 = -w_1(\alpha v_1 + \beta v_1^2 + u_1 w_1) + h_2, \tag{9}$$
$$\dot{w}_1 = u_1^2 - |u_1| + |v_1| + v_1^2 + h_3,$$

where the control is $\mathbf{h} = [h_1, h_2, h_3]^T$.

In order to reveal the difference between the driver system (7) and the response system (9), the state errors can be defined as

$$e_1 = u + u_1,$$
$$e_2 = v + v_1, \tag{10}$$
$$e_3 = w + w_1,$$

and we obtain

$$\dot{e}_1 = \dot{u} + \dot{u}_1,$$
$$\dot{e}_2 = \dot{v} + \dot{v}_1, \tag{11}$$
$$\dot{e}_3 = \dot{w} + \dot{w}_1.$$

Combining (7), (9), (10), and (11), we get the state errors system

$$
\begin{aligned}
\dot{e}_1 &= e_3 + h_1, \\
\dot{e}_2 &= -w(\alpha v + \beta v^2 + uw) - w_1(\alpha v_1 + \beta v_1^2 + u_1 w_1) + h_2, \\
\dot{e}_3 &= u^2 - |u| + |v| + v^2 + u_1^2 - |u_1| + |v_1| + v_1^2 + h_3.
\end{aligned}
\tag{12}
$$

We choose the control proposed by

$$
\begin{aligned}
h_1 &= -e_3 - k_1 e_1, \\
h_2 &= w(\alpha v + \beta v^2 + uw) + w_1(\alpha v_1 + \beta v_1^2 + u_1 w_1) - k_2 e_2, \\
h_3 &= -u^2 + |u| - |v| - v^2 - u_1^2 + |u_1| - |v_1| - v_1^2 - k_3 e_3,
\end{aligned}
\tag{13}
$$

where $k_i > 0$ $(i = 1, 2, 3)$ are the positive gain constants used to control the rate of anti-synchronization. By substituting (12) into (11), we get the state errors system

$$
\begin{aligned}
\dot{e}_1 &= -k_1 e_1, \\
\dot{e}_2 &= -k_2 e_2, \\
\dot{e}_3 &= -k_3 e_3.
\end{aligned}
\tag{14}
$$

Obviously, the eigenvalues $(-k_1, -k_2, -k_3)$ of the Jacobian matrix of the state errors system are negative. Then, the complete anti-synchronization between the driver system (7) and the response system (9) is proved.

In numerical simulation, we assume the initial values of the driver system (7) and the response system (9) to be

$$
\begin{aligned}
u(0) &= 0.06, \\
v(0) &= 0.01, \\
w(0) &= 0.01, \\
u_1(0) &= -0.20, \\
v_1(0) &= -0.09, \\
w_1(0) &= 0.07.
\end{aligned}
\tag{15}
$$

Then, the initial values of the state errors system (12) are

$$
\begin{aligned}
e_1(0) &= 0.40, \\
e_2(0) &= -0.08, \\
e_3(0) &= 0.08.
\end{aligned}
\tag{16}
$$

The positive gain constants here are selected as $k_1 = k_2 = k_3 = 3$. It is obvious in Figure 12 that there exists anti-synchronization of the respective states of the new systems with two closed curve equilibrium (7) and (9). The time history of the synchronization errors e_1, e_2, e_3 is shown in Figure 13 which plots the anti-synchronization of the driver-response system.

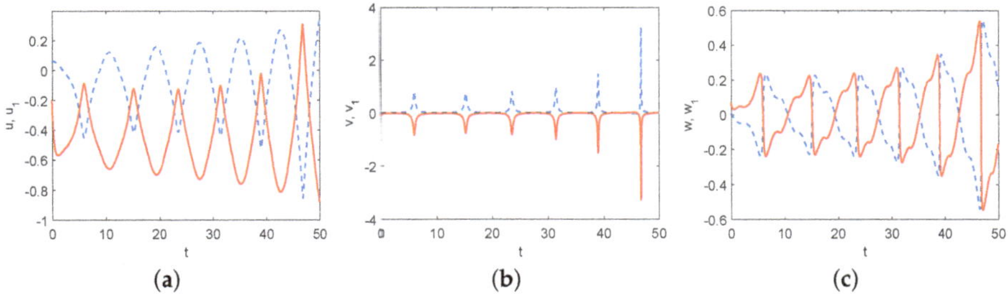

Figure 12. Anti-synchronization of the driver-response system: (**a**) u, u_1, (**b**) v, v_1, (**c**) w, w_1, the driver system (**dashed lines**), the response system (**solid lines**).

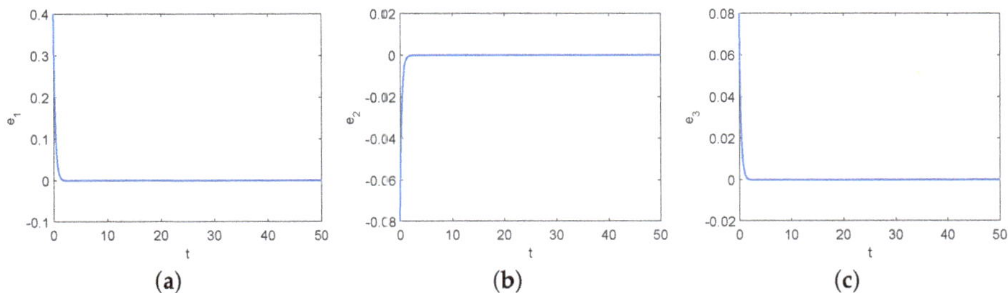

Figure 13. Time history of the anti-synchronization of the state errors system: (**a**) $e_1 - t$, (**b**) $e_2 - t$, (**c**) $e_3 - t$.

4. Conclusions

In this work, we propose and study the following new system:

$$
\begin{aligned}
\dot{u} &= w, \\
\dot{v} &= -w(\alpha v + \beta v^2 + uw), \\
\dot{w} &= u^2 - |u| + |v| + v^2,
\end{aligned}
$$

with eyed-shaped equilibrium points which are located on two closed curves passing the same point. In Section 2, some dynamical properties of the proposed system are presented, which were investigated using bifurcation diagram, phase portrait, maximal Lyapunov exponents, and Kaplan-Yorke dimension. Furthermore, the anti-synchronization of systems is obtained by using active control in Section 3. This study will broaden the current knowledge of chaotic systems with infinitely many equilibria.

Author Contributions: Both authors contributed equally to this work. Both authors read and approved the final manuscript.

Funding: The first author is funded by National Natural Science Foundation of China (Grant No.11672207). The second author is supported by Grant No. MOST 107-2115-M-017-004-MY2 of the Ministry of Science and Technology of the Republic of China.

Acknowledgments: The authors wish to express their hearty thanks to the anonymous referees for their valuable suggestions and comments.

References

1. Lorenz, E.N. Deterministic nonperiodic flow. *J. Atmos. Sci.* **1963**, *20*, 130–141. [CrossRef]
2. Sprott, J.C. Some simple chaotic flows. *Phys. Rev. E* **1994**, *50*, R647–R650. [CrossRef]
3. Sprott, J.C. *Elegant Chaos Algebraically Simple Chaotic Flows*; World Scientific: Singapore, 2010; pp. 10–11.

4. Azar, A.T.; Vaidyanathan, S. *Advances in Chaos Theory and Intelligent Control*; Springer: Berlin, Germany, 2016; pp. 50–61.

5. Pehlivan, I.; Moroz, I.M.; Vaidyanathan, S. Analysis, synchronization and circuit design of a novel butterfly attractor. *J. Sound Vib.* **2014**, *333*, 5077–5096. [CrossRef]

6. Akgul, A.; Moroz, I.; Pehlivan, I.; Vaidyanathan, S. A new four-scroll chaotic attractor and its engineering applications. *Optik* **2016**, *127*, 5491–5499. [CrossRef]

7. Pham, V.T.; Volos, C.; Jafari, S.; Wei, Z.; Wang, X. Constructing a novel no-equilibrium chaotic system. *Int. J. Bifurc. Chaos* **2014**, *24*, 1450073. [CrossRef]

8. Molaie, M.; Jafari, S.; Sprott, J.C.; Golpayegani, S. Simple chaotic flows with one stable equilibrium. *Int. J. Bifurc. Chaos* **2013**, *23*, 1350188. [CrossRef]

9. Jafari, S.; Sprott, J.C. Simple chaotic flows with a line equilibrium. *Chaos Solitons Fractals* **2013**, *57*, 79–84. [CrossRef]

10. Gotthans, T.; Petrzela, J. New class of chaotic systems with circular equilibrium. *Nonlin. Dyn.* **2015**, *81*, 1143–1149. [CrossRef]

11. Gotthans, T.; Sprott, J.C.; Petrzela, J. Simple chaotic flow with circle and square equilibrium. *Int. J. Bifurc. Chaos* **2016**, *26*, 1650137. [CrossRef]

12. Pham, V.T.; Jafari, S.; Volos, C.; Giakoumis, A.; Vaidyanathan, S.; Kapitaniak, T. A chaotic system with equilibria located on the rounded square loop and its circuit implementation. *IEEE Trans. Circuits Syst. II Express Briefs* **2016**, *63*, 878–882. [CrossRef]

13. Zhu, X.; Du, W.-S. A new family of chaotic systems with different closed curve equilibrium. *Mathematics* **2019**, *7*, 94. [CrossRef]

14. Bondarenko, V.E. Information processing, memories, and synchronization in chaotic neural network with the time delay. *Complexity* **2005**, *11*, 39–52. [CrossRef]

15. Cicek, S.; Ferikoglu, A.; Pehlivan, I. A new 3D chaotic system: Dynamical analysis, electronic circuit design, active control synchronization and chaotic masking communication application. *Optik* **2016**, *127*, 4024–4030. [CrossRef]

16. Leonov, G.; Kuznetsov, N.; Seldedzhi, S.; Vagaitsev, V. Hidden oscillations in dynamical systems. *Trans. Syst. Contr.* **2011**, *6*, 54–67.

17. Dudkowski, D.; Jafari, S.; Kapitaniak, T.; Kuznetsov, N.V.; Leonov, G.A.; Prasad, A. Hidden attractors in dynamical systems. *Phys. Rep.* **2016**, *637*, 1–50. [CrossRef]

18. Sprott, J.C. A proposed standard for the publication of new chaotic systems. *Int. J. Bifurc. Chaos* **2011**, *21*, 2391–2394. [CrossRef]

19. Wang, X.; Pham, V.T.; Volos, C. Dynamics, circuit design, and synchronization of a new chaotic system with closed curve equilibrium. *Complexity* **2017**, *2017*, 7138971. [CrossRef]

20. Wolf, A.; Swift, J.B.; Swinney, H.L.; Vastano, J.A. Determining Lyapunov exponents from a time series. *Phys. D Nonlinear Phenom.* **1985**, *16*, 285–317. [CrossRef]

21. Pecora, L.M.; Carroll, T.L. Synchronization in chaotic systems. *Phys. Rev. Lett.* **1990**, *64*, 821–824. [CrossRef]

22. Boccaletti, S.; Kurths, J.; Osipov, G.; Valladares, D.L.; Zhou, C.S. The synchronization of chaotic systems. *Phys. Rep. A* **2002**, *366*, 1–101. [CrossRef]

23. Klein, E.; Mislovaty, R.; Kanter, I.; Kinzel, W. Public-channel cryptography using chaos synchronization. *Phys. Rev. E* **2005**, *72*, 016214. [CrossRef] [PubMed]

24. Li, Y.N.; Chen, L.; Cai, Z.S.; Zhao, X.Z. Experimental study of chaos synchronization in the Belousov-Zhabotinsky chemical system. *Chaos Solitons Fractals* **2004**, *22*, 767–771. [CrossRef]

25. Glass, L. Synchronization and rhythmic processes in physiology. *Nature* **2001**, *410*, 277–284. [CrossRef] [PubMed]

26. Zhang, X.H.; Zhou, S.B. Chaos synchronization for bi-directional coupled two-neuron systems with discrete delays. *Lect. Notes Comput. Sci.* **2005**, *3496*, 351–356.

27. Yousefi, S.; Maistrenko, Y.; Popovych, S. Complex dynamics in a simple model of interdependent open economies. *Discret. Dyn. Nat. Soc.* **2000**, *5*, 161–177. [CrossRef]

Article

Three-Saddle-Foci Chaotic Behavior of a Modified Jerk Circuit with Chua's Diode

Pattrawut Chansangiam

Department of Mathematics, Faculty of Science, King Mongkut's Institute of Technology Ladkrabang, Bangkok 10520, Thailand; pattrawut.ch@kmitl.ac.th; Tel.: +66-9352-666-00

Received: 25 September 2020; Accepted: 28 October 2020; Published: 30 October 2020

Abstract: This paper investigates the chaotic behavior of a modified jerk circuit with Chua's diode. The Chua's diode considered here is a nonlinear resistor having a symmetric piecewise linear voltage-current characteristic. To describe the system, we apply fundamental laws in electrical circuit theory to formulate a mathematical model in terms of a third-order (jerk) nonlinear differential equation, or equivalently, a system of three first-order differential equations. The analysis shows that this system has three collinear equilibrium points. The time waveform and the trajectories about each equilibrium point depend on its associated eigenvalues. We prove that all three equilibrium points are of type saddle focus, meaning that the trajectory of $(x(t), y(t))$ diverges in a spiral form but $z(t)$ converges to the equilibrium point for any initial point $(x(0), y(0), z(0))$. Numerical simulation illustrates that the oscillations are dense, have no period, are highly sensitive to initial conditions, and have a chaotic hidden attractor.

Keywords: chaos theory, electrical circuit analysis, jerk circuit, Chua's diode, system of differential equations, hidden attractor.

PACS: 02.10.Ud; 02.30.Hq; 05.45.Pq; 84.32.-y

---

1. Introduction

Nowadays, chaos theory is an important subject dealing with physics, mathematics, and engineering. A chaos system is a nonlinear dynamical system that has a non-periodic oscillation of waveforms. It is sensitive to initial conditions and has the self-similarity property. A significant development of chaos theory is the discovery of the celebrated Chua's system by L.O. Chua in 1983. This system was described by a set of three first-order ordinary differential equations (ODEs). Chua's discovery has encourged others to look for more chaotic systems, for example, systems of the type Rössler, jerk [1,2], circulant [3,4], hyperjerk [5,6], and hyper chaotic [5,7,8]. In addition, several chaotic circuits have been investigated, for example, Lorenz-based chaotic circuits [9,10], Chua' circuits [11–14], Wien-type chaotic oscillator [15], and chaotic jerk circuits [16–19]. Chaos theory has increasingly attracted much attention due to its wide applications in physical/natural/health sciences and engineering, for example, communication systems, weather forecasting, image encryption [20], celestial mechanics [21], population models [22], hydrology [23], cardiotocography [24], and dynamical disease [25]. Chaos theory as formulated for physical dynamic systems turns out to be useful in social science. For example, chaos theory can be applied to a simple nonlinear model concerning arms race; see, for example [26,27]. The works [28,29] substantiate the chaotic phenomena in dynamic love affair models.

L.O. Chua [14] investigated the chaotic theory for a simple famous circuit in Figure 1, known nowadays as Chua's circuit. The circuit consists of only resistors, capacitors, and a nonlinear resistor. The nonlinear resistor, also called Chua's diode, consists of many op-amps. Many researchers discussed several ways to modify the classical Chua's circuit to a more complicated circuit having chaotic phenomenon. Morgul [30] used an inductorless realization of a Chua's diode consisting of the Wien-bridge oscillator, coupled in parallel with the same nonlinear resistor used in the classical Chua's diode. Numerical experiments illustrated similar chaotic behavior. Aissi and Kazakos [31] modified the Chua's circuit by replacing the op-amps in Chua's diode with RC op-amps. Stouboulos et al. [32] modified the oscillator so that it consists of a nonlinear resistor and a negative conductance, demonstrating the birth and catastrophe of the double-bell strange attractor for different values of frequency. Kyprianidis [33] investigated the anti-monotonicity of the Chua's circuit, which is the creation of forward period-doubling bifurcation sequences followed by reverse period-doubling sequences. The work [34] of Kyprianidis and Fotiadou shows a possible way to replace the piecewise linear characteristic of the Chua's diode with a smooth cubic polynomial. Recently, the work [35] investigates chaotic behavior of the classical Chua's circuit with two nonlinear resistors. The existence of two nonlinear resistors in that case implies that the system has three equilibrium points.

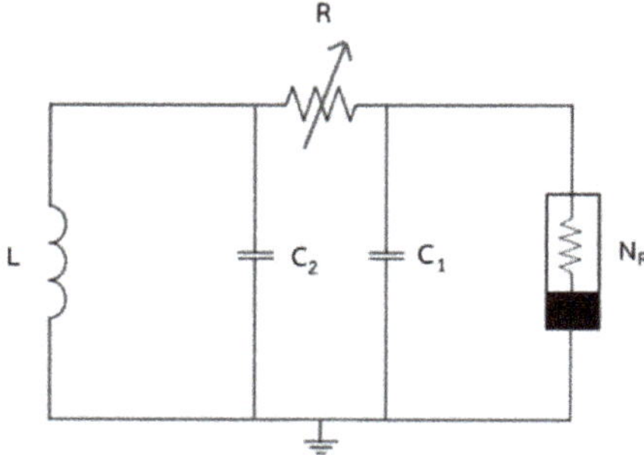

Figure 1. Chua's circuit [14].

In 2011, Sprott [19] studied a simple chaotic jerk circuit, as shown in Figure 2, consisting of only five electronic components: two capacitors, an inductor, an adaptive resistor and a nonlinear resistor. His work shows a chaotic behavior of the trajectories around the equilibriums of the system, and launches a quest for other circuits that chaotically oscillate. Indeed, this circuit can be formulated into a third-order ODE consisting of a nonlinear term, called a "jerk"or the third-order derivative of a variable.

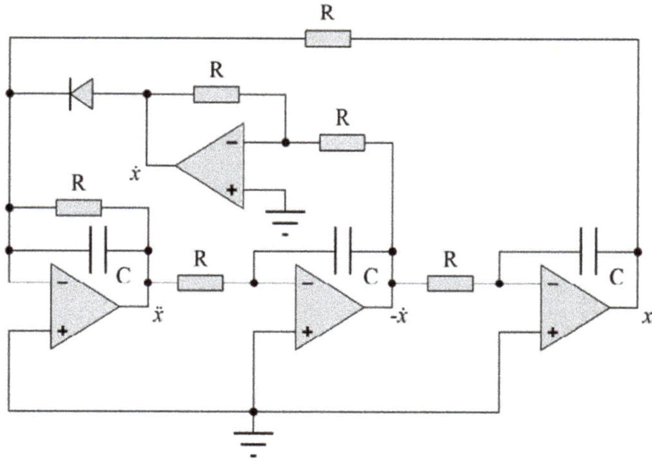

Figure 2. A chaotic jerk circuit [19].

According to much recent interest about chaotic oscillators based on jerk equations, this paper investigates the chaotic behavior of a new chaotic jerk circuit. We modify the chaotic jerk circuit in [19] so that there is a Chua's diode connected parallel to the nonlinear resistor as in Figure 3. The existence of Chua's diode discriminates the proposed system to the system [19]. The voltage-current characteristic of the Chua's diode satisfies a symmetric piecewise linear relation. To describe our system (see Section 2), we apply fundamental laws in electrical circuit theory to formulate a mathematical model in terms of a third-order (jerk) nonlinear ODE, or a system of three first-order ODE. The analysis in Section 3 shows that this system has three collinear symmetric equilibrium points. The time waveform about each equilibrium point depends on its associated eigenvalues. We prove that all three equilibrium points are of type saddle focus node, meaning that the trajectories of $(x(t), y(t))$ diverge in a spiral form but $z(t)$ converges to the equilibrium point for any initial value $(x(0), y(0), z(0))$. Numerical simulation in Section 4 illustrates the chaotic phenomenon, including time waveforms, trajectories about each equilibrium point, effects of changing initial points, and existence of a chaotic hidden attractor. Finally, we summarize the paper in Section 5. In particular, we compare our work to [14,19].

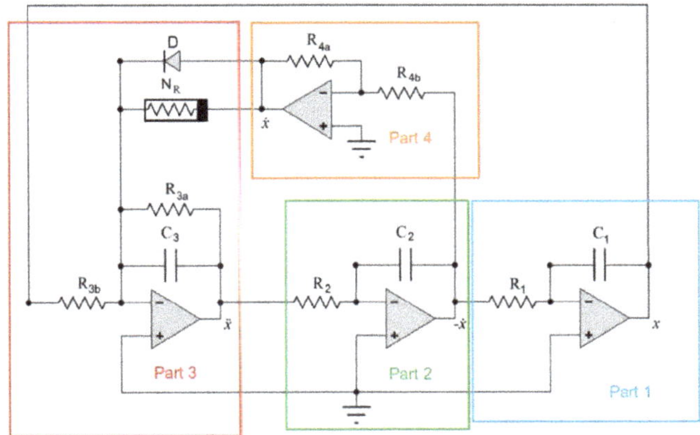

Figure 3. A modified chaotic jerk circuit with Chua's diode.

2. Formulation of a Modified Chaotic Jerk Circuit with Chua's Diode to a System of ODEs

In this section, we formulate a mathematical model for a modified chaotic jerk circuit with Chua's diode in terms of a system of ODEs concerning a piecewise linear function and exponential term. We divide the circuit into four parts, as illustrated in Figure 3. Our analysis is based on fundamental theory of electrical circuit analysis such as Ohm's law, Kirchhoff's current law (KCL) and Kirchhoff's voltage law (KVL).

For Part 1, using KCL and the current-voltage equation for the capacitor, we have

$$\frac{v_{R_1}}{R_1} = i_{R_1} = i_{C_1} = C_1 \frac{dv_{C_1}}{dt} = C_1 \dot{v}_{C_1}.$$

Now, since $v_{R_1} = v_{C_2}$, we obtain $\dot{v}_{C_1} = v_{C_2}/(R_1 C_1)$. Without loss of generality, we may normalize the value of $R_1 C_1$ to be 1 *ms* and we thus have

$$\dot{v}_{C_1} = v_{C_2}. \tag{1}$$

Similarly, for Part 2 we reach $\dot{v}_{C_2} = v_{C_3}/(R_2 C_2)$. Setting the time constant $R_2 C_2 := 1$ yields

$$\dot{v}_{C_2} = v_{C_3}. \tag{2}$$

For Part 3, we have by KCL that $i_{R_{3b}} + i_{N_R} + i_D = i_{R_{3a}} + i_{C_3}$. It follows that

$$\dot{v}_{C_3} = -\frac{v_{C_1}}{R_{3b}C_3} - \frac{v_{C_3}}{R_{3a}C_3} + \frac{i_{N_R}}{C_3} + \frac{i_D}{C_3}.$$

Setting the time constants $R_{3a}C_3$ and $R_{3b}C_3$ to be $1ms$, we get

$$\dot{v}_{C_3} = -v_{C_1} - v_{C_3} + R_{3a}(i_{N_R} + i_D). \tag{3}$$

The circuit in Figure 4 is a more complicated one since it consists of two nonlinear resistors. For the nonlinear resistor on the left, we have by Ohm's law that $v_{N_R} = i_{R_3}R_3$, $v_e = (R_{2c} + R_{3c})i_{R_3}$ and $v_{N_R} - v_e = i_x R_1$, where v_e is the voltage of the op-amp on the left hand side. Combining these three equations to get $v_{N_R} = i_x R_x$ where

$$R_x = -\frac{R_{1c}R_{3c}}{R_{2c}}.$$

Similarly, for the nonlinear resistor on the right, we obtain that $v_{N_R} = i_y R_y$ where

$$R_y = -\frac{R_{4c}R_{6c}}{R_{5c}}.$$

Using KCL at node c, we have $i_{N_R} - i_x - i_y = 0$. Then the current i_{N_R} satisfies the relation

$$v_{N_R} = i_{N_R}(R_x + R_y).$$

However, as pointed out in [19], the behavior of i_{N_R} depends on the voltage v_{C_1}. Indeed, when $v_e < v_f$, the graph of i_{N_R} with respect to v_{C_1} is as follows:

From Figure 5, we have

$$i_{N_R} = \left(\frac{1}{R_x} + \frac{1}{R_{4c}}\right)v_{C_1} + \frac{1}{2}\left(\frac{1}{R_y} - \frac{1}{R_{4c}}\right)\left(\left|v_{C_1} + \frac{v_{f,max}}{v_f}v_{C_1}\right| - \left|v_{C_1} - \frac{v_{f,max}}{v_f}v_{C_1}\right|\right), \tag{4}$$

where $v_{f,max}$ is the maximum voltage at the node f. The current i_D through the diode D depends on the time-derivative of the voltage v_{C_1} (see, e.g., [18]) as follows:

$$i_D = k^2 T^2 e^{\dot{v}_{C_1}/kT},$$

where k is the Boltzmann constant and T is the absolute temperature of the P-N junction. Let us denote $\alpha := kT$. Of particular interest is that the chaos persists when α tends to zero. Since

$$\lim_{\alpha \to 0^+} \alpha^2 e^{y/\alpha} = \infty.$$

At Part 4, we use KCL to analyze this part and we get $i_{R_{4b}} = i_{R_{4a}}$. From Parts 2 and 4, we have by Ohm's law that $v_{C_2}/R_{4b} = v_{R_{4a}}/R_{4a}$ and, thus, the second capacitive voltage is

$$v_{C_2} = \frac{R_{4b}}{R_{4a}}v_{R_{4a}}.$$

For convenience, denote

$$m_0 = R_{3a}\left(\frac{1}{R_x} + \frac{1}{R_y}\right), \quad m_1 = R_{3a}\left(\frac{1}{R_x} + \frac{1}{R_{4c}}\right).$$

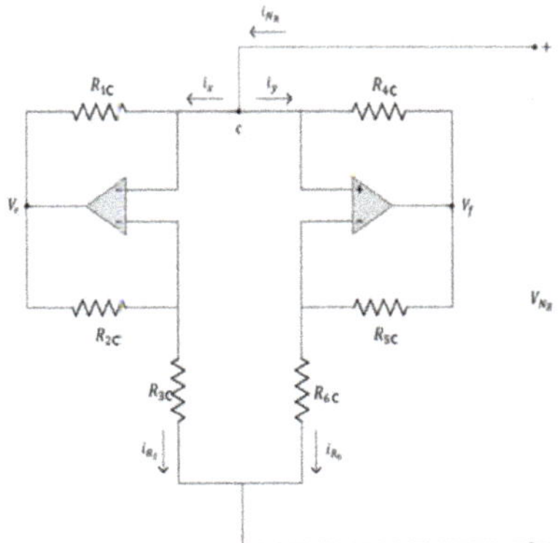

Figure 4. Two nonlinear resistors in Chua's circuit.

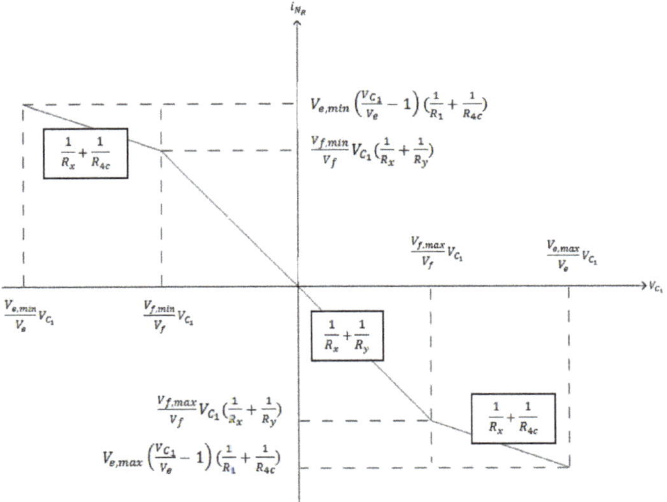

Figure 5. I-V characteristic of nonlinear resistors.

Let us rescale the variables $v_{C_1}, v_{C_2}, v_{C_3}$ to new variables x, y, z so that the current i_{N_R} is reduced to

$$g(x) = m_1 x + 0.5(m_0 - m_1)\left(|x+1| - |x-1|\right),\tag{5}$$

so that the characteristic in Figure 5 becomes that in Figure 6.

Thus, the third-order (jerk) system can be described by the group of Equations (1)–(3), or equivalently, the following system of three first-order ODEs:

$$\begin{aligned}
\dot{x} &= y,\\
\dot{y} &= z,\\
\dot{z} &= -x - z + g(x) + \alpha^2 e^{\frac{y}{\alpha}}.
\end{aligned}\tag{6}$$

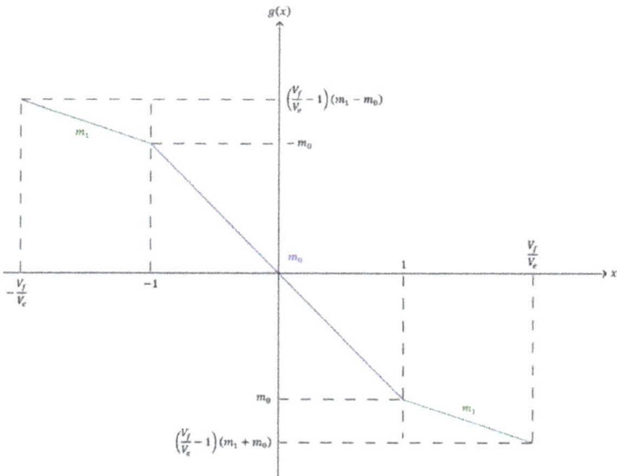

Figure 6. Changing scales of I-V characteristic of nonlinear resistors.

3. Analysis for Chaotic Behavior of the System

In order to analyze the behavior of the dynamical system (6), we need to find all its equilibrium points. Note that Figure 6 illustrates how the three-segment piecewise function $g(x)$ depends on the range of x. The investigation of equilibrium points is thus divided into three cases in Sections 3.1–3.3. We then prove that each equilibrium point is of type saddle focus in Section 3.4. In Section 3.5, we describe how to find an initial point to localize a hidden attractor of the system.

3.1. Case 1: $-1 < x < 1$

From Equation (5), we have $g(x) = m_0 x$. At the equilibrium point, we get $y = \dot{x} = 0$, $z = \dot{y} = 0$, and

$$-x - z + m_0 x + \alpha^2 e^{y/\alpha} = \dot{z} = 0.$$

Thus, the equilibrium for Case 1 is given by $E_1 = (x_1, y_1, z_1) = (\frac{\alpha^2}{1-m_0}, 0, 0)$. When α tends to 0, the equilibrium point reaches the origin $(0, 0, 0)$.

The system (6) can be put in the vector form

$$X'(t) = AX(t) + B(t), \tag{7}$$

where

$$X(t) = \begin{bmatrix} x(t) \\ y(t) \\ z(t) \end{bmatrix}, \quad B(t) = \begin{bmatrix} 0 \\ 0 \\ \alpha^2 e^{y(t)/\alpha} \end{bmatrix}, \quad A = \begin{bmatrix} 0 & 1 & 0 \\ 0 & 0 & 1 \\ m_0 - 1 & 0 & -1 \end{bmatrix}.$$

3.2. Case 2: $-\frac{v_f}{v_e} \leqslant x \leqslant -1$

In this case, we have $g(x) = m_1 x - m_0 + m_1$. At the equilibrium point, we obtain $y = \dot{x} = 0$, $z = \dot{y} = 0$, and

$$-x - z + m_1 x - m_0 + m_1 + \alpha^2 e^{y/\alpha} = \dot{z} = 0.$$

Thus, the equilibrium for Case 2 is given by

$$E_2 = (x_2, y_2, z_2) = (\frac{\alpha^2 - m_0 + m_1}{1 - m_1}, 0, 0).$$

When $\alpha \to 0$, we have (x_2, y_2, z_2) reaches the point $(\frac{-m_0 + m_1}{1 - m_1}, 0, 0)$. The system (6) can be put in the vector form (7) where

$$A = \begin{bmatrix} 0 & 1 & 0 \\ 0 & 0 & 1 \\ m_1 - 1 & 0 & -1 \end{bmatrix}, \quad B(t) = \begin{bmatrix} 0 \\ 0 \\ -m_0 + m_1 + \alpha^2 e^{y(t)/\alpha} \end{bmatrix}.$$

3.3. Case 3: $1 \leqslant x \leqslant \frac{v_f}{v_e}$

We have $g(x) = m_1 x + m_0 - m_1$ and, thus, the equilibrium point is given by

$$E_3 = (x_3, y_3, z_3) = (\frac{\alpha^2 + m_0 - m_1}{1 - m_1}, 0, 0).$$

When $\alpha \to 0$, we have (x_3, y_3, z_3) reaches $(\frac{m_0 - m_1}{1 - m_1}, 0, 0)$. We also have the vector form (7) where the Jacobian matrix A is the same as that in the previous case, and

$$B(t) = \begin{bmatrix} 0 \\ 0 \\ m_0 - m_1 + \alpha^2 e^{y(t)/\alpha} \end{bmatrix}.$$

Thus, the system (6) has three colinear equilibrium points on the X-axis. Note that when $\alpha \to 0$, we have that the points E_2 and E_3 are opposite to each other with respect to the origin E_1. This observation shows the symmetry of the equilibrium points.

3.4. Type of Equilibrium Points

Recall the following theorem:

Theorem 1 (see, e.g., [36]). *Let $A(t) = [a_{ij}(t)] \in \mathbb{R}^{n \times n}$ be a continuous matrix-valued function on an interval I (i.e., each $a_{ij}(t)$ is a real-valued continuous function on I). Let $B(t) \in \mathbb{R}^n$ be a continuous vector-valued function on I. Then the following initial value problem*

$$X'(t) = A(t)X(t) + B(t), \quad X(0) = X_0,$$

has a unique solution $X(t) \in \mathbb{R}^n$ on the interval I.

This theorem guarantees the Equation (7) has a unique solution $X(t)$ on any time interval (note that, in this case, A is a constant matrix). Thus in all cases of x, given an initial point $(x(0), y(0), z(0))$, the trajectory of $(x(t), y(t), z(t))$ is uniquely determined. The trajectory of $(x(t), y(t), z(t))$ in a neighborhood of each equilibrium point depends on the signs of the real/imaginary parts of the eigenvalues of the coefficient matrix A.

For Case 1: $-1 < x < 1$, we have the characteristic equation

$$\det(\lambda I - A) = \lambda^3 + \lambda^2 + 1 - m_0 = 0.$$

Since all parameters of the equation are real and the equation degree is odd, we have that a root (says λ_1) is real and other roots are a conjugate pair of complex numbers. Note that $m_0 < 0$ from Figure 6. Now, the product of all roots (eigenvalues) satisfies

$$\lambda_1 \lambda_2 \lambda_3 = m_0 - 1 < 0.$$

Since (λ_2, λ_3) is a complex conjugate pair, the real root λ_1 must be negative. Write $\lambda_2 = a + ib$ and $\lambda_3 = a - ib$, where $a, b \in \mathbb{R}$. Since the sum of products of two roots of the cubic equation satisfies

$$\lambda_1 \lambda_2 + \lambda_2 \lambda_3 + \lambda_3 \lambda_1 = 1,$$

we get

$$a^2 + 2\lambda_1 a + b^2 = 1.$$

Solving this quadratic equation to obtain

$$a = -\lambda_1 \pm \sqrt{\lambda_1^2 - b^2 - 1}.$$

Since $\lambda_1 < 0$ and $\sqrt{\lambda_1^2 - b^2 - 1} < |\lambda_1|$, we get $a > 0$. Hence, an eigenvalue is a negative real and two other eigenvalues are a conjugate pair of complex numbers having positive real parts. Therefore, this equilibrium is a saddle focus, and the trajectory of $(x(t), y(t))$ diverges in a spiral form, but $z(t)$ converges to the equilibrium point for any initial point $(x(0), y(0), z(0))$.

For Cases 2 and 3, the Jacobian matrices are the same and we have the characteristic equation

$$\det(\lambda I - A) = \lambda^3 + \lambda^2 + 1 - m_1 = 0.$$

Since $m_1 < 0$ (from Figure 6), we obtain the same conclusion as in Case 1, i.e., the equilibrium point is a saddle focus.

We summarize the above discussion in the following theorem:

Theorem 2. *The system* (6) *has three equilibrium points, each of which is of type saddle focus. Moreover, the trajectory of $(x(t), y(t))$ diverges in a spiral form, but $z(t)$ converges to the equilibrium point for any initial point $(x(0), y(0), z(0))$.*

Since the equilibrium points are saddle foci, our system has chaotic behavior.

3.5. Localization of a Hidden Attractor of The System

Recall that an oscillation in a dynamical system can be numerically localized if an initial condition from its neighborhood leads to asymptotic behavior. Such an oscillation is known as an attractor, and its attracting set is called the basin of attraction. If the basin of attraction intersects a small neighborhood of an equilibrium point, then such attractor is said to be self-excited; otherwise it is called a hidden attractor. The hidden attractor was discovered in [37] for a generalized Chua's circuit, and then was discovered in the classical Chua's circuit [38].

In order to find a hidden attractor of the system, we will find a suitable initial point $(x(0), y(0), z(0))$ so that our system will have chaos. First, let us write the system (6) into a first-order vector differential equation

$$X'(t) = AX(t) + \psi(r^T X(t))q \tag{8}$$

where $X(t) = [x(t) \ y(t) \ z(t)]^T \in \mathbb{R}^3$, $A \in \mathbb{R}^{3 \times 3}$, $r \in \mathbb{R}^3$, $q \in \mathbb{R}^3$, and $q : \mathbb{R} \to \mathbb{R}$ is a continuous piecewise-differentiable function. Here, $(\cdot)^T$ denotes the transposition operation. To find a periodic oscillation, we introduce a coefficient k of harmonic linearization so that the matrix $A_0 = A + kqr^T$ of the linear system

$$X'(t) = A_0 X(t)$$

has a pair of pure-imaginary eigenvalues $\pm i\omega_0$ for some $\omega_0 > 0$, and the rest of the eigenvalues have negative real parts. Then the system (8) has a periodic solution $X(t)$ such that

$$\sigma(t) := r^T X(t) \approx a \cos \omega_0 t,$$

where the amplitude a is a solution of the integral equation

$$\int_0^{2\pi/\omega_0} (\psi(a \cos \omega_0 t))a \cos \omega_0 t - k(a \cos \omega_0 t)^2 \, dt = 0.$$

Denoting $\phi(\sigma) = \psi(\sigma) - k\sigma$, we can write Equation (8) to

$$X'(t) = A_0 X(t) + q\phi(r^T x).$$

Let us change $\phi(\sigma)$ to $\epsilon\phi(\sigma)$ where ϵ is a small positive number, and investigate a periodic solution of the system

$$X'(t) = A_0 X(t) + \epsilon q\phi(r^T x). \tag{9}$$

Let us introduce the describing function

$$\Phi(a) = \int_0^{2\pi/\omega_0} \phi(a \cos(\omega_0 t)) \cos(\omega_0 t) \, dt.$$

We make an invertible linear transformation $X(t) = SY(t)$ where $S \in \mathbb{R}^{3\times 3}$ is a nonsingular matrix. The following theorem tells us how to choose an initial point in order to get a hidden attractor of the system.

Theorem 3 ([39]). *If there is a positive number a_0 such that $\Phi(a_0) = 0$ and $b_1\Phi'(a_0) < 0$, then the system (9) has a stable periodic solution with initial point*

$$X(0) = S[y_1(0) \ \ y_2(0) \ \ y_3(0)]^T$$

where $y_1(0) = a_0 + O(\epsilon)$, $y_2(0) = 0$, and $y_3 = O_{n-2}(\epsilon)$ with period $O(\epsilon) + \frac{2\pi}{\omega_0}$.

4. Numerical Experiment

In this section, we provide a numerical experiment to illustrate the chaotic behavior of the proposed circuit via MATLAB. Consider the circuit in Figure 3 with the following parameters: $R_1 = 1$ kΩ, $R_2 = 200$ Ω, $R_{3a} = 500$ Ω, $R_{3b} = 500$ Ω, $R_{4a} = 1$ kΩ, $R_{4b} = 1$ kΩ, $R_{1c} = 250$ Ω, $R_{2c} = 250$ Ω, $R_{3c} = 500$ Ω, $R_{4c} = 750$ Ω, $R_{5c} = 180$ Ω, $R_{6c} = 400$ Ω, $C_1 = 1$ µF, $C_2 = 5$ µF, $C_3 = 2$ µF, $m_0 = -0.1768$, $m_1 = -1.1468$, and $\alpha = 0.026077$. We set the initial condition to be $X(0) = (x(0), y(0), z(0)) = (0, -0.7, 0)$.

Remark 1. *In order to obtain the chaotic phenomenon, one can adjust some parameter values of electronics devices in the circuit so that the eigenvalues of the Jacobian matrix satisfy the condition for the type of equilibrium point (see Section 3.4).*

4.1. Mathematical Analysis of the System

For the case of $-1 < x < 1$, the equilibrium points of the system are given by

$$E_1 = (\frac{\alpha^2}{1 - m_0}, 0, 0) = (5.77838 \times 10^{-4}, 0, 0).$$

In this case, we reach the system $X'(t) = AX(t) + B(t)$ where

$$A = \begin{bmatrix} 0 & 1 & 0 \\ 0 & 0 & 1 \\ -1.1768 & 0 & -1 \end{bmatrix}, \quad B(t) = \begin{bmatrix} 0 \\ 0 \\ \alpha^2 e^{y(t)/\alpha} \end{bmatrix}.$$

The eigenvalues of the system associated with the equilibrium point E_1 are the solutions of the cubic equation

$$\lambda^3 + \lambda^2 + 1.1768 = 0.$$

We get the following eigenvalues

$$\lambda_1 = -1.51364, \quad \lambda_2 = 0.25682 + 0.84351i, \quad \lambda_3 = 0.25682 - 0.84351i.$$

For the case $-v_f/v_e \leqslant x \leqslant -1$, we can obtain the equilibrium point

$$E_2 = (\frac{\alpha^2 - m_0 + m_1}{1 - m_1}, 0, 0) = (-0.969422, 0, 0)$$

associated with eigenvalues

$$\lambda_1 = -1.72307, \quad \lambda_2 = 0.36154 + 1.05603i, \quad \lambda_3 = 0.36154 - 1.05603i.$$

For the case $1 \leqslant x \leqslant v_f/v_e$, the system has the equilibrium point $E_3 = (0.452152, 0, 0)$. Note that E_2 and E_3 have the same eigenvalues since their associated matrices are the same.

From the signs of real/imaginary parts of the associated eigenvalues, we conclude that the three equilibrium points E_1, E_2, E_3 are saddle foci. Hence, the proposed circuit has a chaotic behavior.

4.2. Time Waveforms and Trajectories of The System

The time waveforms of $x(t)$, $y(t)$ and $z(t)$ are reported in Figures 7–9, where the time interval is in *ms*. We see that the oscillations in the figures are non-periodic.

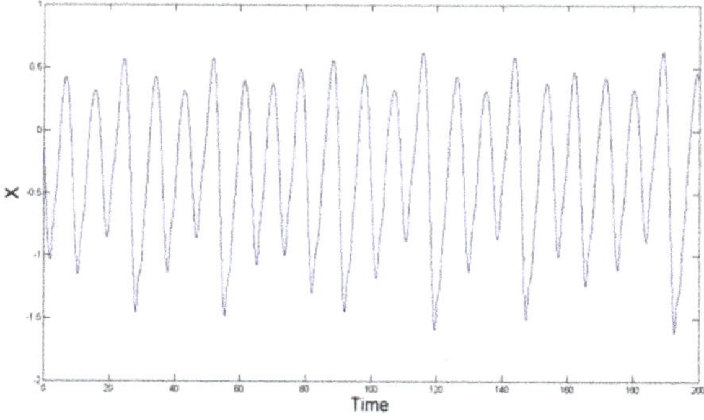

Figure 7. The time waveform of $x(t)$.

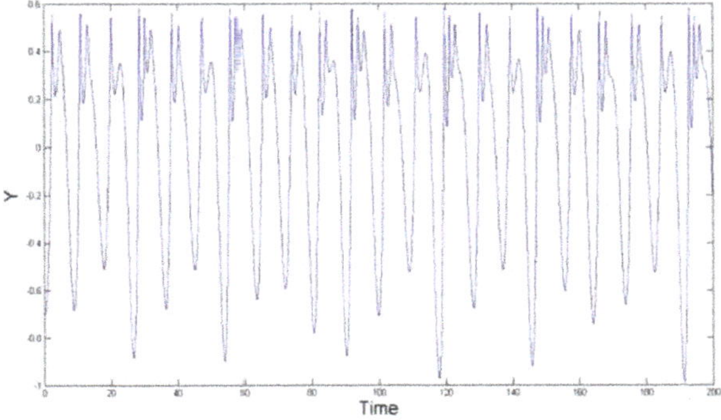

Figure 8. The time waveform of $y(t)$.

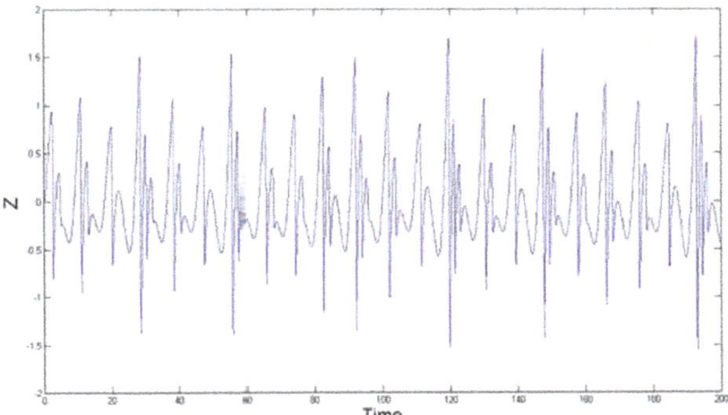

Figure 9. The time waveform of $z(t)$.

The trajectories of $x(t), y(t), z(t)$ in 2D and 3D are numerically simulated in Figures 10–13. We see that the trajectory of $(x(t), y(t))$ diverges in a spiral form, but $z(t)$ converges to the equilibrium point. The trajectories are dense and seem to have no periodic. Thus, chaotic behavior occurs in the modified jerk circuit with Chua's diode. Moreover, the attractor of the system is shown by the blue lines in Figures 10–13. From the 3D plot in Figure 13, we see that the oscillation does not connect with the equilibrium points E_1, E_2, E_3, thus the system has a hidden attractor.

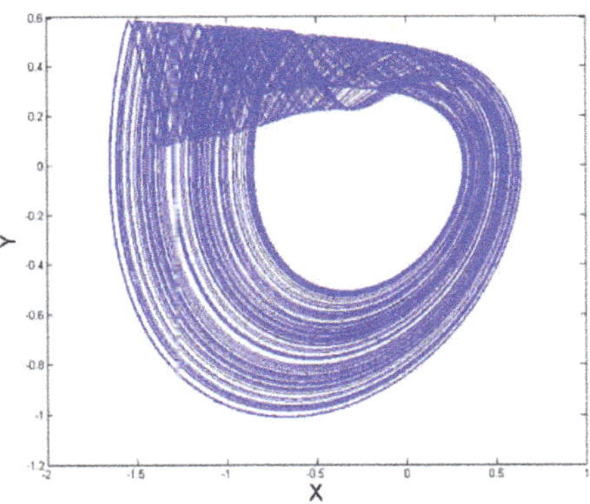

Figure 10. The trajectories of $(x(t), y(t))$ in 2D.

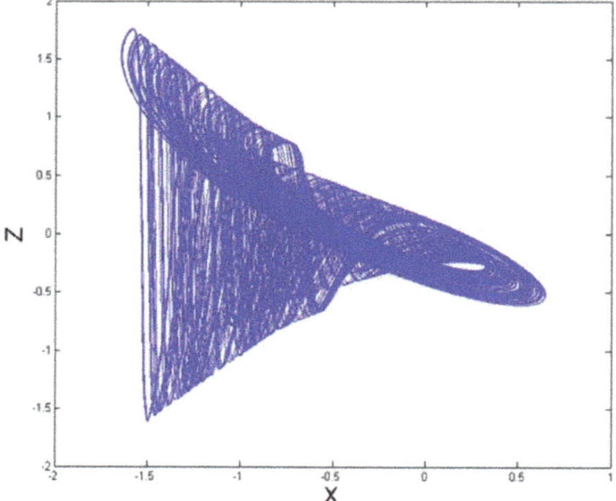

Figure 11. The trajectories of $(x(t), z(t))$ in 2D.

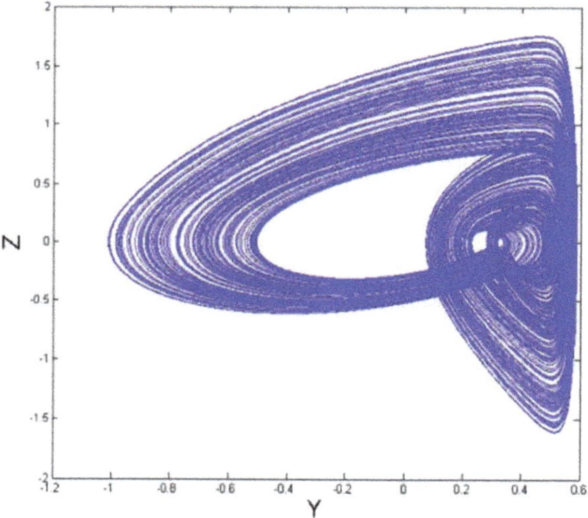

Figure 12. The trajectories of $(y(t), z(t))$ in 2D.

4.3. Effects of Changing Initial Points

Now, we investigate the effect of changing initial points. First, we compare the system behavior when initial values have small changes in the X-axis, namely, $I1 = (0, -0.7, 0)$ and $I2 = (0.0001, -0.7, 0)$; see Figure 14. Next, we consider the case of small changes in the Y-axis, namely, $I1 = (0, -0.7, 0)$ and $I2 = (0, -0.7001, 0)$; the resulting simulation is shown in Figure 15. Finally, the effect of small changes in the Z-axis of the initial point, namely, $I1 = (0, -0.7, 0)$ and $I2 = (0, -0.7, 0.0001)$ is illustrated in Figure 16.

From Figures 14–16, we see that a small difference in initial points leads to a big difference in oscillations of $x(t)$, $y(t)$, $z(t)$. Thus our dynamical system is highly sensitive to initial conditions, a characteristic of a chaotic system.

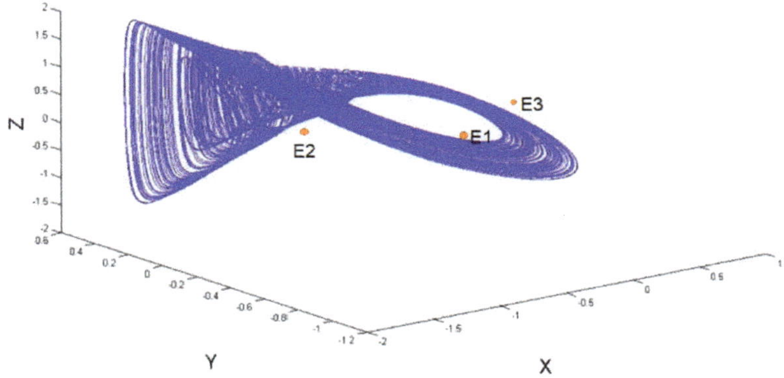

Figure 13. The trajectories of $(x(t), y(t), z(t))$ in 3D.

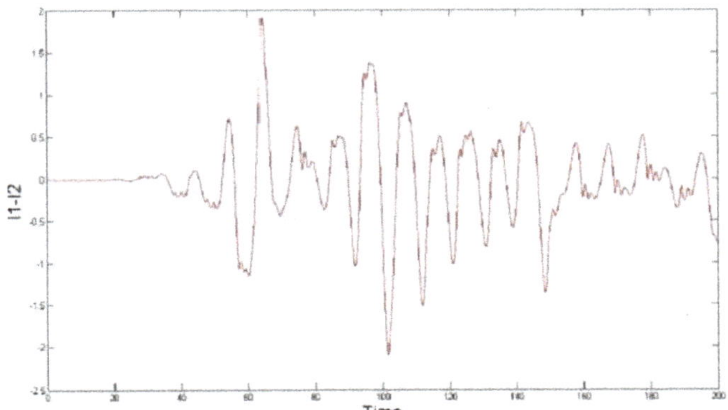

Figure 14. Effects of changing initial points in X-axis from $I1 = (0, -0.7, 0)$ to $I2 = (0.0001, -0.7, 0)$.

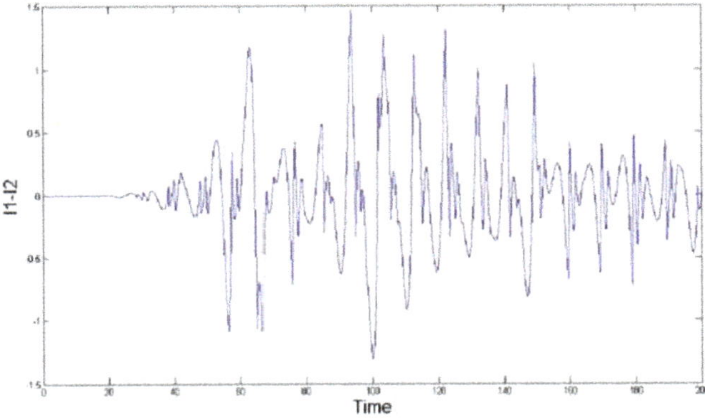

Figure 15. Effects of changing initial points in Y-axis from $I1 = (0, -0.7, 0)$ to $I2 = (0, -0.7001, 0)$.

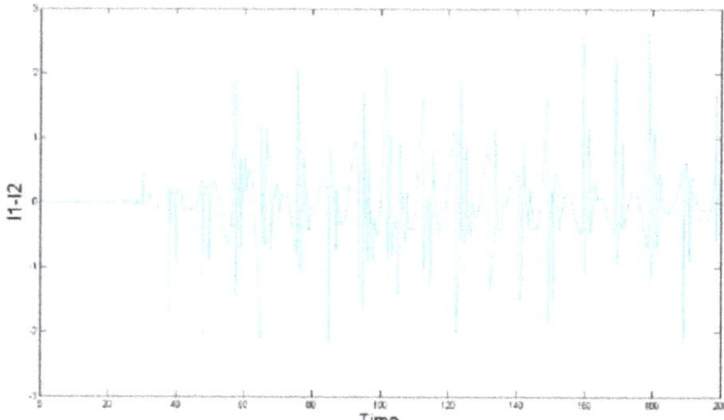

Figure 16. Effects of changing initial points in Z-axis from $I1 = (0, -0.7, 0)$ to $I2 = (0, -0.7, 0.0001)$.

5. Conclusions

We modify a jerk circuit with Chua's diode, and investigate its chaotic properties. This system can be mathematically described by a system of ordinary differential equations with a piecewise linear function and exponential term. The analysis shows that this system has three collinear equilibrium points. The time waveform about each equilibrium point depends on its associated eigenvalues. Indeed, all three equilibrium points are of type saddle focus, meaning that the trajectories of $x(t)$ and $y(t)$ diverge in a spiral form but $z(t)$ converges to the equilibrium point for any initial point $(x(0), y(0), z(0))$. Numerical simulation illustrates that the oscillations are dense, have no period, are highly sensitive to initial conditions, and has a chaotic hidden attractor. Table 1 shows the comparison between three chaotic systems: the proposed system in this paper and the two existing systems in [14,19]. One of the advantages of the proposed system is a higher sensitivity to initial conditions. Therefore, the proposed system enables an alternative model for chaotic theory.

Table 1. The comparisons of a modified chaotic jerk circuit and other related systems.

No.	Terms of Comparison	Ref. [19]	Ref. [14]	This Paper
1	Number of equilibrium points	1	3	3
2	Number of eigenvalues	3	9	9
3	Types of trajectories	1 saddle focus node	1 stable focus node and 2 saddle foci	3 saddle foci
4	Number of components	14	5	15
5	Positions of equilibrium points	a point	3 symmetric points	3 symmetric points
6	Jerk-circuit type	yes	no	yes
7	Existence of Chua's diode	no	yes	yes
8	Existence of chaotic attractors	yes	yes	yes
9	Sensitivity to initial conditions	√√	√	√√√
10	Nonlinear system	yes	yes	yes

Funding: The author would like to thank King Mongkut's Institute of Technology Ladkrabang Research Fund, grant no. KREF046205 for financial supports.

Acknowledgments: This work was supported by King Mongkut's Institute of Technology Ladkrabang.

Conflicts of Interest: The author declares no conflict of interest.

References

1. Sprott, J.C. Some simple chaotic jerk functions. *Am. J. Phys.* **1997**, *65*, 537–543. [CrossRef]
2. Sprott, J.C. *Elegant Chaos: Algebraically Simple Chaotic Flows*; World Scientific: Singapore, 2010.
3. Lorenz, E.N.; Emanuel, K.A. Optimal sites for supplementary weather observations: Simulation with a small model. *J. Atmos. Sci.* **1998**, *55*, 399–414. [CrossRef]
4. Thomas, R. Deterministic chaos seen in terms of feedback circuits: Analysis, synthesis, 'labyrnth chaos. *Int. J. Bifurcat. Chaos Appl. Sci. Eng.* **1999**, *9*, 1889–1905. [CrossRef]
5. Chlouverakis, K.E.; Sprott, J.C. Chaotic hyperjerk systems. *Chaos Solit. Frac.* **2006**, *28*, 739–746. [CrossRef]
6. Mumuangsaen, B.; Srisuchinwong, B. Elementary chaotic snap flows. *Chaos Solit. Frac.* **2011**, *44*, 995–1003. [CrossRef]
7. Rössler, O.E. An equation for hyperchaos. *Phys. Lett. A* **1979**, *71*, 155–157. [CrossRef]
8. Liu, Z.; Lai, Y.C.; Matias, M.A. Universal scaling of Lyapunov exponents in coupled chaotic oscillators. *Phys. Rev. E* **2003**, *67*, 1–4. [CrossRef] [PubMed]
9. Cuomo, K.M.; Oppenheim, A.V. Circuit implementation of synchronized chaos with applications to communications. *Phys. Rev. Lett.* **1993**, *71*, 65–68. [CrossRef]
10. Blakely, J.N.; Eskridge, M.B.; Corron, N.J. A simple Lorenz circuit and its radio frequency implementation. *Chaos* **2007**, *17*, 1–5. [CrossRef]
11. Matsumoto, T.; Chua, L.O.; Komuro, M. The double scroll. *IEEE Trans. Circuits Syst.* **1985**, *32*, 797–818. [CrossRef]
12. Bartissol, P.; Chua, L.O. The double hook. *IEEE Trans. Circuits Syst.* **1988**, *35*, 1512–1522. [CrossRef]
13. Chua, L.O.; Lin, G.-N. Canonical realization of Chua's circuit family. *IEEE Trans. Circuits Syst.* **1990**, *37*, 885–902. [CrossRef]
14. Chua, L.O. The genesis of Chua's circuit. *Archiv. Elektron. Übertragungstechnik* **1992**, *46*, 250–257.
15. Elwakil, A.S.; Kennedy, M.P. High frequency Wien-type chaotic oscillator. *Electron Lett.* **1998**, *34*, 1161–1162. [CrossRef]
16. Srisuchinwong, B.; Treetanakorn, R. Current-tunable chaotic jerk circuit based on only one unity-gain amplifier. *Electron Lett.* **2014**, *50*, 1815–1817. [CrossRef]
17. Srisuchinwong, B.; Nopchinda, D. Current-tunable chaotic jerk oscillator. *Electron Lett.* **2013**, *49*, 587–589. [CrossRef]
18. Sprott, J.C. Simple chaotic systems and circuits. *Am. J. Phys.* **2000**, *68*, 758–763. [CrossRef]
19. Sprott, J.C. A new chaotic jerk circuit. *IEEE Trans. Circuits Syst. II* **2011**, *58*, 240–243. [CrossRef]
20. Xu, M. Cryptanalysis of an image encryption algorithm based on DNA sequence operation and hyper-chaotic system. *3D Res.* **2017**, *8*, 15–24. [CrossRef]
21. Morbidelli, A. Chaotic diffusion in celestial mechanics. *Regul. Chaotic Dyn.* **2001**, *6*, 339–353. [CrossRef]
22. Eduardo, L.; Ruiz, A. Chaos in discrete structured population models. *SIAM J. Appl. Dyn. Syst.* **2012**, *11*, 1200–1214.
23. Sivakumar, B. Chaos theory in hydrology: Important issues and interpretations. *J. Hydrol.* **2000**, *227*, 1–20. [CrossRef]
24. Bozóki, Z. Chaos theory and power spectrum analysis in computerized cardiotocography. *Eur. J. Obstet. Gynecol. Reprod. Biol.* **1997**, *71*, 163–168. [CrossRef]
25. Glass, L. Dynamical Disease: The Impact of Nonlinear Dynamics and Chaos on Cardiology and Medicine. In *The Impact of Chaos on Science and Society*; Grebogi, C., Yorke, J.A., Eds.; United Nations University Press: Tokyo, Japan, 1997.
26. Saperstain, A.M. Chaos–a model for the outbreak of war. *Nature* **1984**, *309*, 303–305. [CrossRef]
27. Grossmann, S.; Mayer-Kress, G. Chaos in the international arms race. *Nature* **1989**, *337*, 702–704. [CrossRef]
28. Huang, L.; Bae, Y. Analysis of chaotic behavior in a novel extended love model considering positive and negative external environment. *Entropy* **2018**, *20*, 365, doi:10.3390/e20050365 [CrossRef]
29. Yoon, J.H.; Bak, G.M. Youngchul Bae, Fuzzy control for chaotic confliction model. *Int. J. Fuzzy Syst.* **2020**, *22*, 1961–1971. [CrossRef]
30. Morgul, O. Inductorless realization of Chua oscillator. *Electron. Lett.* **1995**, *31*, 1403–1404. [CrossRef]

31. Aissi, C.; Kazakos, D. An improved realization of the Chua's circuit using RC-op amps. In Proceedings of the WSEAS International Conference on Signal Processing, Istanbul, Turkey, May 27-30, 2008; Volume 7, pp. 115–118.
32. Stouboulos, I.N.; Kyprianidis, I.M.; Papadopoulou, M.S. Complex chaotic dynamics of the double-bell attractor. *WSEAS Trans. Circuits Syst.* **2008**, *7*, 13–21.
33. Kyprianidis, I.M. New chaotic dynamics in Chua's canonical circuit. *WSEAS Trans. Circuits Circuits Syst.* **2006**, *5*, 1626–1633.
34. Kyprianidis, I.M.; Fotiadou, M.E. Complex dynamics in Chua's canonical circuit with a cubic nonlinearity. *WSEAS Trans. Circuits Syst.* **2006**, *5*, 1036–1043.
35. Limphodaen, L.; Chansangiam, P. Mathematical analysis for classical Chua's circuit with two nonlinear resistors. *Songklanakarin J. Sci. Technol.* **2020**, *42*, 678–687.
36. Goode, S.W. *Differential Equations and Linear Algebra*; Prentice Hall: Englewood Cliffs, NJ, USA, 2000.
37. Kuznetsov, N.V.; Leonov, G.A.; Vagaitsev, V.I. Analytical-numerical method for attractor localization of generalized Chua's system. *Int. Fed. Autom. Control. Proc.* **2010**, *4*, 29–33. [CrossRef]
38. Leonov, G.A.; Kuznetsov, N.V.; Vagaitsev, V. Localization of hidden Chua's attractors. *Phys. Lett. A* **2011**, *375*, 2230–2233. [CrossRef]
39. Leonov, G.A.; Kuznetsov, N.V. Hidden attractors in dynamical systems. From hidden oscillations in Hilbert-Kolmogorov, Aizerman, and Kalman problems to hidden chaotic attractors in Chua circuits. *Int. Bifurc. Chaos* **2013**, *23*. [CrossRef]

Publisher's Note: MDPI stays neutral with regard to jurisdictional claims in published maps and institutional affiliations.

Article

Two New Asymmetric Boolean Chaos Oscillators with No Dependence on Incommensurate Time-Delays and Their Circuit Implementation

Jesus M. Munoz-Pacheco [†], Tonatiuh García-Chávez [†], Victor R. Gonzalez-Diaz [*,†], Gisela de La Fuente-Cortes [†] and Luz del Carmen Gómez-Pavón [†]

Faculty of Electronics Sciences, Benemérita Universidad Autónoma de Puebla, Puebla 72000, Mexico; jesusm.pacheco@correo.buap.mx (J.M.M.-P.); tonaspiuck@gmail.com (T.G.-C.); gise_tiza@hotmail.com (G.d.L.F.-C.); luz.gomez@correo.buap.mx (L.d.C.G.-P.)
* Correspondence: vicrodolfo.gonzalez@correo.buap.mx; Tel.: +52-222-229-5500
† These authors contributed equally to this work.

Received: 17 February 2020; Accepted: 7 March 2020; Published: 1 April 2020

Abstract: This manuscript introduces two new chaotic oscillators based on autonomous Boolean networks (ABN), preserving asymmetrical logic functions. That means that the ABNs require a combination of XOR-XNOR logic functions. We demonstrate analytically that the two ABNs do not have fixed points, and therefore, can evolve to Boolean chaos. Using the Lyapunov exponent's method, we also prove the chaotic behavior, generated by the proposed chaotic oscillators, is insensitive to incommensurate time-delays paths. As a result, they can be implemented using distinct electronic circuits. More specifically, logic-gates–, GAL–, and FPGA–based implementations verify the theoretical findings. An integrated circuit using a CMOS 180nm fabrication technology is also presented to get a compact chaos oscillator with relatively high-frequency. Dynamical behaviors of those implementations are analyzed using time-series, time-lag embedded attractors, frequency spectra, Poincaré maps, and Lyapunov exponents.

Keywords: chaotic oscillator; lyapunov exponents; poincare map; integrated circuit; fpga; time-delay; boolean networks

---

1. Introduction

Chaos behavior is one of the most studied topics in nonlinear dynamics in recent years. Such interest relies mainly on its extreme sensitivity to the initial conditions. From a real-world application point of view, the random-like patterns generated by chaotic oscillators are currently pointed out as the core for obtaining significant engineering applications, for instance, secure-communications schemes [1–7]; radars [8–10]; sonars [11,12]; liquid mixing [13,14]; adaptive logic gates [15,16]; true random number generators (TRNGs) [17,18]; collective phenomena in physics and biology [19,20]; navigation and control of autonomous mobile robots [21–23]; Internet of Things [24–29]; and so forth. Thereupon, the cutting edge chaos-based applications may need reliable, robust, compact, and faster chaos oscillators.

A remarkable solution to obtain chaotic behavior consists of exploiting the delay paths in autonomous Boolean networks (ABNs) [30–34]. Kauffman proposed the Boolean networks in 1969 as a mathematical framework for studying gene regulatory networks. The mathematics describing ABNs has shown that they could display aperiodic patterns if the Boolean functions have instantaneous response times, the link time-delays are incommensurate, and their nodes perform asymmetric Boolean operations, such as the combination of logic exclusive-OR (XOR) and XNOR functions.

In the context of ABNs, deterministic chaos, also known as Boolean chaos, was initially demonstrated by using Boolean functions implemented with electronic logic circuits (logic gates and field-programmable gate arrays FPGAs) [10,35–38]. At circuit level, the basic principle for obtaining Boolean chaos depends on three main characteristics; the asymmetry between the logic states, the short-pulse rejection phenomenon, and, most importantly, the degradation effect [30,31].

As is well-known, the incommensurate delay between two different nodes of an ABN is the critical parameter to obtain chaotic behaviors since it induces the degradation effect [30–34]. Rosin et al. analyzed two ABNs, one composed of a logic XOR function and two delays τ_{n_k} and τ_{n_l}, and the other one with a logic XNOR function and three delays τ_{n_k}, τ_{n_l}, and τ_{n_m} [38]. They showed that Boolean chaos arises in an FPGA-based implementation when the delays for each of the three delay paths are $\tau_{n_k} \geq 2.8\,\text{ns}$, $\tau_{n_l} \geq 1.7\,\text{ns}$, and $\tau_{n_m} \geq 0.56\,\text{ns}$, respectively. To attain the time-delays, they required 18 extra logic NOT gates to connect the nodes of ABN. Besides, if those time-delays reduce below the minimum, the ABN does not show chaos and evolves to periodic oscillations only, as was analyzed in Ref. [35]. Park et al. presented an ABN composed of a logic XOR gate and ring oscillator [39]. The proposed logic circuit synthesized on an application-specific IC (ASIC), but the design is not straightforward because it also demands specific incommensurate delays in the feedback path to observe Boolean chaos.

Based on the discussion mentioned above, we note a possible benefit of using ABNs can be to get Boolean chaos oscillators with relatively high oscillation frequencies and small form factors since they depend on logic functions only. However, we also found that the proposed ABNs have high sensitivity to the time-delay among network nodes for generating chaos behavior. From a practical point of view, that condition is very complicated to satisfy since the time-delays are heavily related to the electronics technology chosen for implementation. Due to the electronic logic gates being heterogeneous, they do not have the same intrinsic time-delay. As a consequence, the dynamical behaviors of the ABN can be affected by placing the oscillator on a different area into an integrated circuit or FPGA. In conclusion, the previously reported Boolean chaos oscillators may not be suitable for physical realizations with multiple hardware approaches.

In this paper, we propose two ABNs with three and two nodes, respectively. The nodes perform the logic XOR and XNOR operations. This asymmetric approach avoids fixed points in the ABNs, and therefore, their dynamics can converge to chaotic oscillations. By applying the Lyapunov exponent method, we experimentally demonstrate that the Boolean chaos oscillators do not require specific incommensurate time-delays to show chaotic behaviors. Indeed, the Boolean chaos was observed under a wide range of the time-delays for the ABNs nodes. We prove our findings by implementing the proposed ABNs using various logic electronic circuits without any modification neither of the proposed networks nor adding additional path delays. Three discrete physical realizations using commercial logic gates, a Generic Array Logic (GAL), and FPGA are presented. Besides, we design an integrated circuit realization at 180nm fabrication technology.

The structure of the manuscript is as follows. Section 2 introduces the two ABNs and gives the mathematical demonstrations of their equilibrium points. Section 3 shows the analysis based on the Lyapunov exponents to determine the insensibility to time-delays. Section 4 presents the Lyapunov exponents for three different discrete implementations to prove that the ABNs are not affected by the technology. Section 5 introduces a straightforward methodology to design an integrated circuit of the Boolean chaos oscillators. Time-series, phase space reconstruction, Lyapunov exponents, and Poincare maps validate the observed chaos behavior. Finally, the last section concludes the paper.

2. Mathematical Preliminaries

A Boolean network consists of a number of logical nodes interconnected through direct or indirect links. These are nonlinear networks requiring a mathematical base for analysis. Among the present models there are: the Kauffman (N-K) networks, Boolean differential equations, and piecewise-linear differential equations [33,34]. This work uses the Boolean differential equations to develop important mathematical considerations.

2.1. Boolean Differential Equations

Let us consider a system with state variables $\{v_1, v_2, \ldots, v_n\}$, $v_i \in \mathbb{R}$, $i = 1, \ldots, n$. If a Boolean variable x_i is related to each state v_i, depending on a set of thresholds $\sigma_i \in \mathbb{R}$. Then, the set of Boolean variables $x = \{x_1, x_2, \ldots, x_n\}$ gives a simple qualitative description of the system with 2^n possible states. By adding the time dependence through a set of delays $\{\tau_{ij}\}$, $i = 1, \ldots, n, j = 1, \ldots, n, \tau_{ij} > 0$, where τ_{ij} is the time it takes for x_j to affect x_i, there is an associated time delay for each pair of state variables not necessarily obeying $\tau_{ij} = \tau_{ji}$. In this manner, the feedbacks among the Boolean variables can be described by a system of Boolean differential equations as follows [33,34]:

$$
\begin{aligned}
x_1(t) &= f_1(x_1(t - \tau_{11}), x_2(t - \tau_{12}), \ldots, x_n(t - \tau_{1n})), \\
x_2(t) &= f_2(x_1(t - \tau_{21}), x_2(t - \tau_{22}), \ldots, x_n(t - \tau_{2n})), \\
&\;\;\vdots \\
x_n(t) &= f_n(x_1(t - \tau_{n1}), x_2(t - \tau_{n2}), \ldots, x_n(t - \tau_{nn})),
\end{aligned}
\tag{1}
$$

with $f_i : \mathbb{B}^n \to \mathbb{B}$, $i = 1, \ldots, n$, being a set of Boolean functions where $\mathbb{B} = \{0, 1\}$. The system (1) determines the dynamics of a Boolean network considering time delays, thereby defining an *Autonomous Boolean Network* (ABN) [33,34]. The dynamics of the ABN given by Equation (1) is numerically solved once the Boolean functions are defined with initial conditions on an interval $x_i(t) = x_{i0}(t)$ for $t_0 - \tau \le t \le t_0, i = 1, \ldots, n$, where $\tau = \max\{\tau_{ij}\}$ is the memory length of the system.

2.2. Boolean Chaos

In an ideal ABN, the transitions of the signals are arbitrarily fast and the number of transitions increases with time, following a power law. These increasingly fast dynamics result in an unlimited growth of frequency over time, referred to as an ultraviolet catastrophe [30]. However, that phenomenon does not occur in nature because the information-transmitting links and the processing nodes (for instance real logic gates) have a maximum operation frequency, which are physically realized. Hence, they cannot transmit or generate signals above a certain frequency [31]. As a result, the nonideal behaviors of real logic devices are responsible for the origin of chaos in ABNs [30,31]. Those behaviors are (i) Short-pulse rejection (SPR), known as pulse filtering, preventing pulses shorter than a minimum duration from passing through the gate (Theorem 1). (ii) The asymmetry between the logic states, making the propagation delay time through the gate depending on whether the transition is a fall or rise. (iii) The degradation effect triggering a change in the events propagation delay time when they appear in rapid succession. Among them, the degradation effect is the main nonideal behavior source of deterministic chaos in an ABN [32], since Boolean chaos originates from a history-dependent delay [30,31], as defined Lemma 1.

Theorem 1. *For a symmetric ABN consisting of a single XOR logic operation with two self-inputs having delays τ_1 and τ_1, and with τ_{spr} sufficiently small not collapsing to the always-off state occurs before $t = \tau_2$, the trajectory will never reach the always-off state.*

Lemma 1. *For a class of experimental ABN containing at least one XOR connective and feedback loop, deterministic chaos may arise if and only if the degradation effect, which is exhibited at some level in any real ABN, is presented.*

On the other hand, if the ABN has equal delays, the links will produce only regular oscillations. In addition, the fixed points caused by using only symmetric logic functions in the network nodes conduct that the dynamics will always collapse into a low or high logic state, respectively. Theorem 2 and Lemma 2 postulates the conditions. As a reference, the complete proofs of Theorems and Lemmas can be found in [31,35].

Theorem 2. *For a symmetric ABN consisting of a single exclusive-OR (XOR) logic operation with two self-inputs having delays τ_1 and τ_2, the attractors are always periodic.*

Lemma 2. *The experimentally realized ABNs should not include a Boolean fixed point, for which all Boolean functions are satisfied simultaneously.*

2.3. Lyapunov Exponents for ABNs

One of the most reliable tools to demonstrate chaotic behavior is computing the Lyapunov exponent's spectrum [1–29]. A positive Lyapunov exponent is a signature of chaos [40]. It is defined as the exponential divergence of trajectories with nearly identical initial conditions. For the ABNs case, since the states are discrete, indicating a phase space composed just by 2^N states, the Lyapunov exponent's needs to be computed from distance measures tailored for Boolean systems [41].

Zhang et al. proposed a method to estimate the largest Lyapunov exponent using the Boolean distance definition [30]. The approach works as follows. (i) Acquire experimentally a long time series from an output voltage of the ABN. (ii) Convert that voltage to a Boolean variable $x(t)$. (iii) Given any two segments of starting at times t_a and t_b, define a Boolean distance with $d(s) = \frac{1}{T} \int_s^{s+T} x(t' + t_a) \oplus x(t' + t_b)dt'$, where T is a fixed parameter, $\oplus$ is the XOR logic operation, and the Boolean distance $d(s)$ evolves as a function of the time s. (iv) Search in $x(t)$ for all the pairs t_a and t_b corresponding to the earliest times in each interval T over which $d(0) < 0.01$. v. Finally, v) compute $\ln\langle d(s)\rangle$, where $\langle\rangle$ means an average over all matching (t_a, t_b) pairs.

As a conclusion, the divergence $\ln\langle d(s)\rangle$ increases exponentially, as expected for an adequate definition of distance between trajectories in a chaotic system [40].

3. The Proposed Boolean Chaos Oscillators (BCOs) and Their Fixed Points

Motivated by Ref. [30], this work introduces two ABNs composed by three and two nodes, respectively. The nodes of ABNs perform asymmetric logic functions, i.e., a combination of Boolean operations XOR and XNOR. We detail the proposed ABNs as follows.

3.1. BCO-1

Figure 1a shows the first Boolean chaos oscillator (BCO). It consists of three nodes where each node has three inputs and one output that propagates to three different nodes. Nodes A and B perform the XOR logic operation while node C executes the XNOR. Expressing BCO-1 in the form of Equation (1), we obtain the following system of Boolean delay equations:

$$
\begin{aligned}
X_a(t) &= X_a(t - \tau_{aa}) \oplus X_b(t - \tau_{ab}) \oplus X_c(t - \tau_{ac}), \\
X_b(t) &= X_a(t - \tau_{ba}) \oplus X_b(t - \tau_{bb}) \oplus X_c(t - \tau_{bc}), \\
X_c(t) &= X_a(t - \tau_{ca}) \oplus X_b(t - \tau_{cb}) \oplus X_c(t - \tau_{cc}) \oplus 1,
\end{aligned}
\tag{2}
$$

with Boolean functions $f_i : \mathbb{B}^3 \to \mathbb{B}$, $i = 1,\ldots,3$, and $\oplus$ the logic XOR operation. The signal propagation time from node j to node i is τ_{ij} for $i, j = a, b, c$.

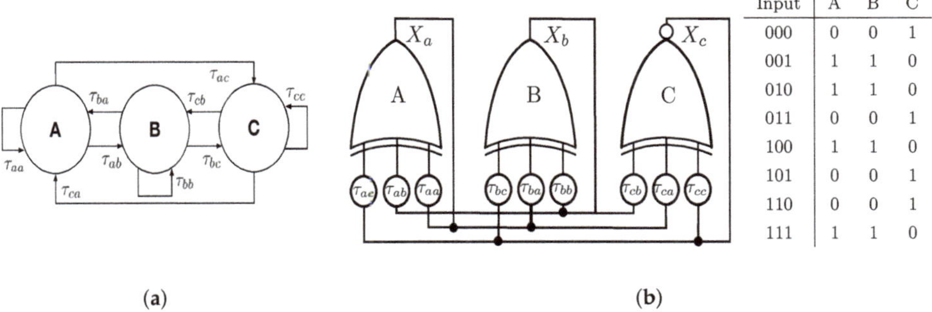

(a)   (b)

Figure 1. (a) Autonomous Boolean networks (ABN) for the proposed Boolean chaos oscillator (BCO-1). (b) An implementation of BCO-1 using electronic logic gates and its look-up table.

Theorem 3. *For an autonomous Boolean network given by the system of Equation* (2), *the orbits are always oscillating* [36]*.*

Proof. A Boolean fixed point provokes nonoscillating dynamics due to some orbits eventually collapsing into the fixed point. To demonstrate the proposed Boolean chaos oscillator converges to sustained oscillations indefinitely, we must prove that there is not a fixed point. By contradiction, we demonstrate this theorem. Let us assume that the BCO-1 has a fixed point (X_a^*, X_b^*, X_c^*), such that:

$$X_a^* = X_a(t - \tau),$$
$$X_b^* = X_b(t - \tau),$$
$$X_c^* = X_c(t - \tau),$$

for $t \gg \tau = \max\{\tau_{aa}, \tau_{ab}, \tau_{ac}, \tau_{ba}, \tau_{bb}, \tau_{bc}, \tau_{ca}, \tau_{cb}, \tau_{cc}\}$. In this manner, the system of Equation (2) recast as:

$$X_a(t) = X_a(t) \oplus X_b(t) \oplus X_c(t), \tag{3}$$
$$X_b(t) = X_a(t) \oplus X_b(t) \oplus X_c(t), \tag{4}$$
$$X_c(t) = X_a(t) \oplus X_b(t) \oplus X_c(t) \oplus 1. \tag{5}$$

By substituting Equations (3) and (4) into (5), we obtain:

$$X_c(t) = X_a(t) \oplus X_b(t) \oplus X_c(t) \oplus X_a(t) \oplus X_b(t) \oplus$$
$$X_c(t) \oplus X_c(t) \oplus 1. \tag{6}$$

Equation (6) implies $X_c(t) = \overline{X_c(t)}$. Since the Boolean space is 2^n, the possible states for $X_c(t)$ are $\{1, 0\}$. Thus, "1" = "0", or vice-versa indicates a contradiction which leads us to conclude that Boolean network (2) does not have fixed points, and therefore, always oscillates. $\square$

3.2. BCO-2

Figure 2a shows the second Boolean chaos oscillator introduced in this work. The proposed topology consists of two nodes, where each node has three inputs and one output connecting to two different nodes. While node A performs the XOR logic operation, node B executes the XNOR. Additionally, the ABN includes two logic NOT operations to obtain the opposed Boolean states for both nodes. The set of Boolean delay equations for the BCO-2 are:

$$X_a(t) = X_a(t - \tau_{aa}) \oplus X_b(t - \tau_{ab}) \oplus \neg X_a(t - \tau_{a\tilde{a}}),$$
$$X_b(t) = X_a(t - \tau_{ba}) \oplus X_b(t - \tau_{bb}) \oplus$$
$$\neg X_b(t - \tau_{b\tilde{b}}) \oplus 1, \tag{7}$$

with $\oplus$ and $\neg$ indicating the logic XOR and NOT operations, respectively. In addition, the Boolean functions $f_i : \mathbb{B}^2 \to \mathbb{B}, i = 1, \ldots, 2$. Similarly to the previous case, it is necessary to prove that the system (7) does not have a Boolean fixed point.

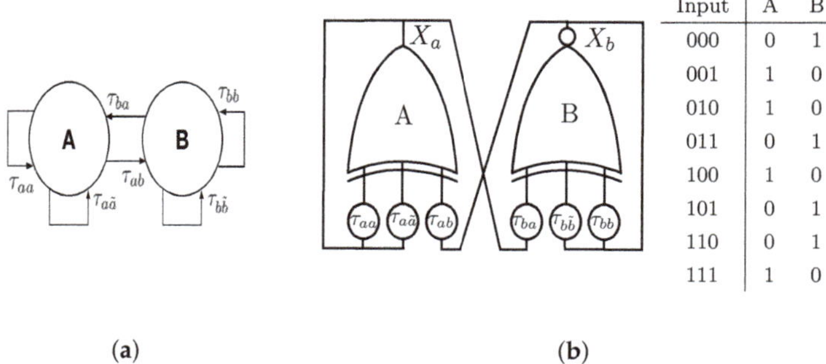

Input	A	B
000	0	1
001	1	0
010	1	0
011	0	1
100	1	0
101	0	1
110	0	1
111	1	0

(a) (b)

Figure 2. (a) ABN for the second Boolean chaos oscillator (BCO-2). (b) An implementation of BCO-2 using electronic logic gates and its look-up table.

Theorem 4. *For an autonomous Boolean network given by the system of Equation (7), the orbits are always oscillating [36].*

Proof. We assume that the BCO-2 has the fixed point (X_a^*, X_b^*), such that $X_a^* = X_a(t - \tau)$ and $X_b^* = X_b(t - \tau)$, for $t \gg \tau = \max\{\tau_{aa}, \tau_{ab}, \tau_{a\tilde{a}}, \tau_{ba}, \tau_{bb}, \tau_{b\tilde{b}}\}$. Therefore, system (7) is rewritten as:

$$X_a(t) = X_a(t) \oplus X_b(t) \oplus \neg X_a(t), \tag{8}$$
$$X_b(t) = X_a(t) \oplus X_b(t) \oplus \neg X_b(t) \oplus 1, \tag{9}$$

By inserting Equation (8) into (9), we obtain:

$$X_b(t) = X_a(t) \oplus X_b(t) \oplus \neg X_a(t) \oplus X_b(t) \oplus$$
$$\neg X_b(t) \oplus 1. \tag{10}$$

Equation (10) means $X_b(t) = \overline{X_b(t)}$. This again implies a contradiction and it is possible to claim that the autonomous Boolean network (7) does not have a fixed point and it will oscillate permanently. $\square$

3.3. Boolean Sensitivity Caused by Asymmetric Logic Functions

As demonstrated in the previous subsection, the presented BCOs do not have fixed points. In this manner, when an autonomous Boolean network is realized experimentally it should include asymmetric Boolean functions to achieve chaotic dynamics [31,35]. As a result, the preference for using logic XOR and XNOR functions in the proposed BCOs lies on the look-up table for these logic operations. Firstly, the idea is considering an equal number of "1"s and "0"s as the output of the XNOR operation to avoid converging into a physical Boolean fixed-point, i.e., where all entries of the look-up table have the same value, and hence inputs and outputs can be the same. From the look-up tables in Figures 1b and 2b, the outputs are different for all inputs, including the cases "000" and "111". On the other hand, since the number of "0"s and "1"s in the look-up tables is equal, the proposed BCOs can have a higher Boolean sensitivity $E = 2K\rho(1 - \rho)$ [30,35]. This is possible with randomly chosen Boolean functions of bias $\rho = 0.5$ (equal number of zeros and ones) and high in-degree K (the number of input connections to a node) or most effectively by using XOR and XNOR Boolean functions as herein.

4. Boolean Chaos Robust to Different Incommensurate Time-Delays

The chaos-based applications demand the exploration of different typologies and implementations to find those that are the most suitable. In addition, the chaotic behavior must be robust. It means that the chaos should be generated consistently for a wide range of parameter values. In the ongoing literature, the reported autonomous Boolean networks only show chaos in certain ranges of the feedback delays [10,30,31,35–39], as was discussed in the Introduction section. At the experimental level, those approaches incorporate an even number of logic NOT gates in the link to act as a time-delay buffer to add extra signal propagation times. Then, the published works need several pairs of NOT gates to satisfy the specific incommensurate time-delays for each one of their links connecting nodes. Otherwise, the chaotic behavior converges to either periodic oscillations or stable dynamics in the Boolean levels high or low.

Conversely, the proposed BCOs in Figures 1 and 2 do not require additional logic NOT gates to generate chaotic oscillations. This means that the chaos behavior depends solely on the incommensurate time-delays, arising only from the intrinsic delay associated with each XOR and XNOR gate. In this manner, we state the following Lemma and Corollary.

Lemma 3. *The Boolean chaotic oscillators of Figures 1 and 2 composed only by logic XOR-XNOR functions evolve to sustained chaotic oscillations not only for different time-delays of the feedback path (additional pairs of logic NOT gates) but also when the time-delays in their links are a function just of the intrinsic delay of each XOR-XNOR gates (no extra logic NOT gates).*

Corollary 1. *As a consequence of Theorems 3 and 4, an autonomous Boolean network without fixed points always presents periodic behavior if its delays are commensurate.*

Proof. To demonstrate Lemma 3 and Corollary 1, we show the physical implementation of the BCOs in Figures 1 and 2 using commercial off-the-shelf logic gates (74HCXXX family), as shown in Figure 3a. The discrete implementation makes it possible to change the time-delay between feedback nodes easily. Then, we study the dynamics of the proposed BCOs using the Lyapunov exponent method.

The scenario is as follows. First, we consider different cases for the incommensurate time-delays of the links. Those time-delays were realized using a pair of two NOT gates wired in series. Thus, from the experimental output signal of nodes C (BCO-1) and B (BCO-2) for each case in Tables 1 and 2, respectively, we collect a long enough time series. For instance, the BCO-1 output signal of node C for cases 1, 3, and 7 is given in Figure 4a,d,g, respectively. On the other hand, Figure 5a,d present the results for the output signal of node B of BCO-2 for cases 1 and 5, respectively.

Next, we compute the largest Lyapunov exponent, λ_{max}, applying the Boolean distance algorithm introduced in Section 2.3. The results in Tables 1 and 2 shows the largest Lyapunov exponent, λ_{max}, is positive for all cases indicating the proposed BCOs generate robust Boolean chaos. Besides, the chaotic behavior was also verified in the BCOs (both cases 1 in Tables 1 and 2, respectively), without extra time-delays with exception from those incommensurate intrinsic delays of the logic XOR and XNOR gates, i.e., the BCO-1 and BCO-2 do not include any logic NOT gates in the feedback paths. □

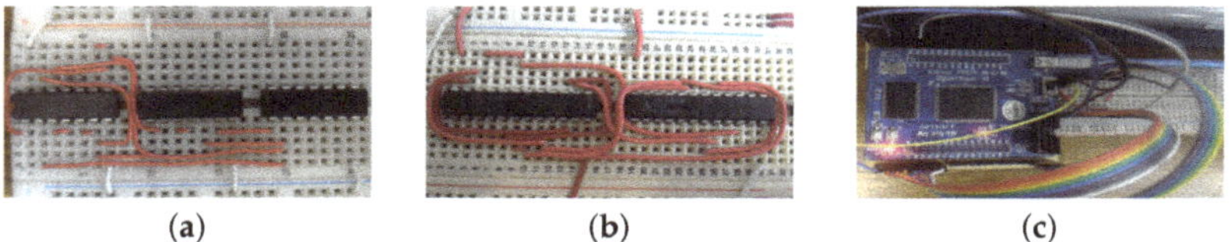

Figure 3. Experimental setup for BCOs in Figures 1 and 2 using (**a**) logic gates 74HCXXX, (**b**) GAL 22V10, and (**c**) FPGA Spartan6.

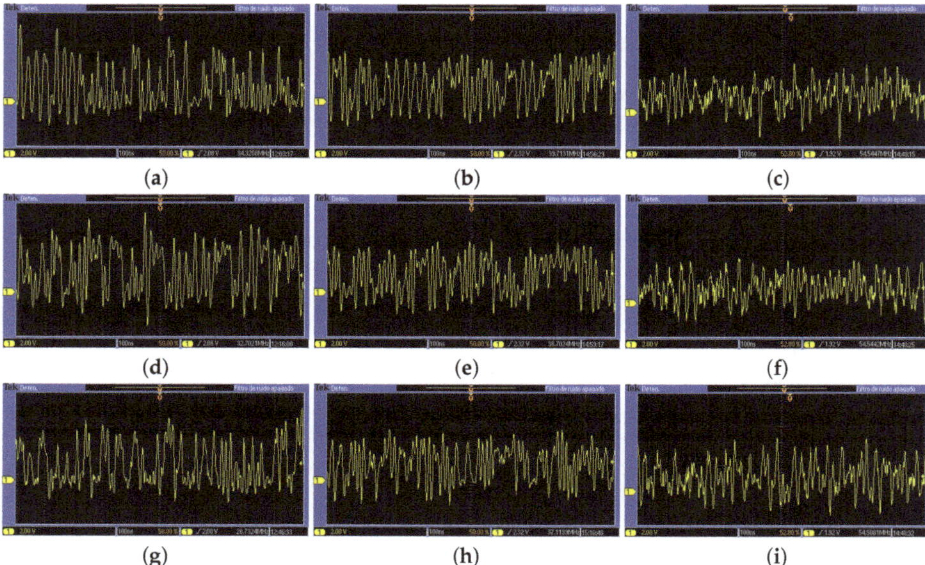

Figure 4. Chaotic oscillations from the output voltage in the node C for BCO-1. The measurements exhibit a 100 ns of time and a 2 V of voltage grid per square. Case 1 in Table 1 with (**a**) logic gates 74HCXXX, (**b**) GAL22V10, and (**c**) FPGA Spartan6. Case 3 in Table 1 with (**d**) logic gates 74HCXXX, (**e**) GAL22V10, and (**f**) FPGA Spartan6. Case 7 in Table 1 with (**g**) logic gates 74HCXXX, (**h**) GAL22V10, and (**i**) FPGA Spartan6.

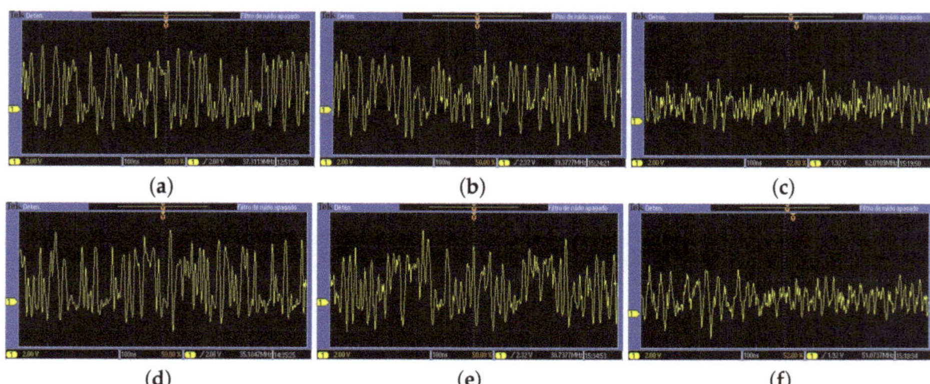

Figure 5. Chaotic oscillations from the output voltage in the node B for BCO-2. The measurements exhibit a 100 ns of time and a 2 V of voltage grid per square. Case 1 in Table 2 with (**a**) logic gates 74HCXXX, (**b**) GAL22V10, and (**c**) FPGA Spartan6. Case 5 in Table 2 with (**d**) logic gates 74HCXXX, (**e**) GAL22V10, and (**f**) FPGA Spartan6.

Tables 1 and 2 suggest the most suitable case for obtaining Boolean chaos is when no other delay paths are incorporated in the links because the higher the time-delays, the lower the magnitude of the largest Lyapunov exponent. From the physical implementation point of view, that is a remarkable feature because we can get a small form factor with the proposed Boolean chaos oscillators. Moreover, since the chaos generation does not depend on determined time-delays, the proposed BCOs can be implemented with several hardware technologies, as demonstrated in the next subsection.

Table 1. Largest Lyapunov exponent (λ_{max}) of the BCO in Figure 1 for different time-delays in the feedback paths. The symbol "-" means no extra time-delay, while "$\sqrt{}$" refers to a time-delay composed of two logic NOT gates.

Case	Time-Delay									Lyapunov Exponent
	τ_{aa}	τ_{ab}	τ_{ac}	τ_{ba}	τ_{bb}	τ_{bc}	τ_{ca}	τ_{cb}	τ_{cc}	λ_{max}
1	-	-	-	-	-	-	-	-	-	0.2306
2	-	-	-	-	-	-	-	-	$\sqrt{}$	0.2079
3	$\sqrt{}$	-	-	-	-	-	-	-	-	0.2275
4	$\sqrt{}$	-	-	-	-	-	-	-	$\sqrt{}$	0.2057
5	$\sqrt{}$	-	-	-	$\sqrt{}$	-	-	-	$\sqrt{}$	0.2076
6	$\sqrt{}$	$\sqrt{}$	$\sqrt{}$	-	-	-	-	-	-	0.2101
7	-	-	-	-	-	-	$\sqrt{}$	$\sqrt{}$	$\sqrt{}$	0.2121
8	$\sqrt{}$	$\sqrt{}$	$\sqrt{}$	-	-	-	$\sqrt{}$	$\sqrt{}$	$\sqrt{}$	0.1774
9	$\sqrt{}$	$\sqrt{}$	$\sqrt{}$	$\sqrt{}$	$\sqrt{}$	$\sqrt{}$	-	-	-	0.1808
10	$\sqrt{}$	$\sqrt{}$	$\sqrt{}$	$\sqrt{}$	$\sqrt{}$	$\sqrt{}$	$\sqrt{}$	$\sqrt{}$	$\sqrt{}$	0.1896
11	-	$\sqrt{}$	$\sqrt{}$	$\sqrt{}$	$\sqrt{}$	$\sqrt{}$	$\sqrt{}$	$\sqrt{}$	$\sqrt{}$	0.1862
12	-	$\sqrt{}$	$\sqrt{}$	$\sqrt{}$	$\sqrt{}$	$\sqrt{}$	$\sqrt{}$	$\sqrt{}$	-	0.1707

Table 2. Lyapunov exponent of BCO in Figure 2 for different time-delays in the feedback paths.

Case	Time-Delay						Lyapunov Exponent
	τ_{aa}	τ_{ab}	$\tau_{a\tilde{a}}$	τ_{bb}	τ_{ba}	$\tau_{b\tilde{b}}$	λ_{max}
1	-	-	-	-	-	-	0.1644
2	-	-	$\sqrt{}$	-	-	-	0.0960
3	-	-	-	-	-	$\sqrt{}$	0.1495
4	$\sqrt{}$	-	-	$\sqrt{}$	-	-	0.1442
5	-	$\sqrt{}$	-	-	-	-	0.1525
6	-	-	-	-	$\sqrt{}$	-	0.1448

Boolean Chaos Robust to Distinct Discrete Physical Implementation

This subsection presents three different physical implementations of the Boolean chaos oscillators. The dynamics are affected by physical constraints and hardware differences. This may lead to time-delay variations where the boolean chaos displays. Therefore, the BCO-1 and BCO-2 are constructed with three different electronic devices (i) commercial-off-the-shelf logic gates (introduced previously), (ii) a GAL, and (iii) an FPGA. The experiments in Figure 3 demonstrate the robust generation of Boolean chaos. From the circuit conception, the implementations show the chaotic behavior source is the degradation effect [32]. In addition, there are no additional procedures to calculate the delay paths to achieve chaotic oscillations.

The implementation considers all cases of Tables 1 and 2, but for the sake of simplicity, the Table 3 displays only the examples where the largest Lyapunov exponent is higher. In particular, case 1 for both BCOs is of particular interest because they do not need extra time-delays for generating chaos. More specifically, we use the GAL22V10 for realizing both BCOs, as given in Figure 3b. The programming of the GAL was performed with VHDL language. Figure 4b,e,h give the experimental results for the cases 1, 3, and 7 of BCO-1; while Figure 5b,e shows the results for cases 1 and 5 of BCO-2. For FPGA implementation, the Spartan 6 was employed (Figure 3c). The experimental results are shown in Figure 4c,f,i for cases 1, 3, 7 in Table 1, respectively. Figure 5c,f display the output signal for cases 1 and 5 in Table 2, respectively.

The measurements exhibit a 100 ns of time and a 2V of voltage grid per square. For all the three presented implementations, the output voltages (Figures 4 and 5) show the cumbersome temporal oscillations without evident periodicity. This continuous-time evolution can be identified as Boolean chaos. To verify the chaotic behavior, we compute the largest Lyapunov exponent for each implemented case of the corresponding technology, as shown in Figure 6. Table 3 also shows that the Lyapunov exponents for the three physical implementations have a similar value. This behavior suggests the Boolean chaos of proposed BCOs is robust to the distinct physical implementations changing the technology.

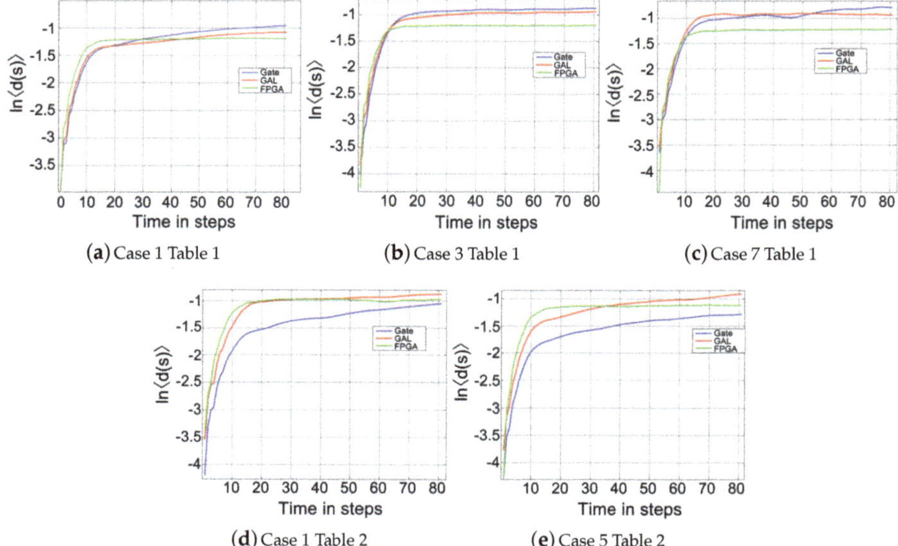

Figure 6. The divergence $\ln\langle d(s)\rangle$ to determine the largest Lyapunov exponent of the attractor for cases in Table 3 from each discrete physical implementation (Logic gates, GAL, FPGA).

Table 3. Largest Lyapunov exponent (λ_{max}) for BCOs in Figures 1 and 2 implemented experimentally with different design technologies considering the time-delays of Tables 1 and 2.

	Logic Gates	GAL	FPGA
BCO-1			
λ_{max} (case 1, Table 1)	0.230	0.224	0.209
λ_{max} (case 3, Table 1)	0.227	0.221	0.194
λ_{max} (case 7, Table 1)	0.212	0.211	0.185
BCO-2			
λ_{max} (case 1, Table 2)	0.164	0.160	0.157
λ_{max} (case 5, Table 2)	0.152	0.150	0.148

It is worth to noting that, although the intrinsic time-delays of the logic XOR and XNOR gates change among the physical realizations, Case-1 for both BCOs continues generating chaotic behavior. In agreement with Lemma 4, the fact that there is Boolean chaos, for various implementations without extra time-delays in the feedback links, demonstrates that the proposed BCOs are not overly sensitive to heterogeneous intrinsic time-delays.

5. An Application Specific Integrated Circuit for the Proposed Boolean Chaos Oscillators

5.1. Chip Design

This section describes the integrated circuit-based implementation of the prospected BCOs in this work. The Boolean chaos generators are described with Verilog, a Hardware Description Language (HDL), using the UMC 180 nm Generic Core Cell Library. The BCO-1 hardware description in Figure 7a uses one XNOR3S and two XOR3S cells from the Generic Core library. That verilogHDL code synthesizes the three logic gates, whereas the Encounter tool (from Cadence Design Systems) executes a generic routing algorithm. Similarly, the BCO-2 of Figure 7b uses one XOR3S and one XNOR3S for the description with the verilogHDL code. The integrated circuit was part of a multiprocess wafer run and is shown in Figure 8 (Left). The size of BCO-1 is 75 μm × 60 μm while the BCO-2 has physical dimensions of 32 μm × 26 μm. The area for biasing rails is considered in both scenarios. In any case, it is possible to reduce the size with routing optimization.

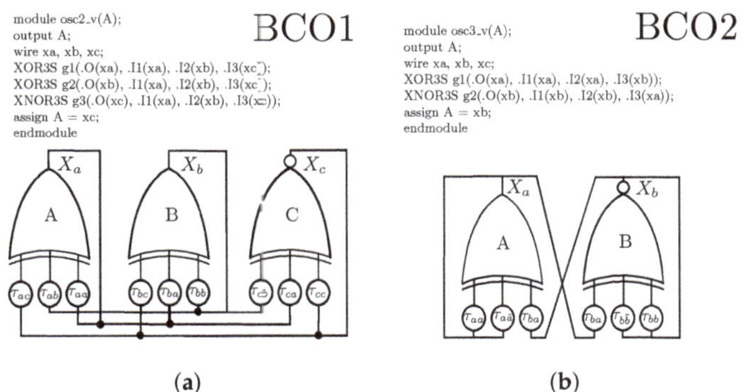

Figure 7. (a) Synthesis codes in VerilogHDL for (a) BCO-1, and (b) BCO-2, respectively.

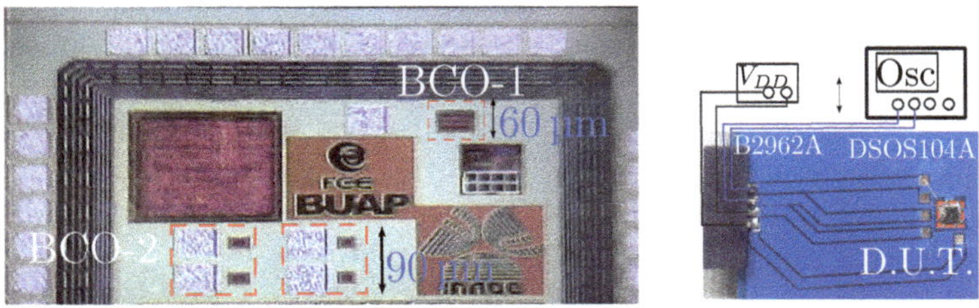

Figure 8. Microphotography of the chip and the test-bench for the integrated circuit.

It is worth noting that the design process of the IC is straightforward, and it does not depend on critical design considerations. However, all the design processes were executed in a semiautomated way using the generic cells and routing tool from Cadence software and UMC 180 nm fabrication technology. Therefore, this demonstrates, once again, the flexibility and robustness of the proposed BCOs.

On the other hand, Figure 8 (Right) shows the test-bench for the integrated circuit. The chip-die is mounted on an FR4 printed circuit board. It is biased with a low-noise $V_{DD} = 1.8\,\text{V}$ voltage source model B2962A while the oscilloscope DSOS104A captures the voltage time-series for further analysis.

5.2. Experimental Results of the Integrated BCO-1 and BCO-2

Now, we present the continuous-time behavior of both BCO-1 and BCO-2 on the integrated circuit. Figure 9 shows the real-time obtained waveforms and the respective dynamical analysis considering:

(i) time-series of the output voltage; (ii) frequency spectra; (iii) time-lag reconstructions of the attractors; (iv) Poincaré mapping set; (v) and largest Lyapunov exponent.

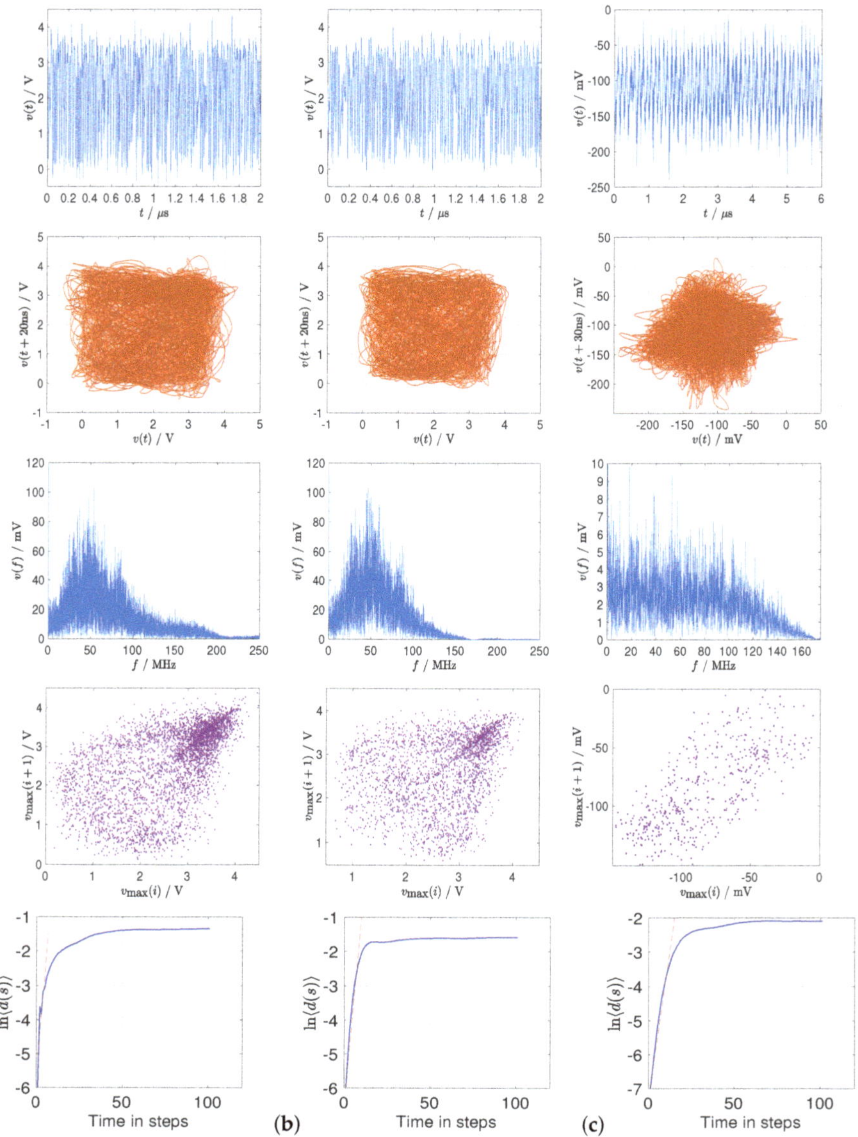

Figure 9. Chaotic dynamics measured experimentally from the integrated circuit of 180 nm at distinct settings for both Boolean chaos oscillators. Top to bottom: Time-series, time-lag embedded attractor, frequency spectrum, Poincaré map, the divergence $\ln\langle d(s)\rangle$ to determine the largest Lyapunov exponent λ_{max} of the attractor. (**a**) Experimental results for BCO-1 @ V_{DD} = 3.3 V with λ_{max} = 0.4496; (**b**) Experimental results for BCO-1 @ V_{DD} = 2.8 V with λ_{max} = 0.4243; (**c**) Experimental results for BCO-2 @ V_{DD} = 3.3 V with λ_{max} = 0.2492.

Figure 9a presents the BCO-1 features for $V_{DD} = 3.3$ V. The time-series shows a random evolution since it has variable cycle amplitudes regarding maxima and minima, and the frequency content is characterized for a predominant broad distribution and with strong content up to 200 MHz, therefore suggesting chaotic oscillations. Besides, the time-lag embedded attractor (lag equal to the first minimum of the time-lag mutual information function) exhibits a chaotic behavior in phase space, whose underlying complexity can be more properly appreciated on the corresponding Poincaré map for amplitudes of successive local maxima. In this manner, the arbitrarily chosen plane sections the attractor in two and thereby enables the visualization of its complex geometry. We found that the Poincaré map has a dense set of points, which has been identified as characteristic dynamics of the chaotic behavior. To quantify these observations, we determine the largest Lyapunov exponent (λ_{max}) of the attractor. The result shows the time evolution of the $\ln\langle d(s)\rangle$. This divergence presents an almost constant slope for the first part of the curve and then it saturates at a maximum value, corresponding to the uncorrelated signals $x(s + T + t_a)$ and $x(s + T + t_b)$. Next, we estimate the value of λ_{max}, assuming that the divergence of the initially similar segments is exponential in the region of constant slope. As a result, the average of all pairs of similar segments is our estimate of the largest Lyapunov exponent for the BCO-1, giving $\lambda_{max} = 0.4496$, which demonstrates that the CMOS Boolean oscillator integrated at 180 nm is chaotic.

Figure 9b shows the dynamical analysis for the same BCO-1 but now with $V_{DD} = 2.8$ V. This BCO-1 displays a clear chaotic attractor in the 0–4 V range and is validated with the Poincaré set. The results for time-series, frequency spectrum up to 150 MHz, Poincaré map, and $\lambda_{max} = 0.4243$ have a similar response to the previous case. Therefore, we can conclude the BCO-1 is robust against bias voltage variations.

The same test is included for the BCO-2 biased to VDD = 3.3 V. The circuit presents a chaotic oscillation but the on-chip pad originates an explicit limitation of the voltage swing. This prototype uses internal pad connections and the reduced swing is a consequence of the extended bonding and absence of I/O cells. Therefore, the on-die probes represent an important load impedance and limit output swing to 200 mV. Figure 9c shows the time evolution, spectral content up to 160 MHz, chaotic attractor, and Poincaré map showing the expected results. Finally, Figure 9c also shows the largest Lyapunov exponent, which has a slope less abrupt but still presents a positive exponent ($\lambda_{max} = 0.2492$) in spite of the small values of the continuous-time sequence.

5.3. Comparison with Similar Implementations

For the sake of reference, Table 4 highlights the principal features of the recent True Random Number Generator (TRNG), systems based on chaotic circuits. The two new boolean chaotic oscillators exhibit competitive numbers compared to the references [20,42]. The comparison includes the most recent works with attempts to fully integrate the system-on-chip. The work in [20] presents a set of inverted-based chaotic oscillators. The area and power consumption are affordable, but the system uses additional off-chip biasing circuits. The chaotic circuit in [42] is fully integrated with the disadvantage of increasing the circuit resources. This is the result of the multiattractor analog system requiring a large bandwidth but the band limitation or frequency centroid is not reported. The two new boolean chaotic oscillators in this work present a reduced circuit size and power dissipation, not considering the chip input-output cells. The proposed boolean chaotic circuits present the most extended frequency span content compared to the recent on-chip implementations.

Table 4. Comparison of the principal features of recent TRNG based on chaotic systems.

	This Work BCO-1	This Work BCO-2	[20]	[42]
Chaos source	Boolean chaos	Boolean chaos	Chaotic oscillation	Multiattractor
Integrated	Fully	Fully	Partially	Fully
Technology	180 nm	180 nm	180 nm	180 nm
Size $(\mu m)^2$	4500	832	28,000	$(315,000 \times 383,000)$
Static power (μw)	0.2	0.09	25	3660
Speed limit (MHz)	200	160	10	NA

6. Conclusions

Two autonomous Boolean networks that generate Boolean chaos have been introduced. From the mathematical model, it was shown that the logical states would never reach a fixed-point, and therefore, will oscillate permanently. The proposed Boolean chaos oscillators exhibited no dependence on incommensurate time-delays, as demonstrated by computing the Lyapunov exponents under various scenarios for the delay paths. The correct physical implementations of the two Boolean chaos oscillators are good evidence of the predicted conditions. Therefore, the BCOs are reliable and robust to be implemented with multiple circuit implementations, both discrete as integrated.

In particular, the synthesis of the chaotic oscillators in an integrated circuit has shown the benefits of a compact CMOS chaos generator with areas $0.0045 \, \text{mm}^2$ and $0.000\,832 \, \text{mm}^2$ for BCO-1 and BCO-2, respectively, as well as high-speed chaotic oscillations with relevant amplitude content up to 200 MHz. Several dynamical analyses such as time-series, chaotic attractors, Poincaré maps, and Lyapunov exponents validated the experimental results.

In this manner, the proposed Boolean chaos oscillators could be useful for various engineering applications, for instance, random number generators, since the design is straightforward, robust, compact, and can be implemented in many options of hardware without needing special considerations.

Author Contributions: Conceptualization, J.M.M.-P. and T.G.-C.; methodology, J.M.M.-P. and V.R.G.-D.; software, J.M.M.-P., V.R.G.-D., G.d.L.F.-C.; validation, V.R.G.-D., T.G.-C. and G.d.L.F.-C.; formal analysis, J.M.M.-P.; investigation, J.M.M.-P., T.G.-C., and L.d.C.G.-P.; writing—original draft preparation, J.M.M.-P. and T.G.-C.; writing—review and editing, J.M.M.-P., V.R.G.-D., L.d.C.G.-P. All authors have read and agreed to the published version of the manuscript.

Funding: This research received no external funding.

Acknowledgments: J.M. Munoz-Pacheco acknowledges the National Council of Science and Technology (CONACYT) MEXICO for the financial support through Project No. 258880 (Proyecto Apoyado por el Fondo Sectorial de Investigacion para la Educacion). The Authors also thank CONACYT for the Lab project INFRA-2013-1-205873. The authors thankfully acknowledge computer resources, technical advice, and support provided by Laboratorio Nacional de Supercómputo del Sureste de México(LNS), a member of the CONACYT national laboratories, with project No. 202001025C.

Conflicts of Interest: The authors declare no conflict of interest.

References

1. Yin, Q.; Wang, C. A new chaotic image encryption scheme using breadth-first search and dynamic diffusion. *Int. J. Bifurc. Chaos* **2018**, *28*, 1850047. [CrossRef]
2. Kaddoum, G. Wireless chaos-based communication systems: A comprehensive survey. *IEEE Access* **2016**, *4*, 2621–2648. [CrossRef]
3. Kocamaz, U.E.; Çiçek, S.; Uyaroğlu, Y. Secure communication with chaos and electronic circuit design using passivity-based synchronization. *J. Circuits Syst. Comput.* **2018**, *27*, 1850057. [CrossRef]
4. Herceg, M.; Miličević, K.; Matić, T. Frequency-translated differential chaos shift keying for chaos-based communications. *J. Frankl. Inst.* **2016**, *353*, 2966–2979.

5. Dmitriev, A.; Efremova, E.; Nikishov, A.Y. Generating dynamic microwave chaos in self-oscillating ring system based on complementary metal-oxide-semiconductor structure. *Tech. Phys. Lett.* **2010**, *36*, 430–432. [CrossRef]

6. Herceg, M.; Vranješ, D.; Grbić, R.; Job, J. Chaos-Based transmitted-reference ultra-wideband communications. *Int. J. Electron.* **2019**, *106*, 160–172. [CrossRef]

7. Gomez-Pavon, L.; Munoz-Pacheco, J.; Luis-Ramos, A. Synchronous Chaos Generation in an Er^{+3}-Doped Fiber Laser System. *IEEE Photonics J.* **2015**, *7*, 1–6. [CrossRef]

8. Liu, Z.; Zhu, X.; Hu, W.; Jiang, F. Principles of chaotic signal radar. *Int. J. Bifurc. Chaos* **2007**, *17*, 1735–1739. [CrossRef]

9. Xu, H.; Li, L.; Li, Y.; Zhang, J.; Han, H.; Liu, L.; Li, J. Chaos-Based Through-Wall Life-Detection Radar. *Int. J. Bifurc. Chaos* **2019**, *29*, 1930020. [CrossRef]

10. Qiao, J.; Xu, H.; Zhang, J.; Han, H.; Wang, B. High-resolution and anti-jamming chaotic guided radar prototype for perimeter intrusion detection. *J. Electromagn. Waves Appl.* **2019**, *33*, 1060–1069. [CrossRef]

11. Fortuna, L.; Frasca, M.; Rizzo, A. Chaotic pulse position modulation to improve the efficiency of sonar sensors. *IEEE Trans. Instrum. Meas.* **2003**, *52*, 1809–1814. [CrossRef]

12. Shin, S.; Kim, M.H.; Choi, S.B. Ultrasonic distance measurement method with crosstalk rejection at high measurement rate. *IEEE Trans. Instrum. Meas.* **2018**, *68*, 972–979. [CrossRef]

13. Honglad, S.; San-Um, W. Automatic stand-alone liquid mixer with chaotic PWM control using diode-based Rössler system. In Proceedings of the IEEE 2014 International Electrical Engineering Congress (iEECON), Chonburi, Thailand, 19–21 March 2014; pp. 1–4.

14. Xie, T.; Chen, M.; Xu, C.; Chen, J. High-throughput extraction and separation of Ce (III) and Pr (III) using a chaotic advection microextractor. *Chem. Eng. J.* **2019**, *356*, 382–392. [CrossRef]

15. Murali, K.; Sinha, S.; Mohamed, I.R. Chaos computing: Experimental realization of NOR gate using a simple chaotic circuit. *Phys. Lett. A* **2005**, *339*, 39–44. [CrossRef]

16. Kia, B.; Mobley, K.; Ditto, W.L. An integrated circuit design for a dynamics-based reconfigurable logic block. *IEEE Trans. Circuits Syst. II Express Briefs* **2017**, *64*, 715–719. [CrossRef]

17. Wannaboon, C.; Tachibana, M.; San-Um, W. A 0.18-μm CMOS high-data-rate true random bit generator through $\Delta\Sigma$ modulation of chaotic jerk circuit signals. *Chaos Interdiscip. J. Nonlinear Sci.* **2018**, *28*, 063126. [CrossRef]

18. Li, B.; Liao, X.; Jiang, Y. A novel image encryption scheme based on improved random number generator and its implementation. *Nonlinear Dyn.* **2019**, *95*, 1781–1805. [CrossRef]

19. Minati, L. Experimental implementation of networked chaotic oscillators based on cross-coupled inverter rings in a CMOS integrated circuit. *J. Circuits Syst. Comput.* **2015**, *24*, 1550144. [CrossRef]

20. Minati, L.; Frasca, M.; Yoshimura, N.; Ricci, L.; Oświecimka, P.; Koike, Y.; Masu, K.; Ito, H. Current-Starved Cross-Coupled CMOS Inverter Rings as Versatile Generators of Chaotic and Neural-Like Dynamics Over Multiple Frequency Decades. *IEEE Access* **2019**, *7*, 54638–54657. [CrossRef]

21. Volos, C.K.; Kyprianidis, I.M.; Stouboulos, I.N. Experimental investigation on coverage performance of a chaotic autonomous mobile robot. *Robot. Auton. Syst.* **2013**, *61*, 1314–1322. [CrossRef]

22. Petavratzis, E.K.; Volos, C.K.; Stouboulos, I.N.; Nistazakis, H.E.; Kyritsi, K.G.; Valavanis, K.P. Coverage Performance of a Chaotic Mobile Robot Using an Inverse Pheromone Model. In Proceedings of the IEEE 2019 8th International Conference on Modern Circuits and Systems Technologies (MOCAST), Thessaloniki, Greece, 13–15 May 2019; pp. 1–4.

23. Ansari, U.; Bajodah, A.H.; Kada, B. Development and experimental investigation of a Quadrotor's robust generalized dynamic inversion control system. *Nonlinear Dyn.* **2019**, *96*, 1541–1557. [CrossRef]

24. Montero-Canela, R.; Zambrano-Serrano, E.; Tamariz-Flores, E.I.; Muñoz-Pacheco, J.M.; Torrealba-Meléndez, R. Fractional chaos based-cryptosystem for generating encryption keys in Ad Hoc networks. *Ad Hoc Netw.* **2020**, *97*, 102005. [CrossRef]

25. Wang, T.; Wang, D.; Wu, K. Chaotic adaptive synchronization control and application in chaotic secure communication for industrial Internet of Things. *IEEE Access* **2018**, *6*, 8584–8590. [CrossRef]

26. Mareca, P.; Bordel, B. Robust hardware-supported chaotic cryptosystems for streaming commutations among reduced computing power nodes. *Analog Integr. Circuits Signal Process.* **2019**, *98*, 11–26. [CrossRef]

27. Muñoz-Pacheco, J.M.; Zambrano-Serrano, E.; Félix-Beltrán, O.; Gómez-Pavón, L.C.; Luis-Ramos, A. Synchronization of PWL function-based 2D and 3D multi-scroll chaotic systems. *Nonlinear Dyn.* **2012**, *70*, 1633–1643. [CrossRef]

28. Flores-Vergara, A.; Garcia-Guerrero, E.; Inzunza-González, E.; López-Bonilla, O.; Rodríguez-Orozco, E.; Cardenas-Valdez, J.; Tlelo-Cuautle, E. Implementing a chaotic cryptosystem in a 64-bit embedded system by using multiple-precision arithmetic. *Nonlinear Dyn.* **2019**, *96*, 497–516. [CrossRef]

29. Raza, S.F.; Satpute, V. A novel bit permutation-based image encryption algorithm. *Nonlinear Dyn.* **2019**, *95*, 859–873. [CrossRef]

30. Zhang, R.; de S.Cavalcante, H.L.D.; Gao, Z.; Gauthier, D.J.; Socolar, J.E.S.; Adams, M.M.; Lathrop, D.P. Boolean chaos. *Phys. Rev. E* **2009**, *80*, 045202. [CrossRef]

31. Cavalcante, H.L.; Gauthier, D.J.; Socolar, J.E.; Zhang, R. On the origin of chaos in autonomous Boolean networks. *Philos. Trans. R. Soc. A Math. Phys. Eng. Sci.* **2010**, *368*, 495–513. [CrossRef]

32. Bellido-Diaz, M.; Juan-Chico, J.; Acosta, A.; Valencia, M.; Huertas, J. Logical modelling of delay degradation effect in static CMOS gates. *IEE Proc. Circuits Devices Syst.* **2000**, *147*, 107–117. [CrossRef]

33. Ghil, M.; Mullhaupt, A. Boolean delay equations. II. Periodic and aperiodic solutions. *J. Stat. Phys.* **1985**, *41*, 125–173. [CrossRef]

34. Dee, D.; Ghil, M. Boolean difference equations, I: Formulation and dynamic behavior. *SIAM J. Appl. Math.* **1984**, *44*, 111–126. [CrossRef]

35. Rosin, D.P. *Dynamics of Complex Autonomous Boolean Networks*; Springer: Berlin, Germany, 2014.

36. Rivera-Durón, R.R.; Campos-Cantón, E.; Campos-Cantón, I.; Gauthier, D.J. Forced synchronization of autonomous dynamical Boolean networks. *Chaos Interdiscip. J. Nonlinear Sci.* **2015**, *25*, 083113. [CrossRef] [PubMed]

37. Rosin, D.P.; Rontani, D.; Gauthier, D.J. Ultrafast physical generation of random numbers using hybrid Boolean networks. *Phys. Rev. E* **2013**, *87*, 040902. [CrossRef]

38. Rosin, D.P.; Rontani, D.; Gauthier, D.J.; Schöll, E. Experiments on autonomous Boolean networks. *Chaos Interdiscip. J. Nonlinear Sci.* **2013**, *23*, 025102. [CrossRef]

39. Park, M.; Rodgers, J.C.; Lathrop, D.P. True random number generation using CMOS Boolean chaotic oscillator. *Microelectron. J.* **2015**, *46*, 1364 – 1370. [CrossRef]

40. Wolf, A.; Swift, J.B.; Swinney, H.L.; Vastano, J.A. Determining Lyapunov exponents from a time series. *Phys. D Nonlinear Phenom.* **1985**, *16*, 285–317. [CrossRef]

41. Luque, B.; Solé, R.V. Lyapunov exponents in random Boolean networks. *Phys. A Stat. Mech. Appl.* **2000**, *284*, 33–45. [CrossRef]

42. Zhang, X.; Wang, C. A Novel Multi-Attractor Period Multi-Scroll Chaotic Integrated Circuit Based on CMOS Wide Adjustable CCCII. *IEEE Access* **2019**, *7*, 16336–16350. [CrossRef]

Article

The Effect of a Non-Local Fractional Operator in an Asymmetrical Glucose-Insulin Regulatory System: Analysis, Synchronization and Electronic Implementation

Jesus M. Munoz-Pacheco [1], Cornelio Posadas-Castillo [2] and Ernesto Zambrano-Serrano [2],*

[1] Faculty of Electronics Sciences, Autonomous University of Puebla, Av. San Claudio y 18 Sur, Edif. FCE2, Puebla C.P. 72570, Mexico; jesusm.pacheco@correo.buap.mx

[2] Facultad de Ingeniería Mecánica y Eléctrica, Universidad Autónoma de Nuevo León, Av. Pedro de Alba S/N, Cd. Universitaria, San Nicolás de los Garza C.P. 66455, Nuevo León, Mexico; cornelio.posadascs@uanl.edu.mx

* Correspondence: ernesto_zambrano1@nemdigital.sep.gob.mx; Tel.: +52-81-8329-4020

Received: 11 July 2020; Accepted: 19 August 2020; Published: 21 August 2020

Abstract: For studying biological conditions with higher precision, the memory characteristics defined by the fractional-order versions of living dynamical systems have been pointed out as a meaningful approach. Therefore, we analyze the dynamics of a glucose-insulin regulatory system by applying a non-local fractional operator in order to represent the memory of the underlying system, and whose state-variables define the population densities of insulin, glucose, and β-cells, respectively. We focus mainly on four parameters that are associated with different disorders (type 1 and type 2 diabetes mellitus, hypoglycemia, and hyperinsulinemia) to determine their observation ranges as a relation to the fractional-order. Like many preceding works in biosystems, the resulting analysis showed chaotic behaviors related to the fractional-order and system parameters. Subsequently, we propose an active control scheme for forcing the chaotic regime (an illness) to follow a periodic oscillatory state, i.e., a disorder-free equilibrium. Finally, we also present the electronic realization of the fractional glucose-insulin regulatory model to prove the conceptual findings.

Keywords: fractional-order; glucose-insulin system; chaotic attractor; active control; synchronization

1. Introduction

Homeostasis is the tendency of organisms to auto-regular and maintain their internal environment in a stable state [1], For instance, an excellent model to describe the homeostatic process in the organism is the glucose-insulin system [1,2]. On one side, when the glucose level is low, arises diverse pathologies (anxiety, tremors, obfuscation, coma, etc.). On the other side, microvascular damages in the retina, kidney, and neuronal injuries, which lead to chronic renal insufficiency and blindness, are originated by high glucose concentrations. The principal pathology of glucose homeostasis is diabetes. Diabetes can be stated as a chronic disorder provoked when the pancreas does not produce insulin in enough quantities, and also if the body cannot successfully process it, resulting in atypical high blood sugar levels. Because of the autoimmune annihilation of β-cells, the insulin released by the pancreas to the human organism is not enough to maintain a certain healthy level, and therefore, the Type 1 Diabetes Mellitus (T1DM) may be manifested. On the contrary, when the beta-cells can produce supranormal, or average concentrations of insulin, but cannot be adequately used in reducing glycemia, the Type 2 Diabetes Mellitus (T2DM) appears [3].

Health agencies [4] demonstrate people, mainly adults, living with diabetes has almost quadrupled from 1980 to 422 million. This rise is primarily due to the rise in T2DM and factors driving it, including overweight and obesity. Diverse mathematical models using ODEs and OdEs have provided a common path to understand multiple complex systems [5,6]. In the last years, because of the long-memory of fractional-order operators, fractional-order systems have gained extensive attention for describing and understanding physical and biological systems [5,7–10]. For instance, the analysis of diverse biological systems, such as bio-impedance, drug diffusion, respiratory tissue, and so forth, was reported in [11] using fractional calculus. In [12], the authors characterize the trade-offs between HIV infection and the tumor-immune system employing fractional-order biological systems. Kheiri and Jafari [13] formulated a fractional-order theory that was focused on the multi-patch HIV/AIDS model to analyze whether human migration has effects on the propagation of the HIV/AIDS outbreak. In [14], the authors reported a fractional-order Izhikevich system to obtain insights on distinct neuronal spike responses, including bursting, fast-, regular-spiking, and chattering, as the no-integer order varies. The functioning at electrical level of a fractional-order system of an isolated β-cell is presented in [15,16]. In [17], was reported the comparison among diverse scenarios, such as integer, constant, no-constant and fractional-order derivatives, in order to explain the memory index. In [18], the transmission issues of a susceptible-infected-recovered model were analyzed. They found a proper yield of memory of the fractional-order systems to forecast the pandemic spread. Finally, [19] discovered that the neuron's firing rate could be emulated with a fractional derivative and a slowly varying of the parameters. Therefore, the unique rat neocortical pyramidal neurons have several time-scales.

Therefore, the study of fractional-order biological models continues been critical for an accurate analysis of several health conditions, as well as being essential to understanding this significant open-topic.

Additionally, the synchronization and collective dynamics play an essential role both in physical [20–22] and biological systems [15,23–26]. By one side, the synchronization is necessary for systems with stable behavior, for instance, the human heartbeat and respiration [27,28]; nevertheless, on the other hand, the synchronized state could provoke severe pathologies such as the Parkinson's disease where the excessive synchronization correlates with a motor deficit. For the glucose-insulin regulatory system, a synchronization state is mandatory, since this condition is closely related to a disorder-free state where the glucose-insulin concentrations synchronize. In [29], the β-cell synchronization is fundamental to effectuate a pulsatile insulin liberation, which carries more substantial hypoglycemic results compared to constant secretion. In [30], the synchronization between gap-junctions and adjacent cells is essential to limit the heterogeneity and biological noise, thence obtaining a robust activation of the β-cells population within the islet. Pecora and Carroll in [31] showed that two nonlinear dynamical systems could be synchronized by introducing appropriate coupling, since then, a variety of approaches were proposed to deal with synchronized states of fractional-order chaos generators, and these include backstepping, adaptive, and active controls [32,33]. Regarding the chaos synchronization applying the active control method [34], it has been demonstrated that an active controller can be straightforwardly designed to achieve synchronization globally if the nonlinearity of the system is known. Thereupon, it is considered to be a promising control strategy due to its straightforwardness. [33].

In addition, the experimental verification of fractional-order models is a topic that has been attracting the attention of researchers [35–39]. In this scenario, ARM-based embedded systems have become in a central block integrated with non-embedded technologies, such as FPGAs, DSPs, and microcontrollers. Subsequently, now it is possible to implement complex software and hardware functionality on a single chip [40]. Several digital hardware have been reported and verified contemplating integer-order chaotic systems [41–43]. Notwithstanding, just a short-list of papers have studied the implementation of fractional chaos oscillators on ARM platforms [44,45].

Motivated by the discussion mentioned above, we analyze the effect of a non-local fractional operator in an asymmetrical glucose-insulin regulatory system. More specifically, the system parameters that were related to hypoglycemia, hyperinsulinemia, T1DM, and T2DM, were studied by using both analytical (stability of equilibrium) and numerical (bifurcation diagrams and basins of attraction) techniques. Additionally, we found that the system presents a chaotic attractor, as demonstrated by its phase portraits and Lyapunov exponent spectrum. Besides, the theoretical insights of the glucose-insulin interaction can be validated by using the proposed electronic implementation based on the embedded SoC ARM Broadcom BCM2837B0. Latter, the synchronization between a chaotic behavior (disorder) and periodic behaviors (average condition) were achieved by applying a simple control strategy.

The sections of the manuscript contain the following. The fractional-order glucose-insulin system, in the sense of Caputo, is presented in Section 2. The stability of the equilibrium points is analyzed in Section 3. Several dynamical analysis for the hypoglycemia, hyperinsulinemia, T1D, and T2D are presented in Section 4. Two synchronization schemes using the active control approach are introduced in Section 5. The ARM implementation is carried out in Section 6, which is followed by the conclusions in Section 7.

2. Fractional-Order Glucose-Insulin Regulatory System

Because diverse biological systems present memory effects, direct generalization maybe by applying fractional-order differential equations, i.e., arbitrary (non-integer) orders. In this manner, we can get a closer insight into the real phenomena. The benefits of fractional-order are mainly to capture the entire time evolution for physical processes, being a kind of memory index, as well as more degrees of freedom for the resulting models. The analysis conducted in this work is inspired by the glucose-insulin regulatory system in [46], as given in Equation (1). This model was derived from the Ackerman, Bajaj, and Rao, and predator-prey Volterra models. As a consequence, the resulting system is formulated as the trade-off relating the glucose and insulin, but including the beta-cells interaction. Apart from normal metabolic conditions described by the classical relation between prey (glucose) and predator (insulin), the model (1) also characterizes the metabolic disorders as chaotic oscillations. Abnormal biological conditions, including glucose-insulin interactions, can be stated as chaotic evolutions of the dynamical systems, as given in [47–50], to mention a few. From experimental data, Ref. [51] also discussed that the T1DM may have chaotic behaviors. In this manner, the proposed analysis of the glucose-insulin regulatory system is in the same line of previous studies, but instead applying fractional-calculus theory.

$$
\begin{aligned}
\dot{x} &= -a_1 x + 0.1xy + 1.09y^2 - 1.08y^3 + 0.03z - 0.06z^2 + a_7 z^3 - 0.19, \\
\dot{y} &= -a_8 xy + 3.84x^2 + 1.2x^3 + 0.3y(1-y) - 1.37z + 0.3z^2 - 0.22z^3 - 0.56, \\
\dot{z} &= a_{15} y - 1.35y^2 + 0.5y^3 + 0.42z + 0.15yz,
\end{aligned}
\tag{1}
$$

we now introduce a fractional-order version [52], but instead applying the Caputo's definition. Let us consider the fractional differential operator $D^q f(t) = J^{n-\gamma} D^n f(t)$, with $q > 0$ and $n \in \mathbb{N}$, where D^n describes the n-order derivative, while J^γ defines the γ-order integral operator in the sense of Riemann-Liouville as

Definition 1. *For a function $f(t)$, the fractional-order (γ) integral, with $\gamma \in \mathbb{R}^+$ can be determined by*

$$
J^\gamma f(t) = \frac{1}{\Gamma(\gamma)} \int_{t_0}^t (t-s)^{\gamma-1} f(s)\,ds,
\tag{2}
$$

where $t \geq t_0$ and $\Gamma(\cdot)$ is the Gamma function, defined as $\Gamma(s) = \int_0^\infty t^{s-1} e^{-t} dt$.

In this manner, the Caputo operator of fractional-order can be established by

Definition 2. *The fractional derivative of Caputo with order q for a function $f(t) \in \mathbb{C}^n([t_0, \infty), \mathbb{R})$ is given as*

$$
D^q f(t) = \frac{1}{\Gamma(n-q)} \int_{t_0}^t \frac{f^{(n)}(s)}{(t-s)^{q-n+1}}\,ds,
\tag{3}
$$

and when $0 < q < 1$ is given, as follows

$$D^q f(t) = \frac{1}{\Gamma(1-q)} \int_{t_0}^{t} \frac{f'(s)}{(t-s)^q} ds. \tag{4}$$

By substituting Equation (4) in Equation (1), we get the fractional-order glucose-insulin regulatory system that is given by

$$\begin{aligned}
D^{q_1} x &= -a_1 x + 0.1xy + 1.09y^2 - 1.08y^3 + 0.03z - 0.06z^2 + a_7 z^3 - 0.19, \\
D^{q_2} y &= -a_8 xy + 3.84x^2 + 1.2x^3 + 0.3y(1-y) - 1.37z + 0.3z^2 - 0.22z^3 - 0.56, \\
D^{q_3} z &= a_{15} y - 1.35y^2 + 0.5y^3 + 0.42z + 0.15yz,
\end{aligned} \tag{5}$$

where $D^{q_i}(x, y, z)$ with $i = 1, 2, 3$ is the the fractional derivative operator in Caputo's sense [53], q_i is the fractional-order satisfying $0 < q_i \leq 1$, (x, y, z) describe the population density of insulin, the population size of glucose, and the population size of β-cells, respectively. The representative reduction of insulin levels for glucose deficiency is described by a_1, whereas the parameter a_8 stands for the impact of insulin on glucose. Additionally, the increment rate of insulin concentrations released by b-cells is provided by a_7. Finally, the increasing rate of the β-cells provoked, when the glucose levels also grow, is given by a_{15} [46]. From the nature of how this system was conceived, the parameters must be nonnegative [46].

The discrete-time version for the proposed system (5) is set as:

$$\begin{aligned}
x_{n+1} &= x_0 + \frac{h^{q_1}}{\Gamma(q_1 + 2)} \left(f_1(x_{n+1}^p, y_{n+1}^p, z_{n+1}^p) + \sum_{k=0}^{n} \alpha_{1,k,n+1} f_1(x_k, y_k, z_k) \right), \\
y_{n+1} &= y_0 + \frac{h^{q_2}}{\Gamma(q_2 + 2)} \left(f_2(x_{n+1}^p, y_{n+1}^p, z_{n+1}^p) + \sum_{k=0}^{n} \alpha_{2,k,n+1} f_2(x_k, y_k, z_k) \right), \\
z_{n+1} &= z_0 + \frac{h^{q_3}}{\Gamma(q_3 + 2)} \left(f_3(x_{n+1}^p, y_{n+1}^p, z_{n+1}^p) + \sum_{k=0}^{n} \alpha_{3,k,n+1} f_3(x_k, y_k, z_k) \right),
\end{aligned} \tag{6}$$

where

$$\begin{aligned}
x_{n+1}^p &= x_0 + \frac{1}{\Gamma(q_1)} \sum_{k=0}^{n} \beta_{1,k,n+1} f_1(x_k, y_k, z_k), \\
y_{n+1}^p &= y_0 + \frac{1}{\Gamma(q_2)} \sum_{k=0}^{n} \beta_{2,k,n+1} f_2(x_k, y_k, z_k), \\
z_{n+1}^p &= z_0 + \frac{1}{\Gamma(q_3)} \sum_{k=0}^{n} \beta_{3,k,n+1} f_3(x_k, y_k, z_k),
\end{aligned} \tag{7}$$

with

$$\alpha_{i,k,n+1} = \begin{cases} n^{q_i+1} - (n - q_i)(n+1)^{q_i}, k = 0, & k = 0 \\ (n - k + 2)^{q_i+1} + (n - k)^{q_i+1} - 2(n - k + 1)^{q+1}, & 1 \leq k \leq n, \\ 1, & k = n + 1, \end{cases} \tag{8}$$

and

$$\beta_{i,k,n+1} \frac{h^{q_i}}{q_i} \left((n + 1 - k)_i^q - (n - k)_i^q \right). \tag{9}$$

note that x_0, y_0, z_0 are the initial values for, $f_1(x, y, z) = -a_1 x + 0.1xy + 1.09y^2 - 1.08y^3 + 0.03z - 0.06z^2 + a_7 z^3 - 0.19$, $f_2(x, y, z) = -a_8 xy + 3.84x^2 + 1.2x^3 + 0.3y(1-y) - 1.37z + 0.3z^2 - 0.22z^3 - 0.56$ and $f_3(x, y, z) = a_{15} y - 1.35y^2 + 0.5y^3 + 0.42z + 0.15yz$, whether $q_i = q$, $i = 1, 2, 3$ the system is said to be commensurate then the convergence order is described as $|y(t_n) - y_n| = O(h^{\min(2,1+q)})$, $h \to 0$ [54–56].

3. Stability Analysis of Fractional-Order Glucose-Insulin Model

A non-local differential equation with fractional-order $q \in (0,1)$, normally has a stability region larger than that of the integer-order version with $q = 1$ [57]. We introduce the following theorems and definitions to discuss the stability of the proposed fractional-order system.

Definition 3. *A system denoted as*

$$D^{q_i} x_i(t) = f_i(x_1(t), x_2(t), \ldots, x_n(t), t), \quad x_i(0) = c_i, \quad i = 1, 2, \ldots, n, \tag{10}$$

with trajectory $x(t) = 0$, where c_i are the starting conditions is t^{-q} asymptotically stable if there is a nonnegative real q, so that:

$$\forall \|x(t)\| \text{ with } t \leq t_0, \exists N(x(t)), \text{ such that } \forall t > t_0, \|x(t)\| \leq N t^{-q}.$$

Subsequently, fractional-order systems are known to have long memory, since they obey a behavior like t^{-q}, which slowly tends to 0 for the solutions $x(t)$. A particular case of Mittag–Leffler stability is stated as power-law stability t^{-q}. [58,59].

Theorem 1. *Let a commensurate order system described by $D^q x = Ax$, with $x(0) = x_0$, $0 < q < 1$, $x \in \mathbb{R}^n$ and $A \in \mathbb{R}^{n \times n}$, is asymptotically stable if $|\arg(\lambda)| > q\pi/2$, is fulfilled for all eigenvalues of A. Besides, the critical eigenvalues fulfilling $|\arg(\lambda)| = q\pi/2$ holding geometric multiplicity of one, where the geometric multiplicity of an eigenvalue is called the dimension of the subspace vectors v for $Av = \lambda v$ [60–65].*

Theorem 2. *Let $D^q x = Ax$, with $x(0) = x_0$ an incommensurate-order system, where $x = (x_1, x_2, \ldots, x_n)^T \in \mathbb{R}^n$, $D^q x = (D^{q_1} x_1, D^{q_2} x_2, \ldots, D^{q_n} x_n)^T$, $q_i \in \mathbb{R}^+$, $i = 1, 2, \ldots, n$, $q_i \in (0,1)$, and $A = (a_{ij}) \in \mathbb{R}^{n \times n}$, $i = j = 1, 2, \ldots, n$. By supposing w as the lowest common multiple of the denominators u_i's of q_i's, with $q_i = v_i / u_i$, $(u_i, v_i) = 1$, $u_i, v_i \in \mathbb{Z}^+$ for $i = 1, 2, \ldots, n$, the matrix of system is given by*

$$\Delta(\lambda) = \begin{bmatrix} \lambda^{wq_1} - a_{11} & -a_{12} & \cdots & -a_{1n} \\ -a_{21} & \lambda^{wq_2} - a_{22} & \cdots & -a_{2n} \\ \vdots & \vdots & \ddots & \vdots \\ -a_{n1} & -a_{n2} & \cdots & \lambda^{wq_n} - a_{nn} \end{bmatrix}. \tag{11}$$

Therefore, the system $D^q x = Ax$ is globally asymptotically stable if the roots λ's of its characteristic equation $\det(\Delta(\lambda)) = 0$ fulfil $|\arg(\lambda)| > \pi/2w$, [60–65].

Theorem 3. *The equilibria E_* with a instability measure defined as*

$$\rho = (\pi/2w) - \min_i \{|\arg(\lambda_i)|\}, \tag{12}$$

is asymptotically stable if and only if (12) is defined nonpositive. Where λ_i's are the roots of characteristic equation: $\det(diag([\lambda^{wq_1} \quad \lambda^{wq_2} \ldots \lambda^{wq_n}]) - \partial f / \partial x \big|_{x=E_*}) = 0, \forall E_* \in \Omega$ [64,65].

Remark 1. *If ρ is nonnegative, the underlying system can show a chaotic evolution since the equilibrium E_* is unstable [64,65].*

The first step consists of finding the equilibrium points of (5), which are calculated via $\mathbf{f}(\mathbf{x}) = 0$, as follows

$$
\begin{aligned}
0 &= -a_1 x + 0.1xy + 1.09y^2 - 1.08y^3 + 0.03z - 0.06z^2 + a_7 z^3 - 0.19, \\
0 &= -a_8 xy + 3.84x^2 + 1.2x^3 + 0.3y(1-y) - 1.37z + 0.3z^2 - 0.22z^3 - 0.56, \\
0 &= a_{15}y - 1.35y^2 + 0.5y^3 + 0.42z + 0.15yz.
\end{aligned}
\tag{13}
$$

Because of the biological interpretation of state variables [46], the stability analysis for $E^* = (x^*, y^*, z^*)$ is only performed for nonnegative fixed points. Equation (5) has the following Jacobian

$$
J = \begin{pmatrix}
-a_1 + 0.1y* & 0.1x^* + 2.18y^* - 3.24y^{*2} & 0.03 - 0.12z^* + 3a_7 z^{*2} \\
-a_8 y^* + 7.68x^* + 3.6x^{*2} & -a_8 x + 0.3(1 - 2y^*) & -1.37 + 0.6z - 0.66z^{*2} \\
0 & a_{15} - 2.7y^* + 1.5y^{*2} + 0.15z & 0.42 + 0.15y^*
\end{pmatrix}.
\tag{14}
$$

By setting $a_1 = 1.3$, $a_7 = 2.01$, $a_8 = 0.22$, $a_{15} = 0.3$, we compute the equilibrium points and eigenvalues, as shown in Table 1. As can be noted, the system (5) has two nonnegative fixed points characterized for saddle points of index-1 and index-2, respectively. According to Theorem 1, chaos behavior may arise in the fractional-order glucose-insulin system when $|\arg(\lambda)| > q\pi/2$, holds. Therefore, the minimum commensurate fractional-order leading to chaotic oscillations is $q > 0.7638$, which is remarkably lower than that reported in [52].

Table 1. Nonnegative equilibrium points and eigenvalues of the system (5).

E_i	Equilibrium Point	Eigenvalues
E_1	$(0.802, 1.853, 1.286)$	$\lambda_1 = 1.4652, \lambda_{2,3} = -1.4353 \pm 7.7140i$
E_2	$(0.584, 0.832, 0.728)$	$\lambda_1 = -2.5327, \lambda_{2,3} = 0.7664 \pm 1.9699i$

The chaotic attractor existence and local stability of the equilibria are presented below. The Jacobian matrix considering the equilibria E_1 and parameter a_1 is

$$
J_{E_1} = \begin{pmatrix}
-a_1 + 0.1853 & -7.0110 & 9.8541 \\
8.0788 & -0.9887 & -1.6903 \\
0 & 0.6420 & 0.6420
\end{pmatrix},
\tag{15}
$$

which generates the characteristic equation given, as follows

$$
P(\lambda) = \lambda^3 + (a_1 + b_1)\lambda^2 + (b_2 a_1 + b_3)\lambda + b_4 a_1 - b_5 = 0,
\tag{16}
$$

where $b_1 = 0.1054$, $b_2 = 0.2907$, $b_3 = 56.9826$, $b_4 = 0.395$, and $b_5 = 90.7232$, when considering the Routh–Hurwitz theory, obtaining the following conditions $a_1 + b_1 > 0$, $b_2 a_1 + b_3 > 0$, and $b_4 a_1 - b_5 > 0$, which become in $a_1 > -b_1$ and $b_3 > -b_2 a_1$, respectively. Due $b_5 > b_4 a_1$, there is a root positive which is unstable and a pair of complex conjugate with the negative real part. Thus E_1 is unstable being a saddle point of index 1.

The presence of a Hopf bifurcation in the system (5) at the equilibrium point E_1 is analyzed by replacing $\lambda = iw$ in (16) obtaining

$$(iw)^3 + (a_1 + b_1)(iw)^2 + (b_2 a_1 + b_3)(iw) + b_4 a_1 - b_5 = 0, \tag{17}$$

then

$$-(iw)^3 - (a_1 + b_1)w^2 + (b_2 a_1 + b_3)(iw) + b_4 a_1 - b_5 = 0, \tag{18}$$

taking into account the real part of (18), therefore,

$$\omega^2 = \frac{b_5 - b_4 a_1}{-(a_1 + b_1)}, \tag{19}$$

because $a_1 + b_1 > 0$, and $b_5, b_4, a_1 > 0$, hence $\omega^2 < 0$, so it is not possible, then, the equilibrium point E_1 does not suffer a Hopf bifurcation.

In order to study the stability of E_1 we consider the discriminant $D(P)$ of characteristic equation $P(\lambda)$, given as follows

$$D(P) = 18 d_1 d_2 d_3 + (d_1 d_2)^2 - 4 d_3 d_1^3 - 4 d_2^3 - 27 d_3^2, \tag{20}$$

where $d_1 = (a_1 + b_1)$, $d_2 = (b_2 a_1 + b_3)$, and $d_3 = (b_4 a_1 - b_5)$. Following the Routh–Hurwitz stability conditions for fractional-order differential equations [66], we establish the following terms:

(i) If $D(P) < 0$, $d_1 > 0$, $d_2 > 0$, $d_1 d_2 = d_3$, then the equilibrium point E_1 is locally asymptotically stable $\forall q \in (0, 1)$.

(ii) $d_3 > 0$ is the necessary condition for the equilibrium point E_1 to be locally asymptotically stable.

Remark 2. *For parameter values d_i, $i = 1, 2, 3$ with $d_1 = 1.4054$, $d_2 = 57.3606$ and $d_3 = -90.2097$ the discriminant $D(P) = -1.098041820 \times 10^6$, but $d_1 d_2 = 80.6144 \neq d_3$, and $d_3 < 0$ thence, the Routh–Hurwitz conditions are unsatisfied. Thus, the E_1 is unstable for the given parameters.*

When $q_1 = q_2 = q_3 \equiv q = 0.9 = 9/10$, with $w = 10$, the characteristic equation of (5) in the equilibrium point $E_1 = (0.802, 1.853, 1.286)$ is given by Theorem 3, as follows

$$\lambda^{27} + (a_1 + b_1)\lambda^{18} + (b_2 a_1 + b_3)\lambda^9 + b_4 a_1 - b_5 = 0, \tag{21}$$

if we set $a_1 = 1.3$ the characteristic equation has an unstable root $\lambda = 1.0434$ and the $|\arg(\lambda)| < \pi/20$, moreover the E_1 is saddle point. While the characteristic equation at the equilibrium point $E_2 = (0.584, 0.832, 0.728)$ is

$$\lambda^{27} + (a_1 - 0.3002)\lambda^{18} + (-0.217 a_1 + 0.868)\lambda^9 - 1.2036 a_1 + 12.88179578 = 0, \tag{22}$$

by considering $a_1 = 1.3$ we obtain unstable roots $\lambda_{1,2} = 1.0771 \pm 0.1444i$. thence, the instability measure of the system is $\rho = (\pi/2w) - 0.1333 > 0$. Therefore, the fractional-order system (5) satisfies the necessary condition for exhibiting a chaotic attractor according to Theorem 3. This can be understood by locating the respective eigenvalues in the complex plane. Figure 1, displays the 27 eigenvalues of the system in the complex plane, the unstable region is delimited by the red lines which is denoted by $\pi/2w$. The eigenvalues for E_1 are given in Figure 1a, we can observe that the system has one eigenvalue in the unstable region, it is a saddle point of index 1. Meanwhile, Figure 1b shows the eigenvalues of equilibrium point E_2; it has two unstable eigenvalues (saddle point of index 2).

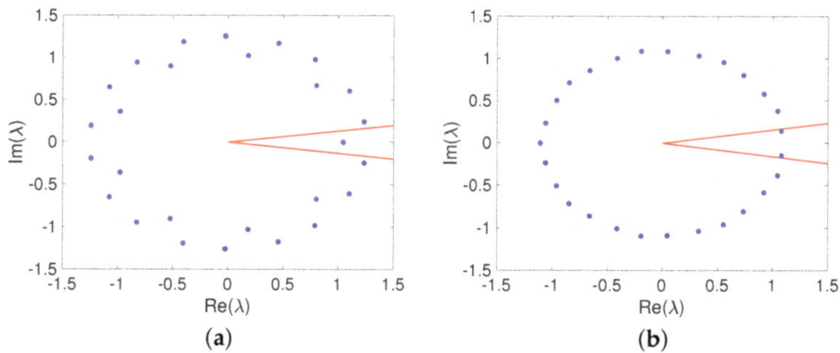

Figure 1. Eigenvalues of the system (5) in the complex plane. (**a**) eigenvalues for equilibria E_1, (**b**) eigenvalues for equilibria E_2.

Figure 2(a)-(c), exhibits the chaotic attractor found in the glucose-insulin model (5) for fractional-orders $q_1 = q_2 = q_3 = 0.9$. The results were computed by applying the predictor-corrector scheme Adams–Bashforth–Moulton (ABM) [54–56]. Due to Caputo's fractional differential operator (4) permits to select both homogeneous and inhomogeneous initial conditions, the ABM algorithm can be executed without particular constraints [56].

As well known, the differential equations that model dynamical systems regularly present symmetries and, therefore, it is rare the solutions do not evolve in a symmetrical orientation. The most common effect is asymmetric attractors for a proper range of parameters, which almost all converges to a singlesymmetric attractor [67]. For dynamical systems with three state variables x, y, and z, there are three types of involuntary symmetries: inversion, rotation, and reflection. We consider the following transformations $(x, y, z) \rightarrow (-x, -y, -z)$, $(x, y, z) \rightarrow (-x, -y, z)$, and $(x, y, z) \rightarrow (-x, y, z)$ corresponding to the invariance of the equations with changes of sign in three, two, and one variable, respectively. As a result, the system (5) does not have a symmetry under the proposed transformations, and its trajectories $(x(t), y(t), z(t))$ cannot cross the $(0, 0, 0)$ coordinate.

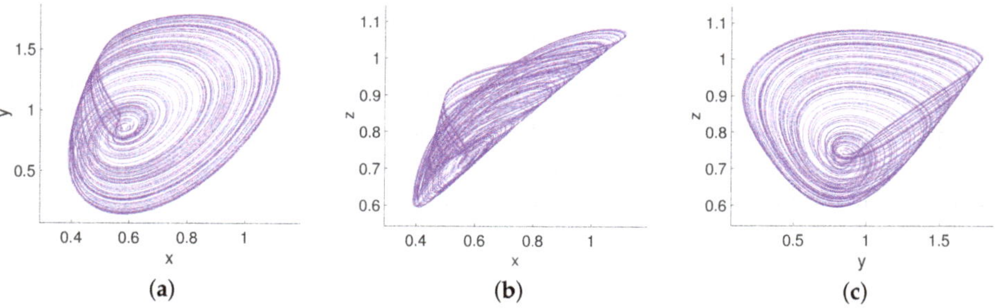

Figure 2. Chaotic attractor of the fractional-order glucose-insulin model (5) for fractional-orders $q = 0.9$, $a_1 = 1.3$, $a_7 = 2.01$, $a_8 = 0.22$, $a_{15} = 0.3$, $(x_0, y_0, z_0) = (0.5, 1.2, 1)$, and integration step-size $h = 0.01$. (**a**) $x - y$ phase plane, (**b**) $x - z$ phase plane, (**c**) $y - z$ phase plane.

4. Numerical Analysis of the Non-Local Fractional Operators on Hypoglycemia, Hyperinsulinemia, T1DM, and T2DM

As reported in [46–48], when the chaotic behavior appears, it could mean the existence of some disorders in the inherent dynamics of a biological system. In this manner, we analyze four health disorders related to fractional-order glucose-insulin model (5) through one- and two-dimensional bifurcation diagrams and Lyapunov exponents. In particular, we construct a map that relates the fractional-order derivative with a specific parameter, such as a_1 (hypoglycemia), a_7 (hyperinsulinemia), a_8 (T1DM), and a_{15} (T2DM).

4.1. Hypoglycemia: Parameter a_1 as a Function of Fractional-Order q

For patients with diabetes, hypoglycemia emerges when the reduction of blood glucose concentration reduces below 185 3.9 mmol/L (70 mg/dL) [68]. This is a critical condition, since hypoglycemia may lead to a life-alarming state. In Equation (5), this complication is analyzed when considering the parameter a_1 and the fractional-order q. It means that, if the rate of insulin decrease, which is represented by a_1 in the system (5), gets low, then the hypoglycemia phenomenon emerge. Therefore, we suppose that the underlying system converges into a chaotic behavior as shown in Figure 2.

Figure 3a exhibits the bifurcation diagram of system (5) with a fixed fractional-order and considering a_1 as a critical parameter. The bifurcation diagram was made when considering the following: when the state-variable x intersects the Poincaré plane provided by $x - p_x = 0$ with $p_x = 0.5$, the measure $r = \sqrt{y^2 + z^2}$ is delineated. It can be observed that system is stable for values of parameter $a_1 > 1.5$ but as the parameter diminishes the behavior turns chaotic.

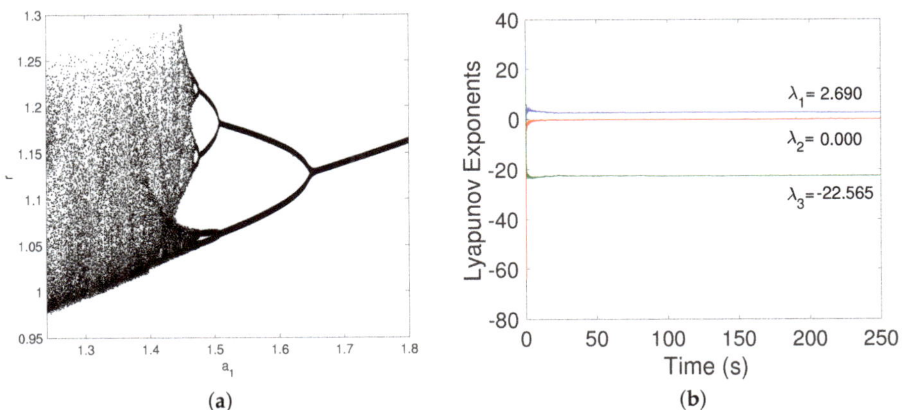

Figure 3. (a) Bifurcation diagram varying the hypoglycemia parameter a_1 and setting $q = 0.9$, and (b) its Lyapunov exponent spectrum when $a_1 = 1.3$.

Additionally, we observed that the fractional-order q produces a shift concerning the bifurcation diagram showed in Ref. [46]. This consideration exemplifies the importance of considering a fractional-order derivative in the dynamical system, i.e., when values lesser than $a_1 = 2.3$ are set in the integer-order system [46], chaotic behavior was observed; however, this limit is different for the fractional-order model ($a_1 \leq 1.45$). It is at this moment when we could mention that a disorder appears. The numerical results of Lyapunov exponents denoted by λ_i with $i = 1, 2, 3$ are shown in Figure 3b for $a_1 = 1.3$ and $q = 0.9$ by applying Wolf's algorithm [69]. The fractional-order glucose-insulin system is chaotic because of the exponents are $\lambda_1 > 0$, $\lambda_2 = 0$ and $\lambda_3 < 0$ with $|\lambda_1| < |\lambda_2 + \lambda_3|$. Those results imply that the system is sensitive to tiny variations of its initial conditions [70,71].

Besides, a two-dimensional bifurcation diagram between the hypoglycemia parameter a_1 and q is presented in Figure 4. The unbounded behavior is represented by green regions, whereas chaos regions are denoted by red color. The black regions lead to healthy behavior (free of hypoglycemia). We found the lower the fractional-order, the lower the effect of a_1. The basin of attraction in the plane $x(0) - y(0)$ for $z(0) = 1$, $q = 0.9$ and $a_1 = 1.3$, is plotted in Figure 4b, the yellow region stand for a chaotic attractor shown in Figure 2, whereas initial conditions from blue region converge into a unbounded behavior. Finally, Figure 5a–c and Figure 6, presents the phase portraits and Lyapunov exponents, respectively, of healthy behavior for system (5) obtaining a Lyapunov exponent with magnitude zero and two negatives.

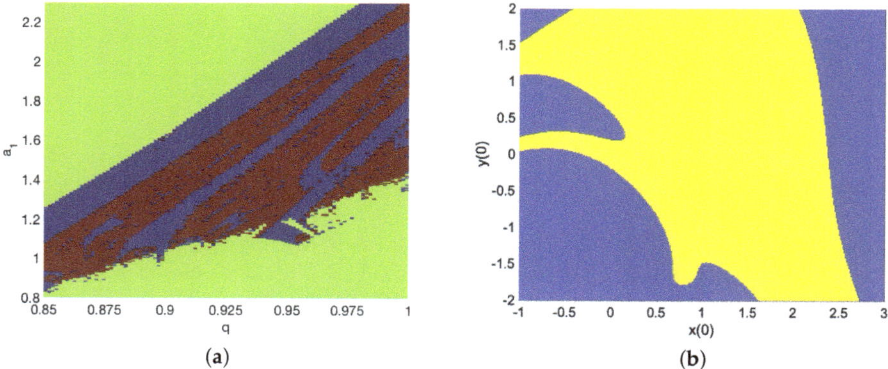

(a)　　　　　(b)

Figure 4. (**a**) Two-dimensional bifurcation diagram for the hypoglycemia parameter a_1 and fractional-order q, where green region stands for unbounded behavior, red for chaotic behavior, and blue regions lead to healthy (periodic) behavior. (**b**) Basin of attraction on the $x(0) - y(0)$ plane with $z(0) = 1$, $q = 0.9$, and $a_1 = 1.3$ showing the chaotic behavior. The initial conditions marked in the yellow color lead into a bounded chaotic attractor, whereas the initial conditions in blue region converge into unbounded behavior.

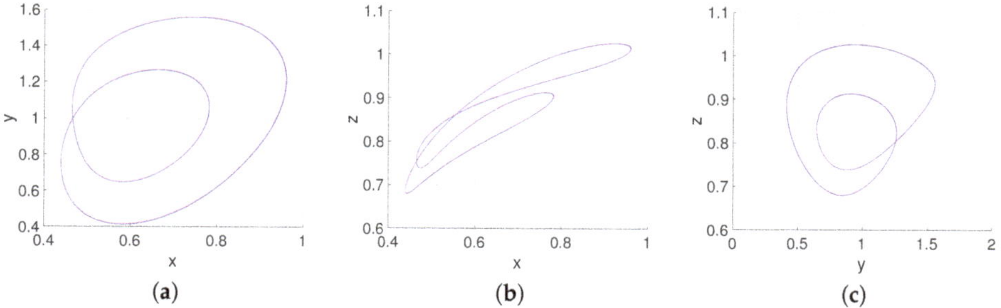

(a)　　　　　(b)　　　　　(c)

Figure 5. Stable behavior of the fractional-order system (5) considering $q = 0.9$, $a_1 = 1.55$, and initial conditions $(x_0, y_0, z_0) = (0.5, 1.2, 1)$, with a integration step-size $h = 0.01$. (**a**) $x - y$ phase plane, (**b**) $x - z$ phase plane, (**c**) $y - z$ phase plane.

4.2. Hyperinsulinemia: Parameter a_7 as a Function of Fractional-Order q

Hyperinsulinemia means the quantity of insulin in the blood is higher than normal levels. Hyperinsulinemia is most often caused by insulin resistance, both humans and animals [72]. A condition in which the body is not capable of acts in the right form to the effects of insulin. Consequently, in order to compensate the high blood glucose levels, the pancreatic β-cells irrigate more insulin [73–76]. Hyperinsulinemia condition is analyzed in the fractional-order glucose-insulin regulatory system (5) by the parameter a_7. Figure 7a shows the bifurcation diagram for different values of a_7 as a function of fractional order q.

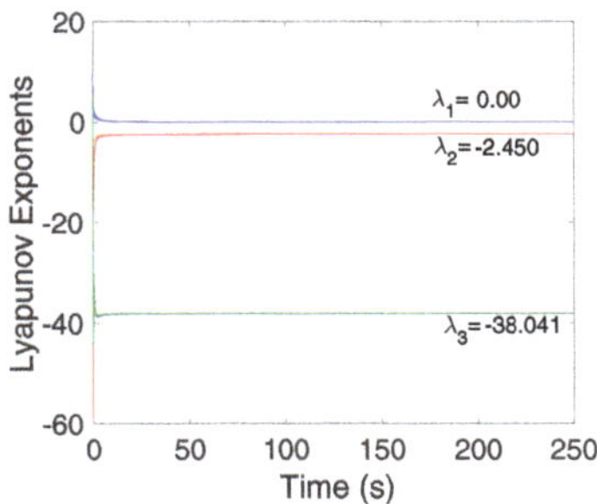

Figure 6. Lyapunov spectrum of (5): $a_1 = 1.55$ and $q = 0.9$.

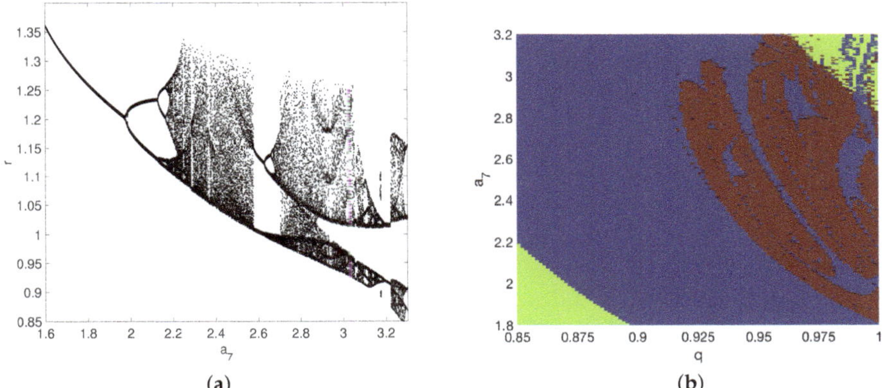

(**a**)

(**b**)

Figure 7. (**a**) Bifurcation diagram varying the hyperinsulinemia parameter a_7 and setting $q = 0.95$ and (**b**) Two-dimensional bifurcation diagram for a_7 and fractional-order q where the unbounded behavior is represented by the green regions; chaotic behavior is denoted by red regions, and the healthy behavior (free of hyperinsulinemia) is given by the blue regions.

From Figure 7a, we can observe that the system is stable when the values for a_7 are small, which describes the increased rate of insulin. If a_7 increases, the system becomes in a chaotic manner, which can be proved by Proposition 1 as follows

Proposition 1. *When* $q_1 = q_2 = q_3 \equiv q = 0.95$ *and* $a_1 = 2.04$, $a_7 = 2.4$, $a_8 = 0.22$, $a_{15} = 0.3$, *the system* (5) *exhibits a chaotic attractor.*

Proof. To demonstrate the nonlinear behavior (chaotic behavior) in (5), it is mandatory that the instability measure ρ defined in (12) is nonnegative. When considering $q = 0.95$, $a_7 = 2.4$, and $w = 100$, the characteristic equation at the equilibrium point $E_1 = (0.802, 1.866, 1.273)$ is

$$\lambda^{285} + 2.149\lambda^{190} + 58.565\lambda^{95} - 102.703, \tag{23}$$

with a unstable root $\lambda = 1.0049$, whereas the characteristic polynomial at $E_2 = (0.606, 0.889, 0.812)$ is

$$\lambda^{285} + 1.764\lambda^{190} + 1.658\lambda^{95} + 17.26, \tag{24}$$

with unstable roots $\lambda_{1,2} = 1.0090 \pm 0.0137i$, then $\rho = (\pi/2w) - 0.0136 > 0$. This result implies system (5) could generate a chaotic attractor when $q = 0.95$ and $a_1 = 2.04$, $a_7 = 2.4$, $a_8 = 0.22$, $a_{15} = 0.3$. $\square$

Besides, the phenomenon antimonotonicity is stated in Figure 7a, which refers to the creation of period orbits followed by their nullification with reverse bifurcation sequences [77]. This phenomenon is one of the most common paths to chaos [78,79]. Antimonotonicity was found in Equation (5) by sweeping a_7 in the interval $2.6 \leq a_7 \leq 3.2$ with $q = 0.95$. Additionally, we obtain the Lyapunov exponents $\lambda_1 = 1.6617$, $\lambda_2 = 0$, $\lambda_3 = -24.7646$) indicating chaos.

On the other hand, Figure 7b gives the two-dimensional bifurcation diagram between the hyperinsulinemia parameter a_7 and the fractional-order q. The unbounded behavior is represented by the green regions; chaotic behavior is denoted by red regions and the healthy behavior (free of hyperinsulinemia) is given by the blue regions. We found that hyperinsulinemia disorder depends on the value of fractional-order. For values $q < 0.92$, the hyperinsulinemia tends to periodic oscillations.

4.3. Type-2 Diabetes Mellitus: Parameter a_7 as a Function of Fractional-Order q

The abnormal insulin secretion of the pancreatic β-cells is commonly related to T2DM or non-insulin-dependent diabetes mellitus, which is known to be a disorder with insulin resistance [80–82]. The interconnection among T2DM, insulin resistance, and obesity relies on the β-cell dysfunction [80,83]. T2DM condition is characterized by the parameter a_8 in (5). Figure 8a shows the bifurcation diagram for the parameter a_8 with a fractional-order $q = 0.98$. The parameter a_8 is appropriate to understand the insulin resistance of the human body since it describes the effect of emitted insulin on glucose level [46]. In the bifurcation diagram, that phenomenon is detected when $a_8 < 0.37$, which is associated with chaotic behavior, as demonstrated by Proposition 2.

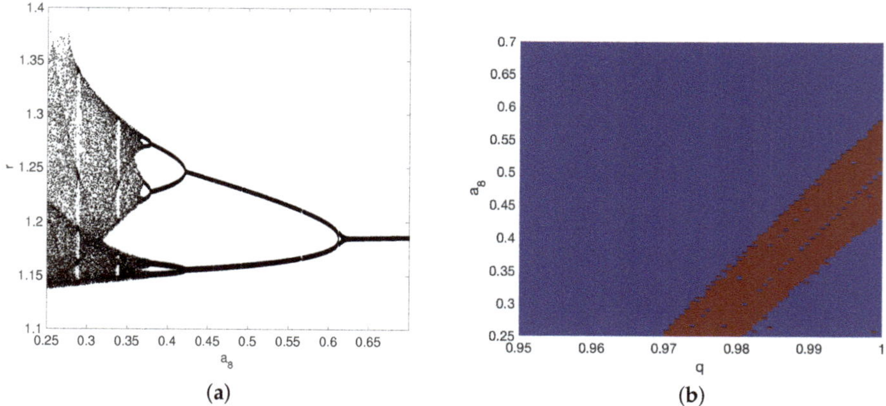

Figure 8. (a) Bifurcation diagram varying the T2DM parameter a_8 and setting $q = 0.98$ and (b) Two-dimensional bifurcation diagram for a_8 and fractional-order q, the chaotic behavior is denoted by red regions, and the periodic behavior (healthy behavior) is given by the blue regions.

Proposition 2. *When $q_1 = q_2 = q_3 \equiv q = 0.98$, and $a_1 = 2.04$, $a_7 = 2.01$, $a_8 = 0.27$, $a_{15} = 0.3$, the system (5) exhibits a chaotic attractor.*

Proof. By applying Theorem 3, we can determine the instability measure ρ. When ρ is strictly positive, a chaos condition could be established. By selecting $q = 0.98$, $a_8 = 0.27$, and $w = 100$ the characteristic polynomial at $E_1 = (0.814, 1.813, 1.320)$ is

$$\lambda^{294} + 2.174\lambda^{196} + 54.782\lambda^{98} - 82.2, \tag{25}$$

with unstable root $\lambda = 1.0033$, while at the equilibrium point $E_2 = (0.63, 0.937, 0.879)$ is

$$\lambda^{294} + 1.818\lambda^{196} + 2.903\lambda^{98} + 16.537, \tag{26}$$

with unstable roots $\lambda_{1,2} = 1.0089 \pm 0.0140i$, where the instability measure of the system is $\rho = (\pi/2w) - 0.0138 > 0$. Thus, the system (5) fulfills the essential requirement for getting chaos when $q = 0.98$ and $a_1 = 2.04$, $a_7 = 2.01$, $a_8 = 0.27$, $a_{15} = 0.3$. $\square$

Additionally, we compute the Lyapunov exponents $\lambda_1 = 0.56$, $\lambda_2 = 0$, $\lambda_3 = -21.03$. Figure 8b sketches the two-dimensional bifurcation diagram for the T2DM parameter a_8 and the fractional-order q. Analogous previous cases, the red areas evolve to chaos, whereas the blue regions converge to a stable behavior (healthy condition). There is a linear fit between the fractional-order and T2DB. The lower the fractional-order, the lower the value for a_8, where the T2DM disorder is observed. Besides, for $q < 0.97$ the T2DM disappear for $0.25 < a_8 < 0.7$. These results suggest that the T2DM is not presented in the glucose-insulin system (5) for lowers fractional-orders.

4.4. Type-1 Diabetes Mellitus: Parameter c_{15} as a Function of Fractional-Order q

T1DM is a common autoimmune disease that originates when the pancreatic β-cells cannot produce insulin at normal levels, and patients will require hormone dosage for their entire life. [84]. The fractional-order system (5) exhibits this condition when the density of β-cells distinguished by a_{15} reduces and, therefore, the pancreas may not secrete sufficient insulin to stabilize the glucose concentration.

The bifurcation diagram of Equation (5) is shown in Figure 9a when considering a_{15} as critical parameter with $q = 0.95$. As can be seen, the system (5) exhibits different types of steady behaviors for specific values of a_{15}. However, whether this parameter decreases, the system behaves chaotically, as is demonstrated in Proposition 3.

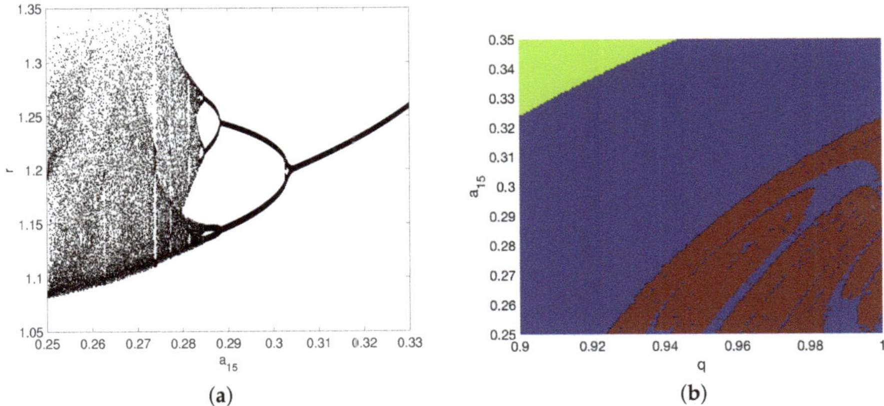

(a)   (b)

Figure 9. (a) Bifurcation diagram varying the T1DM parameter a_{15} and setting $q = 0.95$, and (b) Two-dimensional bifurcation diagram for a_{15} and fractional-order q. The unbounded behavior is represented by the green regions; chaotic behavior is denoted by red regions, and the periodic behavior (healthy behavior) is given by the blue regions.

Proposition 3. *When $q_1 = q_2 = q_3 \equiv q = 0.95$ and $a_1 = 2.04$, $a_7 = 2.01$, $a_8 = 0.22$, $a_{15} = 0.26$, the system (5) exhibits a chaotic attractor.*

Proof. By selecting $q = 0.95$, $a_{15} = 0.26$, and $w = 100$, we attain the characteristic polynomial at $E_1 = (0.823, 1.881, 1.366)$ as

$$\lambda^{285} + 2.159\lambda^{190} + 61.954\lambda^{95} - 106.369, \tag{27}$$

being $\lambda = 1.0048$ the unstable root. At the equilibrium point $E_2 = (0.618, 0.883, 0.867)$, we obtain

$$\lambda^{285} + 1.765\lambda^{190} + 1.535\lambda^{95} + 17.503, \tag{28}$$

with unstable roots $\lambda_{1,2} = 1.0091 \pm 0.0137i$. Therefore, $\rho = (\pi/2w) - 0.0135 > 0$. In this manner, the proposed system (1) fulfills Theorem 3 for generating a chaotic attractor. $\quad\square$

The Lyapunov exponents when $q = 0.95$ and $a_1 = 2.04$, $a_7 = 2.01$, $a_8 = 0.22$, $a_{15} = 0.26$ are $\lambda_1 = 1.7733$, $\lambda_2 = 0$, and $\lambda_3 = -24.3966$. Similarly previous case, Figure 9b presents the two-dimensional bifurcation diagram relating a_{15} and the fractional-order q. The green, red, and blue colors denote unbounded, chaotic, and steady-state behaviors, respectively. A healthy behavior, free of T1DM, is found for fractional-orders lowers than $q < 0.925$. Those results may imply that lower fractional-orders mitigate the effect of the reduction of population density of β-cells for T1DM.

5. Synchronization between Fractional-Order Glucose Insulin Systems

Synchronization is a nonlinear phenomenon that was observed in biological systems; it is seen on isolated cells [15], clusters of cells as in organisms, and even in collective dynamics of populations [25]. Regarding the glucose-insulin system, it has been shown pancreatic β-cells also present a collective behavior whose synchronization underlies a small-world functional organization [24–26]. Thus, the synchronization is crucial to effectuate a pulsatile insulin liberation in cells, which guarantees more substantial hypoglycemic effects. Hence, we study the synchronization between fractional-order glucose-insulin regulatory systems. We expect that the synchronization state converges into a periodic behavior, because it is the typical response in a suband blood glucose concentrations. We define thject with normal metabolic conditions, allowing with this, the synchronization between the insulin e drive and response system, as follows

$$
\begin{aligned}
D^{q_1} x_1 &= -a_1 x_1 + 0.1 x_1 y_1 + 1.09 y_1^2 - 1.08 y_1^3 + 0.03 z_1 - 0.06 z_1^2 + a_7 z_1^3 - 0.19, \\
D^{q_2} y_1 &= -a_8 x_1 y_1 + 3.84 x_1^2 + 1.2 x_1^3 + 0.3 y_1 (1 - y_1) - 1.37 z_1 + 0.3 z_1^2 - 0.22 z_1^3 - 0.56, \\
D^{q_3} z_1 &= a_{15} y_1 - 1.35 y_1^2 + 0.5 y_1^3 + 0.42 z_1 + 0.15 y_1 z_1,
\end{aligned}
\tag{29}
$$

and

$$
\begin{aligned}
D^{q_1} x_2 &= -\hat{a}_1 x_2 + 0.1 x_2 y_2 + 1.09 y_2^2 - 1.08 y_2^3 + 0.03 z_2 - 0.06 z_2^2 + \hat{a}_7 z_2^3 - 0.19 + u_1, \\
D^{q_2} y_2 &= -\hat{a}_8 x_2 y_2 + 3.84 x_2^2 + 1.2 x_2^3 + 0.3 y_2 (1 - y_2) - 1.37 z_2 + 0.3 z_2^2 - 0.22 z_2^3 - 0.56 + u_2, \\
D^{q_3} z_2 &= \hat{a}_{15} y_2 - 1.35 y_2^2 + 0.5 y_2^3 + 0.42 z_2 + 0.15 y_2 z_2 + u_3,
\end{aligned}
\tag{30}
$$

where u_1, u_2, u_3 in (30) represents the unknown control terms, and the error can be defined by

$$
\begin{aligned}
e_1 &= x_2 - x_1, \\
e_2 &= y_2 - y_1, \\
e_3 &= z_2 - z_1.
\end{aligned}
\tag{31}
$$

To achieve the synchronization, it is essential that the errors $e_i \to 0$ as $t \to \infty$ with $i = 1, 2, 3$. Equation (31), together with (29) and (30), yield the error system

$$
\begin{aligned}
D^{q_1} e_1 &= -\hat{a}_1 x_2 + 0.1 x_2 y_2 + 1.09 y_2^2 - 1.08 y_2^3 + 0.03 z_2 - 0.06 z_2^2 + \hat{a}_7 z_2^3 + \\
&\quad + a_1 x_1 - 0.1 x_1 y_1 - 1.09 y_1^2 + 1.08 y_1^3 - 0.03 z_1 + 0.06 z_1^2 - a_7 z_1^3 + u_1, \\
D^{q_2} e_2 &= -\hat{a}_8 x_2 y_2 + 3.84 x_2^2 + 1.2 x_2^3 + 0.3 y_2 (1 - y_2) - 1.37 z_2 + 0.3 z_2^2 - 0.22 z_2^3 \\
&\quad + a_8 x_1 y_1 - 3.84 x_1^2 - 1.2 x_1^3 - 0.3 y_1 (1 - y_1) + 1.37 z_1 - 0.3 z_1^2 + 0.22 z_1^3 + u_2, \\
D^{q_3} e_3 &= \hat{a}_{15} y_2 - 1.35 y_2^2 + 0.5 y_2^3 + 0.42 z_2 + 0.15 y_2 z_2 \\
&\quad - a_{15} y_1 + 1.35 y_1^2 - 0.5 y_1^3 - 0.42 z_1 - 0.15 y_1 z_1 + u_3.
\end{aligned}
\tag{32}
$$

Let us define the active control functions u_i with $i = 1, 2, 3$

$$
\begin{aligned}
u_1 &= V_1 + \hat{a}_1 x_2 - 0.1 x_2 y_2 - 1.09 y_2^2 + 1.08 y_2^3 - 0.03 z_2 + 0.06 z_2^2 - \hat{a}_7 z_2^3 - \\
&\quad a_1 x_1 + 0.1 x_1 y_1 + 1.09 y_1^2 - 1.08 y_1^3 + 0.03 z_1 - 0.06 z_1^2 + a_7 z_1^3, \\
u_2 &= V_2 + \hat{a}_8 x_2 y_2 - 3.84 x_2^2 - 1.2 x_2^3 - 0.3 y_2 (1 - y_2) + 1.37 z_2 - 0.3 z_2^2 + 0.22 z_2^3 \\
&\quad - a_8 x_1 y_1 + 3.84 x_1^2 + 1.2 x_1^3 + 0.3 y_1 (1 - y_1) - 1.37 z_1 + 0.3 z_1^2 - 0.22 z_1^3, \\
u_3 &= V_3 - \hat{a}_{15} y_2 + 1.35 y_2^2 - 0.5 y_2^3 - 0.42 z_2 - 0.15 y_2 z_2 \\
&\quad + a_{15} y_1 - 1.35 y_1^2 + 0.5 y_1^3 + 0.42 z_1 + 0.15 y_1 z_1,
\end{aligned}
\tag{33}
$$

where the linear functions V_1, V_2, V_3 are given by

$$
\begin{aligned}
V_1 &= -e_1, \\
V_2 &= -e_2, \\
V_3 &= -e_3.
\end{aligned}
\tag{34}
$$

By using (33) and (34), the error system (32) becomes

$$
\begin{bmatrix} D^{q_1} e_1 \\ D^{q_2} e_2 \\ D^{q_3} e_3 \end{bmatrix} = \begin{bmatrix} -1 & 0 & 0 \\ 0 & -1 & 0 \\ 0 & 0 & -1 \end{bmatrix} \begin{bmatrix} e_1 \\ e_2 \\ e_3 \end{bmatrix}.
\tag{35}
$$

The synchronization error vanishes eventually because of the eigenvalues are $-1, -1, -1$ in Equation (35).

The synchronization scenario is as follows. The drive system has a periodic behavior, while the response system is in a chaotic state. We study the Type-1 Diabetes Mellitus (parameter a_8), since it is the most common disorder, and affects most world population as well as it is correlated with obesity. For this case, $a_8 = 0.5$ and $\hat{a}_8 = 0.27$ for drive and response systems, respectively, while $a_1 = \hat{a}_1$, $a_7 = \hat{a}_7$, and $a_{15} = \hat{a}_{15}$. The fractional-order are $q_1 = q_2 = q_3 = q = 0.95$ in both systems with $x_1(0) = 0.53, y_1(0) = 1.31, z_1(0) = 1.03$ and $x_2(0) = 0.5, y_2(0) = 1.1, z_2(0) = 1.3$, for drive and response systems, respectively. Figure 10a–c show the phase planes between the periodic (free of T1DM) and chaotic (with T1DM) systems. Additionally, the synchronization error by considering (36) is given in Figure 11. Due to the error tends to zero as time evolves, we infer that the proposed control strategy is suitable for forcing the system with the disorder to a state free of T1DM. It is worth noting that the control strategy can be extended to incorporate uncertainties and improve the robustness of the synchronization using other approaches, as shown in [22,85]. From a practical biological point of view, for instance, recent works have employed optical-based control using a light-activated Na+ channel, to attain insulin from β-cells both in-vitro and vivo [86,87]. Therefore, our results could be useful for future works where the glucose-insulin system could be controlled with an artificial control signal.

$$\ln(error_1(t)) = \ln\left([x_2 - x_1]\right)^2 ; \quad \ln(error_2(t)) = \ln\left([y_2 - y_1]\right)^2 ; \quad \ln(error_3(t)) = \ln\left([z_2 - z_1]\right)^2 . \tag{36}$$

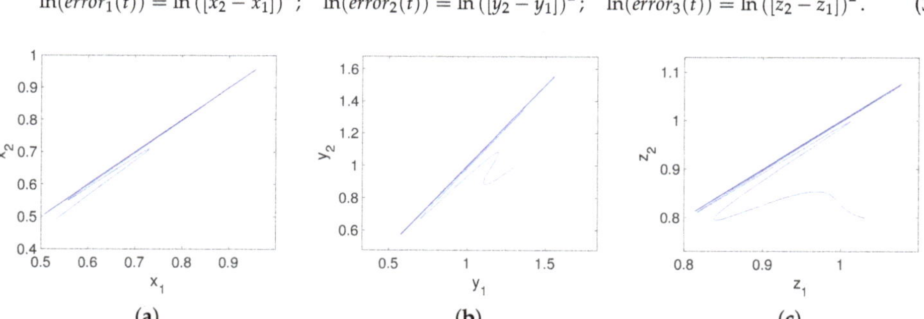

(a)   (b)   (c)

Figure 10. Synchronization planes for the fractional-order glucose-insulin systems (29) and (30) with $a_8 = 0.5$, $\hat{a}_8 = 0.27$, and $q = 0.95$, respectively. (**a**) $x_1 - x_2$ phase plane, (**b**) $y_1 - y_2$ phase plane, (**c**) $z_1 - z_2$ phase plane.

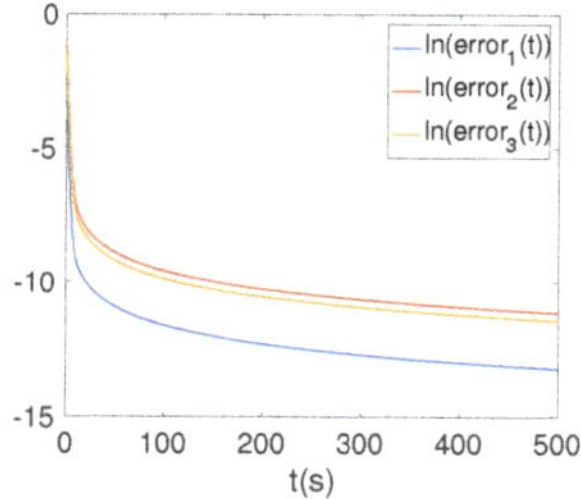

Figure 11. Synchronization error between the fractional-order glucose-insulin systems (29) and (30) when $a_8 = 0.5$, $\hat{a}_8 = 0.27$, $q = 0.95$, and $(x_1, y_1, z_1, x_2, y_2, z_2) = (0.53, 1.31, 1.03, 0.49, 1, 0.8)$, respectively.

6. Physical Realization of the Fractional-Order Glucose-Insulin System Based on an ARM Processor

As well known, the experimental realization of fractional-order dynamical systems is a hot topic that has been attracting the attention of researchers since it is a path for demonstrating the complex dynamics, including chaos [35–39,44,45]. For fractional-order systems, there three typical approaches for getting electronic circuits: frequency-domain approximation, numerical algorithms, and the Adomian decomposition method [35–39,44,45]. The first-mentioned is not recommended for chaos detection, since it may induce incorrect results [62,63]. On the other hand, the second and latter approaches are good options for physically implementing fractional-order systems in re-programmable digital hardware [44,45]. Therefore, we chose the numerical algorithm approach for programming the ABM method. Subsequently, we select herein the ARM SoC Broadcom BCM2837B0 for the experimental verification of the fractional-order glucose-insulin regulatory system. The SoC contains an ARM core with 64-bit. An SDRAM LPDDR2 with 1GB. The ARM cores are capable of running at up to 1.4 GHz. It's possible to create an interface by using the GPIO port with a 16-bit monotonic voltage output D/A converter AD569. Figure 12a,b present the block diagram of the working principle and the main instructions of the pseudo-code, respectively.

After initializing the ARM processor, we set the initial values, h, q, x_0, y_0, z_0. Because of the negative values of the system (1), a positive integer ϕ is needed to offset the time-series of the state-variables to avoid losing data. In this manner, all computed data are now positive. Next, we multiply the data by a positive integer γ to fit them to the DAC resolution of 2^{14} bits. Finally, the obtained results visualize in an oscilloscope, as shown in Figures 13 and 14. We analyze the scenario related to hypoglycemia.

First, we implement the case where system (1) presents the hypoglycemia condition, as given in Figure 13. As can be seen, the experimental phase portraits are pretty similar to those that are shown in Figure 1. Finally, the case when the fractional-order system (1) is free of hypoglycemia, i.e., an steady-state and, therefore, convergers to a periodic attractor, is given in Figure 14. Similar to the previous case, the experimental results are in good agreement with Figure 4. Subsequently, it indicates that the fractional-order glucose-insulin regulatory system was successfully realized on an ARM digital platform.

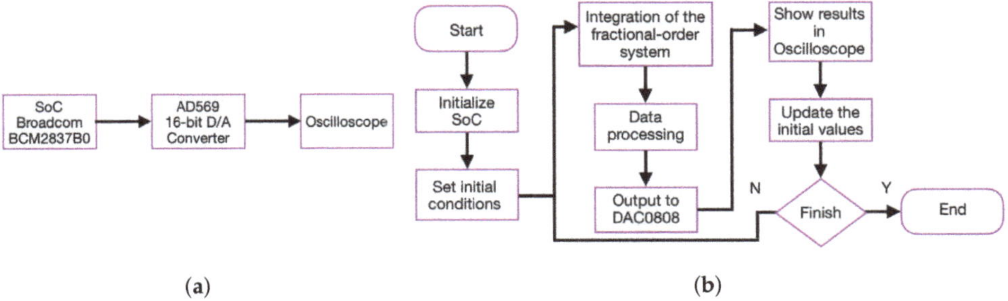

(a)   (b)

Figure 12. (**a**) Simplified diagram of the implementation of the fractional-order glucose-insulin regulatory system (1), and (**b**) the main steps of the proposed algorithm for implementing it on an ARM digital platform.

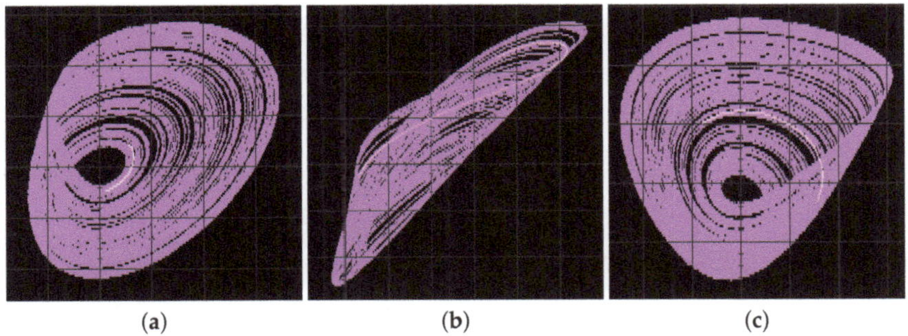

(a)   (b)   (c)

Figure 13. Experimental phase portraits of the fractional-order glucose-insulin regulatory system (1) showing hypoglycemia (chaos behavior) with $h = 0.01$, $a_1 = 1.3$, $a_7 = 2.01$, $a_8 = 0.22$, $a_{15} = 0.3$, $q_1 = q_2 = q_3 = 0.9$ and $(x_0, y_0, z_0) = (0.5, 1.2, 1)$. (**a**) $x - y$ plane, (**b**) $x - z$ plane, (**c**) $y - z$ plane.

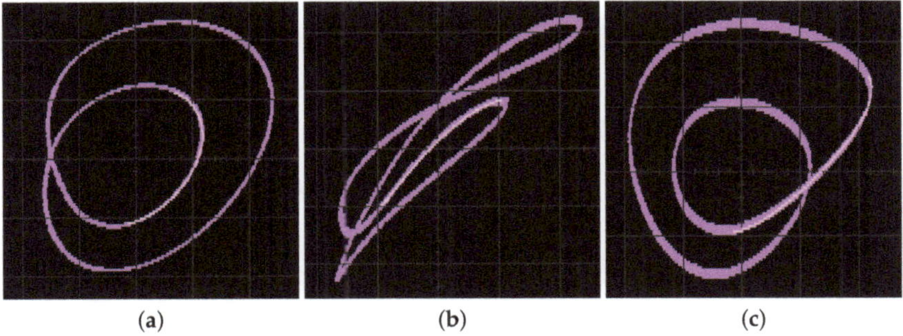

(a)   (b)   (c)

Figure 14. Experimental phase portraits of the fractional-order glucose-insulin regulatory system (1) depicting a steady-state free of hypoglycemia with $h = 0.01$, $a_1 = 1.55$, $a_7 = 2.01$, $a_8 = 0.22$, $a_{15} = 0.3$, $q_1 = q_2 = q_3 = 0.9$ and $(x_0, y_0, z_0) = (0.5, 1.2, 1)$. (**a**) $x - y$ plane, (**b**) $x - z$ plane, (**c**) $y - z$ plane.

7. Conclusions

The dynamical analysis, synchronization, and physical realization of a glucose-insulin regulatory system has been presented by using Caputo's non-local fractional-order operator. In particular, we studied four common disorders, such as T1DM, T2DM, Hypoglycemia, and Hyperinsulinemia. We found that the fractional-order system switches between a chaotic behavior (a health disorder) and a disorder-free state, not only for the values of systems parameters, but also as a function of the fractional-order, due it adds more degrees of freedom in the model. To understand that insight, we computed two-dimensional bifurcations diagrams, which demonstrated the importance of considering the fractional-order (memory index) for getting a higher approximation of the observed behavior because fractional-order systems describe the whole-time domain in the solution, while the integer-order model is related to the local properties. Additionally, a phenomenon of antimonotonicity was observed in the parameter related to the hyperinsulinemia case. Besides, by applying the straightforward active control method, we showed that stable behavior in the fractional glucose-insulin system under the T1DM condition could be attained when it synchronizes with a disorder-free system. We remark that the synchronization can be extended to the remaining conditions. Finally, the electronics approach-based validation of chaotic and periodic behaviors was shown using an ARM digital platform. The experimental observations were in good agreement with the theoretical findings.

In this manner, the system-of-a-chip circuit designs are the right candidate for exploiting the advantages of the fractional-order models due to their simplicity, programmable characteristics, and portability, therefore increasing the fractional-order-based oncoming applications. As future work, an analysis related to robustness of the synchronization scheme will be developed.

Author Contributions: Conceptualization, J.M.M.-P.; Formal analysis, J.M.M.-P. and E.Z.-S; Methodology, E.Z.-S; Supervision, C.P.-C.; Validation, C.P.-C.; Writing—original draft, E.Z.-S.; Writing—review and editing, J.M.M.-P. All authors have read and agreed to the published version of the manuscript.

Funding: This research received no external funding.

Acknowledgments: The authors thankfully acknowledge the computer resources, technical expertise and support provided by the Laboratorio Nacional de Supercómputo del Sureste de México, CONACYT member of the network of national laboratories. J.M. Muñoz-Pacheco thanks CONACyT/MEXICO for the financial support to through Project No. 258880 (Proyecto Apoyado por el Fondo Sectorial de Investigación para la Educación) ; VIEP-BUAP [2020]; and Plan de Trabajo CA [BUAP-CA-276]. C. Posadas-Castillo acknowledges CONACYT/Mexico under Research Grant No. 166654, A1-5-31628 and "Facultad de Ingeniería Mecánica y Eléctrica" (FIME-UANL). E. Zambrano-Serrano acknowledges CONACYT/Mexico (350385) for the financial support to complete a postdoctoral visit to "Facultad de Ingeniería Mecánica y Eléctrica" at UANL.

Conflicts of Interest: The authors declare no conflict of interest.

References

1. Röder, P.V.; Wu, B.; Liu, Y.; Han, W. Pancreatic regulation of glucose homeostasis. *Exp. Mol. Med.* **2016**, *48*, e219. [CrossRef] [PubMed]

2. Puglianiello, A.; Cianfarani, S. Central control of glucose homeostasis. *Rev. Diabet. Stud.* **2006**, *3*, 54. [CrossRef] [PubMed]

3. Palumbo, P.; Ditlevsen, S.; Bertuzzi, A.; De Gaetano, A. Mathematical modeling of the glucose–insulin system: A review. *Math. Biosci.* **2013**, *244*, 69–81. [CrossRef] [PubMed]

4. Roglic, G. WHO Global report on diabetes: A summary. *Int. J. Noncommun. Dis.* **2016**, *1*, 3. [CrossRef]

5. Andrianov, I.; Starushenko, G.; Kvitka, S.; Khajiyeva, L. The Verhulst-Like Equations: Integrable O∆E and ODE with Chaotic Behavior. *Symmetry* **2019**, *11*, 1446. [CrossRef]

6. Rathee, S. ODE models for the management of diabetes: A review. *Int. J. Diabetes Dev. Ctries.* **2017**, *37*, 4–15. [CrossRef]

7. Cruz-Duarte, J.M.; Rosales-García, J.J.; Correa-Cely, C.R. Entropy Generation in a Mass-Spring-Damper System Using a Conformable Model. *Symmetry* **2020**, *12*, 395. [CrossRef]

8. Solís-Pérez, J.E.; Gómez-Aguilar, J.F. Novel Fractional Operators with Three Orders and Power-Law, Exponential Decay and Mittag–Leffler Memories Involving the Truncated M-Derivative. *Symmetry* **2020**, *12*, 626. [CrossRef]

9. Echenausía-Monroy, J.L.; Huerta-Cuellar, G.; Jaimes-Reátegui, R.; García-López, J.H.; Aboites, V.; Cassal-Quiroga, B.B.; Gilardi-Velázquez, H.E. Multistability Emergence through Fractional-Order-Derivatives in a PWL Multi-Scroll System. *Electronics* **2020**, *9*, 880. [CrossRef]

10. Danca, M.F. Puu system of fractional order and its chaos suppression. *Symmetry* **2020**, *12*, 340. [CrossRef]

11. Ionescu, C.; Lopes, A.; Copot, D.; Machado, J.T.; Bates, J. The role of fractional calculus in modeling biological phenomena: A review. *Commun. Nonlinear Sci. Numer. Simul.* **2017**, *51*, 141–159. [CrossRef]

12. Rihan, F.A. Numerical modeling of fractional-Order biological systems. *Abstr. Appl. Anal.* **2013**, *2013*, 1–12. [CrossRef]

13. Kheiri, H.; Jafari, M. Stability analysis of a fractional order model for the HIV/AIDS epidemic in a patchy environment. *J. Comput. Appl. Math.* **2019**, *346*, 323–339. [CrossRef]

14. Teka, W.W.; Upadhyay, R.K.; Mondal, A. Spiking and bursting patterns of fractional-order Izhikevich model. *Commun. Nonlinear Sci. Numer. Simul.* **2018**, *56*, 161–176. [CrossRef]

15. Zambrano-Serrano, E.; Munoz-Pacheco, J.; Gomez-Pavon, L.; Luis-Ramos, A.; Chen, G. Synchronization in a fractional-order model of pancreatic β-cells. *Eur. Phys. J. Spec. Top.* **2018**, *227*, 907–919. [CrossRef]

16. Bodo, B.; Mvogo, A.; Morfu, S. Fractional dynamical behavior of electrical activity in a model of pancreatic β-cells. *Chaos Solitons Fractals* **2017**, *102*, 426–432. [CrossRef]

17. Sun, H.; Chen, W.; Wei, H.; Chen, Y. A comparative study of constant-order and variable-order fractional models in characterizing memory property of systems. *Eur. Phys. J. Spec. Top.* **2011**, *193*, 185. [CrossRef]

18. Saeedian, M.; Khalighi, M.; Azimi-Tafreshi, N.; Jafari, G.; Ausloos, M. Memory effects on epidemic evolution: The susceptible-infected-recovered epidemic model. *Phys. Rev. E* **2017**, *95*, 022409. [CrossRef]

19. Lundstrom, B.N.; Higgs, M.H.; Spain, W.J.; Fairhall, A.L. Fractional differentiation by neocortical pyramidal neurons. *Nat. Neurosci.* **2008**, *11*, 1335. [CrossRef]

20. Lifshitz, R.; Cross, M. Response of parametrically driven nonlinear coupled oscillators with application to micromechanical and nanomechanical resonator arrays. *Phys. Rev. B* **2003**, *67*, 134302. [CrossRef]

21. Bitar, D.; Kacem, N.; Bouhaddi, N. Investigation of modal interactions and their effects on the nonlinear dynamics of a periodic coupled pendulums chain. *Int. J. Mech. Sci.* **2017**, *127*, 130–141. [CrossRef]

22. Chikhaoui, K.; Bitar, D.; Kacem, N.; Bouhaddi, N. Robustness analysis of the collective nonlinear dynamics of a periodic coupled pendulums chain. *Appl. Sci.* **2017**, *7*, 684. [CrossRef]

23. Rosenblum, M.G.; Pikovsky, A.S.; Kurths, J. Synchronization approach to analysis of biological systems. *Fluct. Noise Lett.* **2004**, *4*, L53–L62. [CrossRef]

24. Stožer, A.; Gosak, M.; Dolenšek, J.; Perc, M.; Marhl, M.; Rupnik, M.S.; Korošak, D. Functional connectivity in islets of Langerhans from mouse pancreas tissue slices. *PLoS Comput. Biol.* **2013**, *9*, e1002923. [CrossRef] [PubMed]

25. Loppini, A.; Cherubini, C.; Filippi, S. On the emergent dynamics and synchronization of β-cells networks in response to space-time varying glucose stimuli. *Chaos Solitons Fractals* **2018**, *109*, 269–279. [CrossRef]

26. Barua, A.K.; Goel, P. Isles within islets: The lattice origin of small-world networks in pancreatic tissues. *Phys. D Nonlinear Phenom.* **2016**, *315*, 49–57. [CrossRef]

27. Kotani, K.; Takamasu, K.; Ashkenazy, Y.; Stanley, H.E.; Yamamoto, Y. Model for cardiorespiratory synchronization in humans. *Phys. Rev. E* **2002**, *65*, 051923. [CrossRef]

28. Bartsch, R.; Kantelhardt, J.W.; Penzel, T.; Havlin, S. Experimental evidence for phase synchronization transitions in the human cardiorespiratory system. *Phys. Rev. Lett.* **2007**, *98*, 054102. [CrossRef]

29. Satin, L.S.; Butler, P.C.; Ha, J.; Sherman, A.S. Pulsatile insulin secretion, impaired glucose tolerance and type 2 diabetes. *Mol. Asp. Med.* **2015**, *42*, 61–77. [CrossRef]

30. Ravier, M.A.; Güldenagel, M.; Charollais, A.; Gjinovci, A.; Caille, D.; Söhl, G.; Wollheim, C.B.; Willecke, K.; Henquin, J.C.; Meda, P. Loss of connexin36 channels alters β-cell coupling, islet synchronization of glucose-induced Ca2+ and insulin oscillations, and basal insulin release. *Diabetes* **2005**, *54*, 1798–1807. [CrossRef]

31. Pecora, L.M.; Carroll, T.L. Synchronization in chaotic systems. *Phys. Rev. Lett.* **1990**, *64*, 821. [CrossRef]

32. Shukla, M.K.; Sharma, B. Backstepping based stabilization and synchronization of a class of fractional order chaotic systems. *Chaos Solitons Fractals* **2017**, *102*, 274–284. [CrossRef]

33. Singh, A.K.; Yadav, V.K.; Das, S. Synchronization between fractional order complex chaotic systems with uncertainty. *Optik* **2017**, *133*, 98–107. [CrossRef]

34. Bai, E.W.; Lonngren, K.E. Synchronization of two Lorenz systems using active control. *Chaos Solitons Fractals* **1997**, *8*, 51–58. [CrossRef]

35. Shah, D.K.; Chaurasiya, R.B.; Vyawahare, V.A.; Pichhode, K.; Patil, M.D. FPGA implementation of fractional-order chaotic systems. *AEU-Int. J. Electron. Commun.* **2017**, *78*, 245–257. [CrossRef]

36. Soriano-Sánchez, A.; Posadas-Castillo, C.; Platas-Garza, M.A.; Arellano-Delgado, A. Synchronization and FPGA realization of complex networks with fractional–order Liu chaotic oscillators. *Appl. Math. Comput.* **2018**, *332*, 250–262.

37. Rajagopal, K.; Akgul, A.; Jafari, S.; Karthikeyan, A.; Koyuncu, I. Chaotic chameleon: Dynamic analyses, circuit implementation, FPGA design and fractional-order form with basic analyses. *Chaos Solitons Fractals* **2017**, *103*, 476–487. [CrossRef]

38. He, S.; Sun, K.; Wang, H. Complexity analysis and DSP implementation of the fractional-order Lorenz hyperchaotic system. *Entropy* **2015**, *17*, 8299–8311. [CrossRef]

39. Wang, H.; Sun, K.; He, S. Characteristic analysis and DSP realization of fractional-order simplified Lorenz system based on Adomian decomposition method. *Int. J. Bifurc. Chaos* **2015**, *25*, 1550085. [CrossRef]

40. Evans, J.R.; Arslan, T. Enhanced image detection on an ARM based embedded system. *Des. Autom. Embed. Syst.* **2002**, *6*, 477–487. [CrossRef]

41. Tlelo-Cuautle, E.; Rangel-Magdaleno, J.; Pano-Azucena, A.; Obeso-Rodelo, P.; Nuñez-Perez, J.C. FPGA realization of multi-scroll chaotic oscillators. *Commun. Nonlinear Sci. Numer. Simul.* **2015**, *27*, 66–80. [CrossRef]

42. Chen, H.; He, S.; Azucena, A.D.P.; Yousefpour, A.; Jahanshahi, H.; López, M.A.; Alcaraz, R. A Multistable Chaotic Jerk System with Coexisting and Hidden Attractors: Dynamical and Complexity Analysis, FPGA-Based Realization, and Chaos Stabilization Using a Robust Controller. *Symmetry* **2020**, *12*, 569. [CrossRef]

43. Munoz-Pacheco, J.M.; García-Chávez, T.; Gonzalez-Diaz, V.R.; de La Fuente-Cortes, G.; del Carmen Gómez-Pavón, L. Two new asymmetric Boolean chaos oscillators with no dependence on incommensurate time-delays and their circuit implementation. *Symmetry* **2020**, *12*, 506. [CrossRef]

44. Montero-Canela, R.; Zambrano-Serrano, E.; Tamariz-Flores, E.I.; Muñoz-Pacheco, J.M.; Torrealba-Meléndez, R. Fractional chaos based-cryptosystem for generating encryption keys in Ad Hoc networks. *Ad Hoc Netw.* **2020**, *97*, 102005. [CrossRef]

45. Zambrano-Serrano, E.; Munoz-Pacheco, J.; Campos-Cantón, E. Chaos generation in fractional-order switched systems and its digital implementation. *AEU-Int. J. Electron. Commun.* **2017**, *79*, 43–52. [CrossRef]

46. Shabestari, P.S.; Panahi, S.; Hatef, B.; Jafari, S.; Sprott, J.C. A new chaotic model for glucose-insulin regulatory system. *Chaos Solitons Fractals* **2018**, *112*, 44–51. [CrossRef]

47. Letellier, C.; Denis, F.; Aguirre, L.A. What can be learned from a chaotic cancer model? *J. Theor. Biol.* **2013**, *322*, 7–16. [CrossRef]

48. Jafari, S.; Baghdadi, G.; Golpayegani, S.R.H.; Towhidkhah, F.; Gharibzadeh, S. Is attention deficit hyperactivity disorder a kind of intermittent chaos? *J. Neuropsychiatry Clin. Neurosci.* **2013**, *25*, E02. [CrossRef]

49. Khajanchi, S.; Perc, M.; Ghosh, D. The influence of time delay in a chaotic cancer model. *Chaos Interdiscip. J. Nonlinear Sci.* **2018**, *28*, 103101. [CrossRef]

50. Shabestari, P.S.; Rajagopal, K.; Safarbali, B.; Jafari, S.; Duraisamy, P. A Novel Approach to Numerical Modeling of Metabolic System: Investigation of Chaotic Behavior in Diabetes Mellitus. *Complexity* **2018**, *2018*, 6815190.

51. Ginoux, J.M.; Ruskeepää, H.; Perc, M.; Naeck, R.; Di Costanzo, V.; Bouchouicha, M.; Fnaiech, F.; Sayadi, M.; Hamdi, T. Is type 1 diabetes a chaotic phenomenon? *Chaos Solitons Fractals* **2018**, *111*, 198–205. [CrossRef]

52. Rajagopal, K.; Bayani, A.; Jafari, S.; Karthikeyan, A.; Hussain, I. Chaotic dynamics of a fractional order glucose-insulin regulatory system. *Front. Inf. Technol. Electron. Eng.* **2019**, *21*, 1108–1118. [CrossRef]

53. Ortigueira, M.D.; Machado, J.T. What is a fractional derivative? *J. Comput. Phys.* **2015**, *293*, 4–13. [CrossRef]

54. Diethelm, K.; Ford, N.J.; Freed, A.D. Detailed error analysis for a fractional Adams method. *Numer. Algorithms* **2004**, *36*, 31–52. [CrossRef]

55. Diethelm, K.; Ford, N.J. Analysis of fractional differential equations. *J. Math. Anal. Appl.* **2002**, *265*, 229–248. [CrossRef]

56. Garrappa, R. Numerical solution of fractional differential equations: A survey and a software tutorial. *Mathematics* **2018**, *6*, 16. [CrossRef]

57. Zambrano-Serrano, E.; Campos-Cantón, E.; Muñoz-Pacheco, J.M. Strange attractors generated by a fractional order switching system and its topological horseshoe. *Nonlinear Dyn.* **2016**, *83*, 1629–1641. [CrossRef]

58. Li, Y.; Chen, Y.; Podlubny, I. Mittag–Leffler stability of fractional order nonlinear dynamic systems. *Automatica* **2009**, *45*, 1965–1969. [CrossRef]

59. Petráš, I. *Fractional-Order Nonlinear Systems: Modeling, Analysis and Simulation*; Springer Science & Business Media: Berlin, Germany, 2011.

60. Odibat, Z.; Corson, N.; Aziz-Alaoui, M.; Alsaedi, A. Chaos in fractional order cubic Chua system and synchronization. *Int. J. Bifurc. Chaos* **2017**, *27*, 1750161. [CrossRef]

61. Ahmed, E.; El-Sayed, A.; El-Saka, H.A. Equilibrium points, stability and numerical solutions of fractional-order predator–prey and rabies models. *J. Math. Anal. Appl.* **2007**, *325*, 542–553. [CrossRef]

62. Tavazoei, M.; Haeri, M. Unreliability of frequency-domain approximation in recognising chaos in fractional-order systems. *IET Signal Process.* **2007**, *1*, 171–181. [CrossRef]

63. Tavazoei, M.S.; Haeri, M. A necessary condition for double scroll attractor existence in fractional-order systems. *Phys. Lett. A* **2007**, *367*, 102–113. [CrossRef]

64. Danca, M.F. Hidden chaotic attractors in fractional-order systems. *Nonlinear Dyn.* **2017**, *89*, 577–586. [CrossRef]

65. Tavazoei, M.S.; Haeri, M. Chaotic attractors in incommensurate fractional order systems. *Phys. D Nonlinear Phenom.* **2008**, *237*, 2628–2637. [CrossRef]

66. Ahmed, E.; El-Sayed, A.; El-Saka, H.A. On some Routh–Hurwitz conditions for fractional order differential equations and their applications in Lorenz, Rössler, Chua and Chen systems. *Phys. Lett. A* **2006**, *358*, 1–4. [CrossRef]

67. Sprott, J.C. Simplest chaotic flows with involutional symmetries. *Int. J. Bifurc. Chaos* **2014**, *24*, 1450009. [CrossRef]

68. Tourkmani, A.M.; Alharbi, T.J.; Rsheed, A.M.B.; AlRasheed, A.N.; AlBattal, S.M.; Abdelhay, O.; Hassali, M.A.; Alrasheedy, A.A.; Al Harbi, N.G.; Alqahtani, A. Hypoglycemia in Type 2 Diabetes Mellitus patients: A review article. *Diabetes Metab. Syndr. Clin. Res. Rev.* **2018**, *12*, 791–794. [CrossRef]

69. Wolf, A.; Swift, J.B.; Swinney, H.L.; Vastano, J.A. Determining Lyapunov exponents from a time series. *Phys. D Nonlinear Phenom.* **1985**, *16*, 285–317. [CrossRef]

70. Effah-Poku, S.; Obeng-Denteh, W.; Dontwi, I. A Study of Chaos in Dynamical Systems. *J. Math.* **2018**, *2018*, 1808953. [CrossRef]

71. Leonov, G.A.; Kuznetsov, N.V. Time-varying linearization and the Perron effects. *Int. J. Bifurc. Chaos* **2007**, *17*, 1079–1107. [CrossRef]

72. Lee, Y.; Fluckey, J.D.; Chakraborty, S.; Muthuchamy, M. Hyperglycemia-and hyperinsulinemia-induced insulin resistance causes alterations in cellular bioenergetics and activation of inflammatory signaling in lymphatic muscle. *FASEB J.* **2017**, *31*, 2744–2759. [CrossRef] [PubMed]

73. Shanik, M.H.; Xu, Y.; Škrha, J.; Dankner, R.; Zick, Y.; Roth, J. Insulin resistance and hyperinsulinemia: Is hyperinsulinemia the cart or the horse? *Diabetes Care* **2008**, *31*, S262–S268. [CrossRef] [PubMed]

74. Erion, K.A.; Corkey, B.E. Hyperinsulinemia: A cause of obesity? *Curr. Obes. Rep.* **2017**, *6*, 178–186. [CrossRef] [PubMed]

75. Corkey, B.E. Banting lecture 2011: Hyperinsulinemia: cause or consequence? *Diabetes* **2012**, *61*, 4–13. [CrossRef] [PubMed]

76. Glaser, B. Type 2 diabetes: Hypoinsulinism, hyperinsulinism, or both? *PLoS Med.* **2007**, *4*, e148. [CrossRef] [PubMed]

77. Dawson, S.P.; Grebogi, C.; Yorke, J.A.; Kan, I.; Koçak, H. Antimonotonicity: Inevitable reversals of period-doubling cascades. *Phys. Lett. A* **1992**, *162*, 249–254. [CrossRef]

78. Kengne, J.; Jafari, S.; Njitacke, Z.; Khanian, M.Y.A.; Cheukem, A. Dynamic analysis and electronic circuit implementation of a novel 3D autonomous system without linear terms. *Commun. Nonlinear Sci. Numer. Simul.* **2017**, *52*, 62–76. [CrossRef]

79. Signing, V.F.; Kengne, J.; Pone, J.M. Antimonotonicity, chaos, quasi-periodicity and coexistence of hidden attractors in a new simple 4-D chaotic system with hyperbolic cosine nonlinearity. *Chaos Solitons Fractals* **2019**, *118*, 187–198. [CrossRef]

80. Pandey, A.; Chawla, S.; Guchhait, P. Type-2 diabetes: Current understanding and future perspectives. *IUBMB Life* **2015**, *67*, 506–513. [CrossRef]

81. Bhattacharya, S.; Dey, D.; Roy, S.S. Molecular mechanism of insulin resistance. *J. Biosci.* **2007**, *32*, 405–413. [CrossRef]

82. Quan, W.; Jo, E.K.; Lee, M.S. Role of pancreatic β-cell death and inflammation in diabetes. *Diabetes Obes. Metab.* **2013**, *15*, 141–151. [CrossRef] [PubMed]

83. Weng, J.; Li, Y.; Xu, W.; Shi, L.; Zhang, Q.; Zhu, D.; Hu, Y.; Zhou, Z.; Yan, X.; Tian, H.; et al. Effect of intensive insulin therapy on β-cell function and glycaemic control in patients with newly diagnosed type 2 diabetes: A multicentre randomised parallel-group trial. *Lancet* **2008**, *371*, 1753–1760. [CrossRef]

84. Pearson, J.A.; Agriantonis, A.; Wong, F.S.; Wen, L. Modulation of the immune system by the gut microbiota in the development of type 1 diabetes. *Hum. Vaccines Immunother.* **2018**, *14*, 2580–2596. [CrossRef] [PubMed]

85. Jahanshahi, H.; Yousefpour, A.; Munoz-Pacheco, J.M.; Kacar, S.; Pham, V.T.; Alsaadi, F.E. A new fractional-order hyperchaotic memristor oscillator: Dynamic analysis, robust adaptive synchronization, and its application to voice encryption. *Appl. Math. Comput.* **2020**, *383*, 125310. [CrossRef]

86. Nasteska, D.; Hodson, D.J. The role of beta cell heterogeneity in islet function and insulin release. *J. Mol. Endocrinol.* **2018**, *61*, R43–R60. [CrossRef]

87. Reinbothe, T.M.; Safi, F.; Axelsson, A.S.; Mollet, I.G.; Rosengren, A.H. Optogenetic control of insulin secretion in intact pancreatic islets with β-cell-specific expression of Channelrhodopsin-2. *Islets* **2014**, *6*, e28095. [CrossRef]

 *symmetry*

Article

A New Hyperchaotic Map for a Secure Communication Scheme with an Experimental Realization

Nadia M. G. Al-Saidi [1], Dhurgham Younus [1], Hayder Natiq [2,3], M. R. K. Ariffin [3,4,*], M. A. Asbullah [3,5] and Z. Mahad [3]

1   Department of Applied Sciences, University of Technology, 10001 Baghdad, Iraq; nadia.m.ghanim@uotechnology.edu.iq (N.M.G.A.-S.); as.18.52@grad.uotechnology.edu.iq (D.Y.)
2   Information Technology Collage, Imam Ja'afar Al-Sadiq University, 10001 Baghdad, Iraq; hayder.natiq@sadiq.edu.iq
3   Institute for Mathematical Research, Universiti Putra Malaysia, 43400 UPM Serdang, Selangor, Malaysia; ma_asyraf@upm.edu.my (M.A.A.); zaharimahad@upm.edu.my (Z.M.)
4   Department of Mathematics, Faculty of Science, Universiti Putra Malaysia, 43400 UPM Serdang, Selangor, Malaysia
5   Centre of Foundation Studies for Agriculture Science, Universiti Putra Malaysia, 43400 UPM Serdang, Selangor, Malaysia
*   Correspondence: rezal@upm.edu.my

Received: 27 September 2020; Accepted: 3 November 2020; Published: 15 November 2020

Abstract: Using different chaotic systems in secure communication, nonlinear control, and many other applications has revealed that these systems have several drawbacks in different aspects. This can cause unfavorable effects to chaos-based applications. Therefore, presenting a chaotic map with complex behaviors is considered important. In this paper, we introduce a new 2D chaotic map, namely, the 2D infinite-collapse-Sine model (2D-ICSM). Various metrics including Lyapunov exponents and bifurcation diagrams are used to demonstrate the complex dynamics and robust hyperchaotic behavior of the 2D-ICSM. Furthermore, the cross-correlation coefficient, phase space diagram, and Sample Entropy algorithm prove that the 2D-ICSM has a high sensitivity to initial values and parameters, extreme complexity performance, and a much larger hyperchaotic range than existing maps. To empirically verify the efficiency and simplicity of the 2D-ICSM in practical applications, we propose a symmetric secure communication system using the 2D-ICSM. Experimental results are presented to demonstrate the validity of the proposed system.

Keywords: hyperchaotic behavior; symmetric encryption; arduino microcontrollers; optical channel

---

1. Introduction

Numerous phenomena have been comprehended by studying the complex behaviors in many natural and non-natural dynamical systems. Understanding the chaotic behavior, which is a kind of nonlinear complex dynamical behavior, has provided a significant description of these systems. Although dynamical systems with chaotic behaviors are deterministic, long-term prediction of their behaviors is impossible [1]. Moreover, the sensitivity, topological mixing, and orbits density are the main characteristics of the chaotic systems [2–4]. Therefore, chaotic systems have valuable applications in various fields including computer science, telecommunication, physics, engineering, etc. [5–10]. In particular, due to the similarity between the characteristics of chaotic systems and the diffusion and confusion properties of cryptography [11], a wide body of chaos-based cryptographic applications has been presented in the last few years [12–17].

For the time being, discrete-time systems and continuous-time systems are the major types of chaotic systems. The former type is described by a difference equation, and it can be implemented through an iterative procedure, while the latter one is usually represented by a partial and/or ordinary differential equation. Edward Lorenz was the first to present a chaotic system with continuous-time [18]. Subsequently, several well-known continuous-time chaotic and hyperchaotic systems have been proposed such as Rössler [19], Sprott [20], Chen [21], and Lü [22] systems. On the other hand, the Logistic map, which was presented by Robert May, is the first clear example of a discrete-time system with chaotic behavior [23]. Since then various discrete-time chaotic and hyperchaotic systems have been presented in the literature [24–28].

During the past recent years, significant efforts in the prediction of chaotic systems' behaviors have been devoted through determining their parameters [29], or estimating their states [30]. Predicting the behavior of a chaotic system can render chaos-based cryptosystem insecure [31]. This has raised the need for measuring the complexity of the employed chaotic systems [32,33]. Therefore, numerous algorithms have emerged to measure the complexity of the systems' time series such as Fuzzy Entropy [34], Modified Permutation-Entropy [35], and Sample Entropy [36].

Due to the performance drawbacks of many existing chaotic systems in some attributes, for instance, frail chaos (i.e., chaotic behavior appears only in insulated zones of the system' parameters), it motivated researchers to propose systems with robust chaos that can encourage chaos-based cryptographic applications. An example of such weakness is that through a slight perturbation to a single parameter, it could make the system collapse into a non-chaotic zone of the system [37].

Based on the aforementioned description, this paper proposes a new chaotic discrete-time system, called the 2D infinite-collapse-Sine model (2D-ICSM). The 2D-ICSM exhibits a wide hyperchaotic range, good ergodicity, high complexity, and sensitivity. Therefore, 2D-ICSM could be an ideal source for chaos-based cryptographic applications. The main contributions of this work are as follows.

1. We introduce an analytical framework to understand the dynamical behavior of the 2D-ICSM including stability of its fixed points, bifurcation diagram, and Lyapunov Exponents.
2. We experimentally evaluate the complexity, sensitivity, and randomness of the 2D-ICSM using Sample Entropy, cross-correlation coefficient, and NIST-800-22 statistical test, respectively.
3. To demonstrate the efficiency and simplicity of the 2D-ICSM in practical applications, we design a secure communication system, and then experimented tested it on an optical channel with Arduino microcontrollers.

This paper is organized as follows. Section 2 introduces the 2D-ICSM and studies the stability of its equilibria. In Section 3, we analyze the dynamics of the 2D-ICSM. Section 4 demonstrates the high sensitivity and randomness of the 2D-ICSM. Section 5 demonstrates the detailed complexity performance of the 2D-ICSM. In Section 6, we introduce the proposed secure communication system. Section 7 illustrates the implementation of the communication system. Conclusions are presented in Section 8.

2. The 2D Infinite-Collapse-Sine Model

In this section, we introduce a new 2D chaotic map, called the 2D infinite-collapse-Sine model (2D-ICSM), and then discusses its stability analysis.

2.1. Definition of 2D-ICSM

Among existing 1D discrete-time dynamical systems, the infinite collapse model is considered as one of the best maps that show robust chaotic performance [38]. Mathematically, its dynamical equation is given by

$$x_{n+1} = \sin\left(\frac{\beta}{x_n}\right), \tag{1}$$

where β is the control parameter and x is the state variable. The dynamical behavior of this map can be illustrated by depicting its bifurcation diagram and trajectory in the phase plane, as shown in Figure 1. It can be seen that the numerical solution of this map is in the range of $[-1, 1]$. Besides that, the bifurcation diagram of this map shows that the chaotic attractor appears in limited regions of its parameter. Meanwhile, several non-chaotic regions can be observed as the parameter α increasing. Furthermore, Figure 1b demonstrates that its trajectory only occupies a small space in the phase plane. Such behaviors are widely observed in the existing 1D chaotic maps such as Logistic, Tent, and Sine maps.

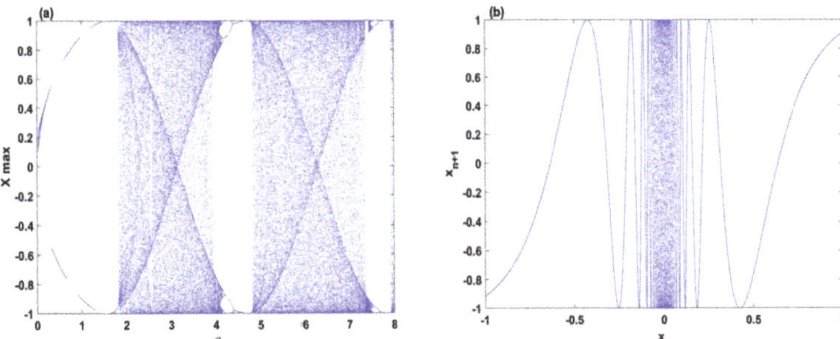

Figure 1. Dynamical behavior of the infinite collapse map (1) with $x_0 = 0.5$: (**a**) bifurcation diagram for $\beta \in [0, 8]$ and (**b**) chaotic attractor for $\beta = 2$.

To tackle the aforementioned issues, we propose a new 2D chaotic map, which consists of four terms with two control parameters. Mathematically, it is defined as follows,

$$\begin{cases} x_{n+1} = \sin\left(\beta y_n\right), \\ y_{n+1} = (\alpha + 2)x_n + \sin\left(\dfrac{\beta}{y_n}\right), \end{cases} \tag{2}$$

where α is the amplitude parameter and β is the internal frequency parameter. It can be seen that the proposed 2D infinite-collapse-Sine model (2D-ICSM) is mainly designed by using three components including a linear variable, 1D Sine map, and 1D infinite collapse map. The linear state variable x_n is used to modulate the output of a 1D infinite collapse map. Therefore, it can enhance the chaotic behavior of the state variable y_{n+1}. Meanwhile, the 1D Sine map is employed to boost randomness to the state variable x_{n+1}.

2.2. Stability Analysis

For discrete-time systems, the fixed point of a function form a graphical point of view is an element in the domain that maps to itself by the function. For instance, P is a fixed point of the function $F(x)$ only when $F^n(P) = P$. To simplify the calculation of the fixed points of 2D-ICSM, we reduce its dimension to become 1D as follows,

$$y^{(v)} = (\alpha + 2)\sin\left(\beta y^{(v)}\right) + \sin\left(\dfrac{\beta}{y^{(v)}}\right), \tag{3}$$

The fixed points of the 2D-ICSM are calculated for two different sets of system parameters. For each set of parameters, we obtain the fixed points of the variable y by Equation (3), and subsequently, the corresponding points of the variable x can be easily obtained by the first equation of the system (2). Figure 2 illustrates how the fixed points of the variable y can be obtained using the graphical method. From this figure, one can notice that the number of fixed points is increased as the values of the amplitude and internal frequency parameters increase. Now, let us collect some fixed points from each set of system parameters to investigate their stability. First, we have extracted the following fixed point from the first set of the system parameters,

$$\left\{ P_1 = \left(x^{(1)}, y^{(1)} \right) = (0.6687, 2.4092). \right.$$

Second, we have extracted three different fixed points from the second set as follows,

$$\begin{cases} P_1 = \left(x^{(1)}, y^{(1)} \right) = (0.9168, 4.1324), \\ P_2 = \left(x^{(2)}, y^{(2)} \right) = (0.7607, 3.5738), \\ P_3 = \left(x^{(3)}, y^{(3)} \right) = (0.1334, 1.5039). \end{cases}$$

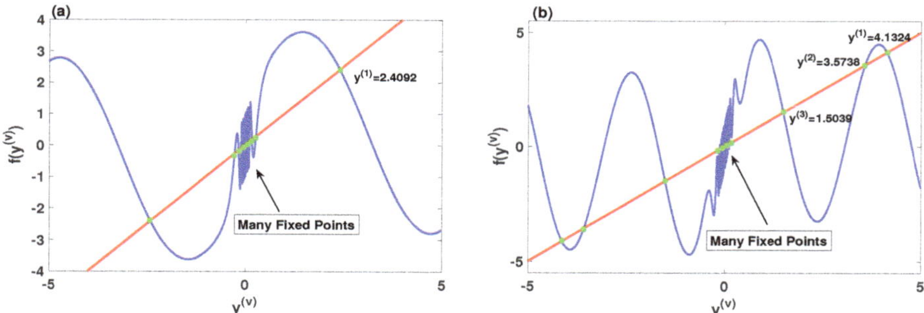

Figure 2. Fixed points of the 2D-ICSM: (**a**) for the parameters $\alpha = \beta = 1$; (**b**) for the parameters $\alpha = \beta = 2$.

The stability of the above-fixed points can be determined by obtaining the Jacobian matrix, which is given by

$$J = \begin{pmatrix} \frac{\partial f_1}{\partial x} & \frac{\partial f_1}{\partial y} \\[2mm] \frac{\partial f_2}{\partial x} & \frac{\partial f_2}{\partial y} \end{pmatrix}.$$

Using the above matrix, the 2D-ICSM is Linearized at any arbitrary fixed point $P_i = (x^*, y^*)$ as follows,

$$J_{P_i} = \begin{pmatrix} 0 & \beta\cos(\beta y) \\[2mm] 2+\alpha & -\frac{\beta}{y^2}\cos\left(\frac{\beta}{y}\right) \end{pmatrix}.$$

Thus, the eigenvalues are obtained by solving the following equation,

$$\lambda^2 + \left(\frac{\beta}{y^2}\cos\left(\frac{\beta}{y}\right) \right)\lambda - (2+\alpha)\beta\cos(\beta y) = 0.$$

It is well-known that the stability of fixed points is dependent on the eigenvalues. When an eigenvalue is within the interval $[-1, 1]$, then the fixed point exhibits a stable state. Otherwise, it shows an unstable state. Moreover, the stability of the obtained fixed points is as illustrated in Table 1. All the selected fixed points of the 2D-ICSM are unstable.

Table 1. The fixed points of the 2D infinite-collapse-Sine model (2D-ICSM) and their stability analysis.

Parameters	Fixed Points	λ_1	λ_2	Stability Analysis
$\alpha = 1, \beta = 1$	P_1	1.53297	-1.53297	unstable
	P_1	1.8162	-1.8162	unstable
$\alpha = 2, \beta = 2$	P_2	2.17218	-2.17218	unstable
	P_3	2.8715	-2.8715	unstable

3. Dynamical Behaviors

This section investigates the dynamical behaviors of 2D-ICSM through the bifurcation diagram, Lyapunov Exponents, and phase space.

3.1. Bifurcation Diagram and Lyapunov Exponents

Typically, the bifurcation diagram and Lyapunov Exponents are used to determined the non-chaotic and chaotic regions of a dynamical system when one of its parameters varies. Furthermore, the Lyapunov exponent is used to evaluate the chaotic properties of a dynamical system. In other words, it could recognize the chaotic and hyperchaotic behaviors of the system. A system is recognized as chaotic when there is one positive Lyapunov Exponent value for each parameter value, whereas the hyperchaotic system has more than one positive Lyapunov Exponent value. The hyperchaotic system exhibits a higher level of randomness, and the generated sequences by the hyperchaotic system show extreme unpredictability.

To investigate the dynamics of 2D-ICSM, we depict its bifurcation diagram and Lyapunov Exponents with the initial values $(0.5, 0.5)$ and for the parameters $0 \leq \alpha \leq 8$ and $\beta = 12$, as shown in Figure 3. It can be seen that 2D-ICSM is hyperchaotic among the whole parameter range, which indicates that its sequences are extremely hard to be predicted.

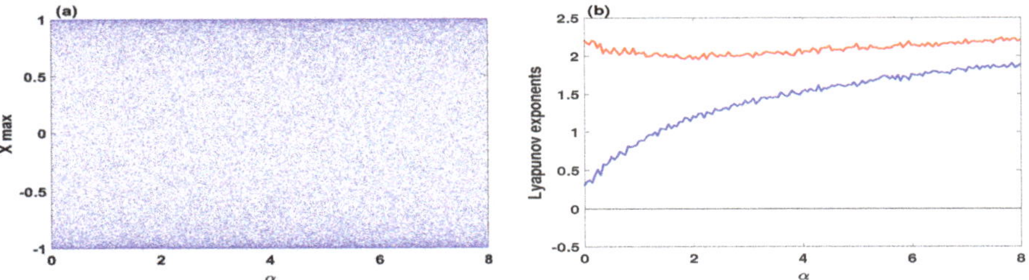

Figure 3. Dynamics of the 2D-ICSM with the initial values $(0.5, 0.5)$ and for the parameter $\beta = 12$: (**a**) bifurcation diagram; (**b**) Lyapunov Exponents.

3.2. Hyperchaotic Attractor

The set of numerical values, which is generated by a chaotic/hyperchaotic map with specific sets of initial values and control parameters, is called chaotic/hyperchaotic attractor. For a 2D map, its attractor can be described by a group of points that occupies a particular region in the phase space. A chaotic/hyperchaotic model has better performance when its attractor is geometrically complicated and occupies a larger range in the phase space. To illustrate the hyperchaotic range of the 2D-ICSM, Figure 4f depicts its attractor in the 2D phase space with the parameters $\alpha = 6$ and $\beta = 12$. Besides that, this figure plots the attractors of several existing chaotic and hyperchaotic models to demonstrate the complicated behavior of the 2D-ICSM. It can be observed that the hyperchaotic attractor of 2D-ICSM fully occupies a 2D phase space ranging $x \in [-1, 1]$ and $y \in [-9, 9]$. This means that 2D-ICSM can generate more unpredictable hyperchaotic sequences and it has a better competitive ergodicity property than existing models.

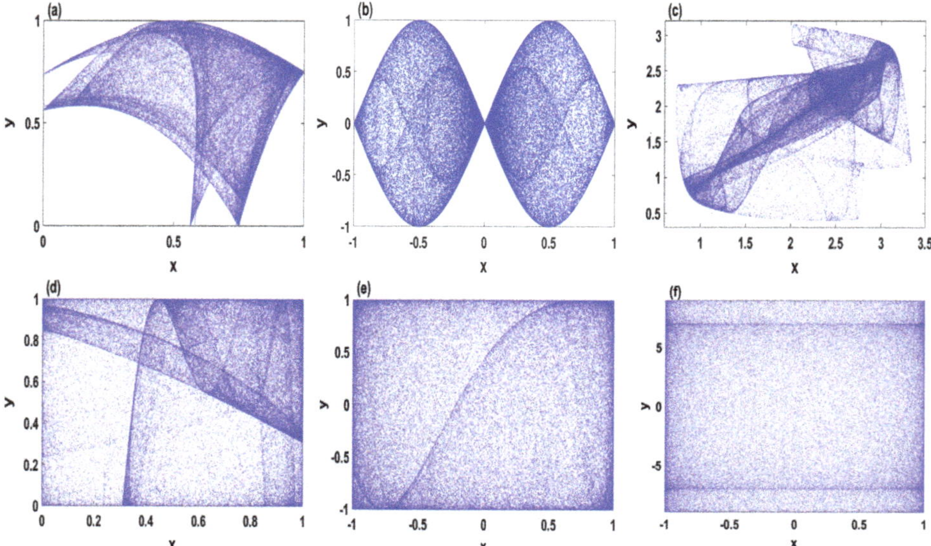

Figure 4. Chaotic and hyperchaotic attractors of different 2D maps: (**a**) 2D-SLMM [12]; (**b**) 2D-SIMM [13]; (**c**) 2D Ushiki map [39]; (**d**) 2D-LASM [14]; (**e**) 2D-LICM [15]; (**f**) the 2D-ICSM.

4. Performance Evaluations

In this section, the sensitivity of the initial conditions and the control parameters is measured by the cross-correlation coefficient. Furthermore, the quantitative values of the randomness of sequences generated by the 2D-ICSM are determined using NIST-800-22 randomness tests.

4.1. Cross-Correlation Coefficient

To estimate the sensitivity of the initial conditions and the control parameters of the 2D-ICSM, we use the cross-correlation coefficient (CCF); its equation is given by

$$CCF(\alpha_t, \beta_t) = \frac{\sum_{t=1}^{N}(\alpha_t - A(\alpha))(\beta_t - A(\beta))}{\sqrt{\sum_{t=1}^{N}(\alpha_t - A(\alpha))^2 \sum_{t=1}^{N}(\beta - A(\beta))^2}}, \tag{4}$$

where $A(\alpha)$ represents the mean value of the time series α_t, meanwhile $A(\beta)$ represents the mean value of the time series β_t. When $CCF(\alpha_t, \beta_t)$ is close to 0, then it can be indicated that these two-time series are diverging.

Figure 5 presents the sensitivity of the 2D-ICSM with the parameters $\alpha \in [0, 8]$ and $\beta = 12$. In this figure, the sensitivity is estimated by calculating the CCF between the original time series and the modified time series. It is important to mention here that the modified time series was generated by the 2D-ICSM using a very small error, $e = 5 \times 10^{-5}$, which was added to the initial value x_0 and the parameter α, as shown in Figure 5a,b, respectively. It can be observed that the 2D-ICSM has a high level of sensitivity to its initial values and control parameters.

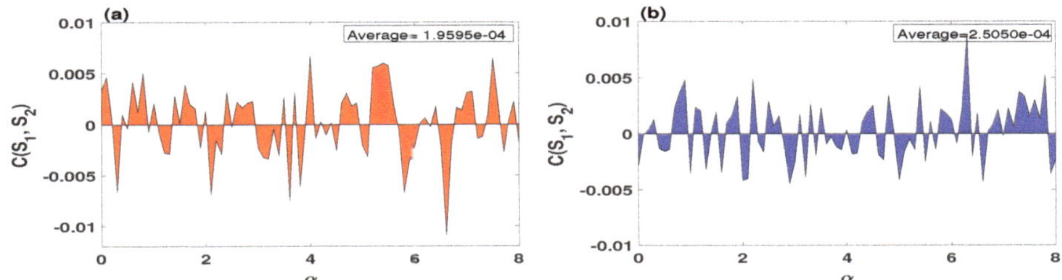

Figure 5. The CCF analysis of the 2D-ICSM with $0 \leq \alpha \leq 8$ and $\beta = 12$: (**a**) the CCF with the initial values $(0.1 + e, 0.1)$; (**b**) the CCF with the initial values $(0.1, 0.1)$ and for $\alpha + e$.

4.2. Chaos-Based Pseudorandom Number Generator

A chaotic map could be a suitable source to generate pseudorandom numbers when it has high sensitivity, good ergodicity, and extreme unpredictability. The existence of these features in a chaotic map can be determined by the NIST-800-22 randomness tests.

It is, therefore, crucial to determine the existence of such features in the 2D-ICSM to examine its ability to be a PRNG. In this regard, we propose a simple strategy, which directly employs the chaotic sequences as pseudorandom numbers by converting each of their values to a 32-bit binary stream using the IEEE 754 float standard. Figure 6 displays the NIST SP800-22 test results of pseudorandom numbers generated by the 2D-ICSM. In this figure, the generated sequence by 2D-ICSM has a length of $100,000,000$ binary bits. It is important to state here that a chaotic map can pass the statistical tests of NIST-800-22 only when the corresponding p-values are greater than the experimental significance level [40]. Consequently, the results in Figure 6 demonstrates the high randomness of the generated pseudo random numbers by the 2D-ICSM.

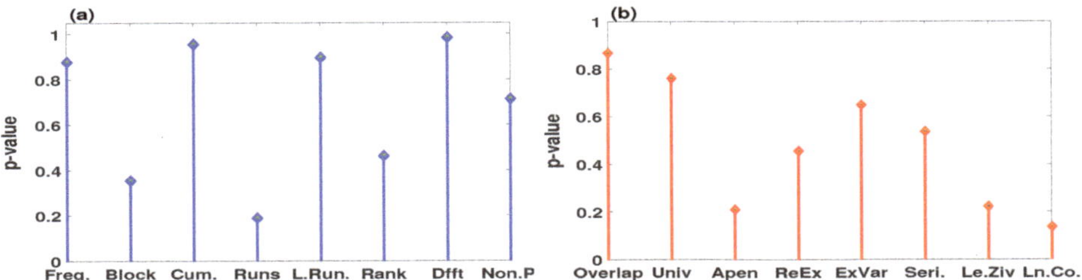

Figure 6. The p-values of the binary sequence generated by PRNG of the 2D-ICSM with the parameters $\alpha = 6$ and $\beta = 12$.

5. Complexity-Based Sample Entropy

In this section, the complexity of 2D-ICSM is investigated through a fundamental algorithm, namely, Sample Entropy (SamEn). The authors of [36] presented SamEn to calculate how much extra information is required to predict the $(t + 1)$th output of a trajectory using its previous (t) outputs. SamEn with larger values indicate a lower degree of regularity of a chaotic map. In other words, the chaotic map exhibits a high level of complexity and unpredictability.

The SamEn algorithm for a given time series $\{x(i), i = 0, 1, 2, \ldots, N - 1\}$ is outlined as follows:

1. Reconstruction: the time series can be reconstructed as follows,

$$X_i = \{x_i, x_{i+\tau}, \ldots, x_{i+(m-1)\tau}\}, \quad X_i \in R^m \tag{5}$$

where m is embedding dimension and τ is time delay.

2. Counting the vector pairs: For a given tolerance parameter r, let B_i be the number of vectors X_j such that

$$d[X_i, X_j] \leq r, \quad i \neq j \tag{6}$$

here, $d[X_i, X_j]$ is the distance between X_i and X_j, which is defined as

$$d[X_i, X_j] = \max\{|x(i+k) - \quad x(j+k)| : \\ 0 \leq k \leq m-1\}. \tag{7}$$

3. Calculating $\theta^m(r)$: According to the obtained number of vector pairs, we can get

$$C_i^m(r) \quad = \frac{B_i}{N-(m-1)\tau}, \tag{8}$$

then calculate $\theta^m(r)$ by

$$\theta^m(r) \quad = \frac{\sum_{i=1}^{N-(m-1)\tau} ln C_i^m(r)}{[N-(m-1)\tau]}. \tag{9}$$

4. Calculating SamEn: Repeating the above steps we can get $\theta^{m+1}(r)$, then SamEn is given by

$$SamEn(m, r, N) = \theta^m(r) - \theta^{m+1}(r). \tag{10}$$

Figure 7a plots SamEn results of the 2D-ICSM when the two parameters α and β are varying simultaneously. This figure provides a more clear vision of the complexity of 2D-ICSM. It can be seen from this figure that the 2D-ICSM exhibits high complexity in most of its parameters setting. However, the highest SamEn values appear whenever the α and β are increasing.

Moreover, Figure 7b depicts the SamEn results of the 2D-ICSM and different chaotic and hyperchaotic maps. It is quite clear that the 2D-ICSM has the largest SamEn values, which indicates that one needs more information to predict the generated sequences by this map.

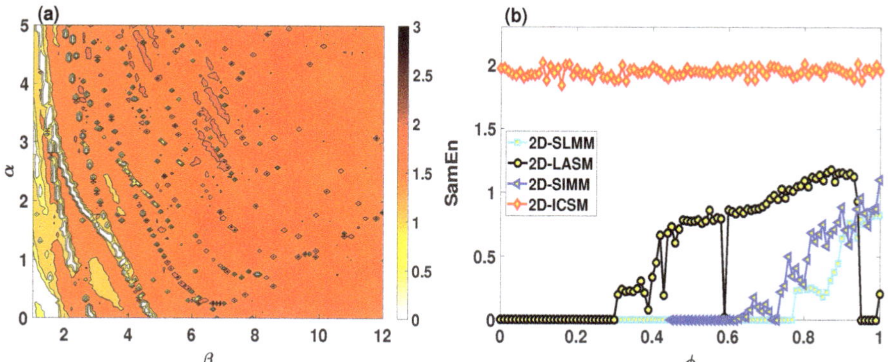

Figure 7. SamEn simulation: (**a**) SamEn values of the 2D-ICSM when its parameters vary; (**b**) SamEn results of different chaotic and hyperchaotic maps, where parameter ϕ represents α, a, a_2, α_1, for the 2D-ICSM, 2D-LASM [14], 2D-SIMM [13], and 2D-SLMM [12], respectively.

6. Chaos Based Cryptography

This section investigates the performance of 2D-ICSM in cryptography applications by designing a symmetric secure communication system. Figure 8 displays the schematic diagram of the proposed symmetric secure communication scheme. As can be observed in this figure, the proposed communication system is designed to transmit a message $M(s)$ between two points in which the 2D-ICSM is employed to encrypt the information.

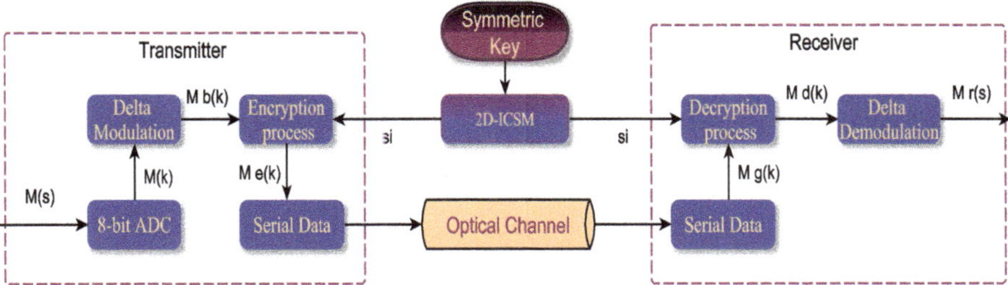

Figure 8. Schematic diagram of the proposed secure communication scheme using the 2D-ICSM.

6.1. Arduino Transmitter

In the proposed communication system, the Arduino is considered as the core of transmission. Here, we employ the Arduino Uno R3 microcontroller boards, which are simple and implemented at a low cost. It has 14 digital input/output pins (6 of them can be used as PWM outputs), 6 analog inputs, a 16 MHz crystal oscillator, a USB connection, and a reset button [41].

First, we convert the signal from the bipolar form to the unipolar form, and this is due to the fact that the Arduino analog inputs only accept unipolar signals in the range from 0 V to 5 V. The communication starts with a message $M(s)$, which is sent to the analog input A0 of the Arduino transmitter, as shown in Figure 9. Second, the input signal is converted from analog to digital using an embedded 8-bit ADC at a maximum rate of 8000 samples per second. Finally, this signal is encrypted by the 2D-ICSM and Delta modulator.

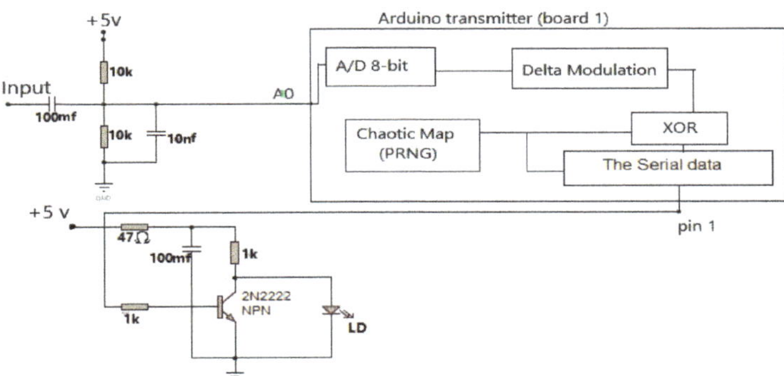

Figure 9. The transmitter circuit.

6.2. Delta Modulation

The Delta modulation is a simple and robust A / D conversion method [42]. It has a comparator in the forward path and an integrator in the feedback path of a simple control loop, as shown in Figure 10. The signal $M(k)$ is the input of the comparator. Meanwhile, $U(n)$ is the integrated output, which has a binary form. The value of Delta modulation depends on the current sample, if it is less than the previous sample, then zero is transmitted as a signal, whereas if the current sample is greater than the previous one, then number one is sent as a transmitted signal. Algorithm 1 illustrates the pseudocode of the Delta modulation process.

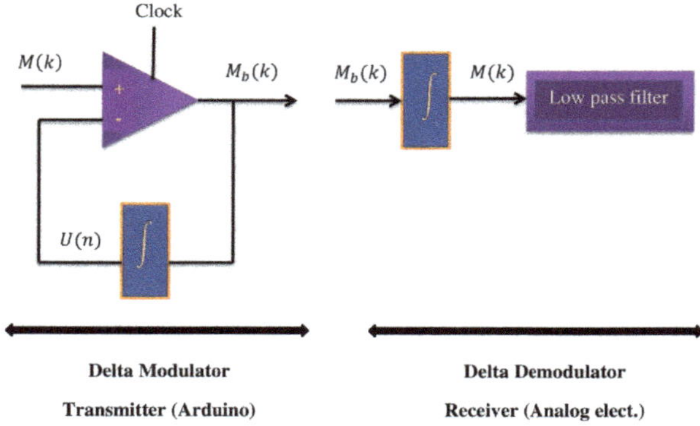

Figure 10. Diagram of the Delta Modulator.

Algorithm 1: Delta Modulation

Input: $M(k)$.
Output: $M_b(k)$.
1 **if** $M(k) > \sum U(n)$, *where* $\sum U(n) = U(n-1) + U(n-2) + \cdots + U(0)$, **then**
2 $\quad\mid\quad M_b(k) = +v$;
3 **else**
4 $\quad\mid\quad M_b(k) = -v$;
5 **end**

6.3. Encryption Process

The encryption process begins after obtaining the signal $M_b(k)$ from the Delta modulator. First, the secret key of the proposed encryption scheme is mainly generated from the initial values and a control parameter of the 2D-ICSM, as illustrated in Figure 11. As can be seen from this figure, the secrete key consists of 5 parts with 232 bits in which (x_0, y_0) and α are the initial values control parameter of the 2D-ICSM. Meanwhile, s and z are added to increase the security by increasing the key space and changing the initial values and control parameters. It is crucial to state here that the proposed key is symmetric, which means that it can be used for both encryption and decryption processes.

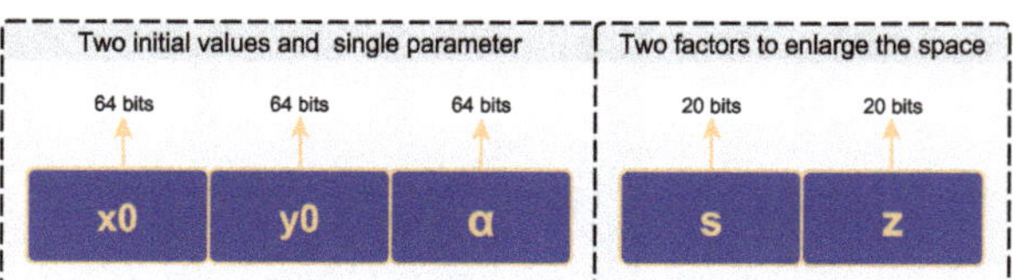

Figure 11. The secret key structure.

Now, the obtained signal $M_b(k)$ can be encrypted using the hyperchaotic sequence generated by the 2D-ICSM. Algorithm 2 illustrates the pseudo-code of generating the secrete key and the encryption process.

Algorithm 2: The encryption process.

Input: The secret key of size 232 bits, $M_b(k)$, and the length of input data Q.

Output: The serial data $M_e(k)$ and s_i.

1   $x_0 = \left(\sum_{j=1}^{64} k[j] \times 2^{64-j} \right) / 2^{64}$;

2   $y_0 = \left(\sum_{j=65}^{128} k[j] \times 2^{128-j} \right) / 2^{64}$;

3   $\alpha = \left(\sum_{j=129}^{192} k[i] \times 2^{192-j} \right) / 2^{64}$;

4   $s = \left(\sum_{j=193}^{212} k[j] \times 2^{212-j} \right) / 2^{20}$;

5   $z = \left(\sum_{j=213}^{232} k[j] \times 2^{232-j} \right)$;

6   $x = (x_0 + s \times z) \mod 1$;

7   $y = (y_0 + s \times z) \mod 1$;

8   $\alpha = (\alpha_0 + s \times z) \mod 8$;

9   Generate the hyperchaotic sequence c_i using the 2D-ICSM;

10   **for** $i=1$ to Q **do**

11   Obtain N-bit key vectors s_i from the chaotic sequence c_i;

12      $M_e(k) = (!M_b(k)\,AND\,s_i)\,OR\,(M_b(k)\,AND\,!s_i)$.

13   **end**

It is crucial to mention here that the encrypted signal will be the pulses (1's on and 0's off) that is sent from pin 1 in Arduino (board 1) to the electronic laser circuit as shown in Figure 9. The laser diode output depends on the diode injected current instead of voltage. The Arduino pin 1 is used to power the diode directly. Its processor draws 30 mA with 40 mA outputs. When the current enters into the laser diode circuit (LDC), it should be controlled by a modulated data stream. This current has high speed, where this circuit inverts the signal phase before the Laser diode (LD) is injected. Figure 12 shows a simulation example of the LDC. The inputs of the data stream are modulated through the LDC directly. The output of LD as an emitted light represents the reaction of "one" or "zero" logic. The direct modulation is considered as the most commonly used. It is utilized to modulate the light intensity for the transmission of information through free space.

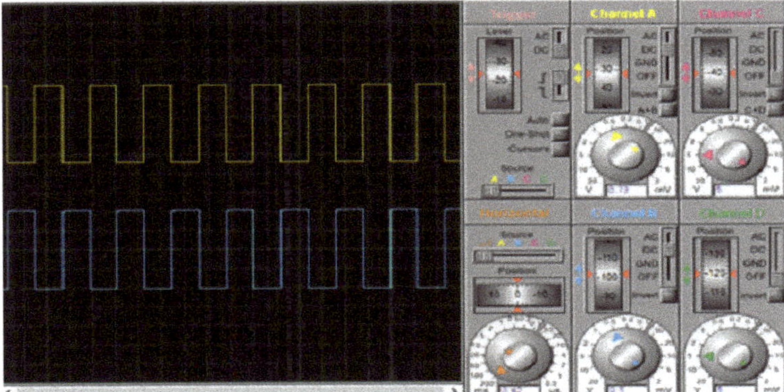

Figure 12. The input signal (yellow) and the output signal of the laser diode circuit (LDC) (blue).

Besides that, the laser beam transmitted by the photodiode is received in the form of light pulses. The photodiode acts as a semiconductor device used to convert light into a current that converts the received beam into an electrical signal within a voltage range between two volts (0–0.5). The encoded signal received by the photodiode passes various stages to amplify the signal voltage and return it to

5 volts, as can be seen in Figure 13. Furthermore, the binary signal is inverted and then sent it to pin 0 of Arduino (Board 2), as shown in Figure 12.

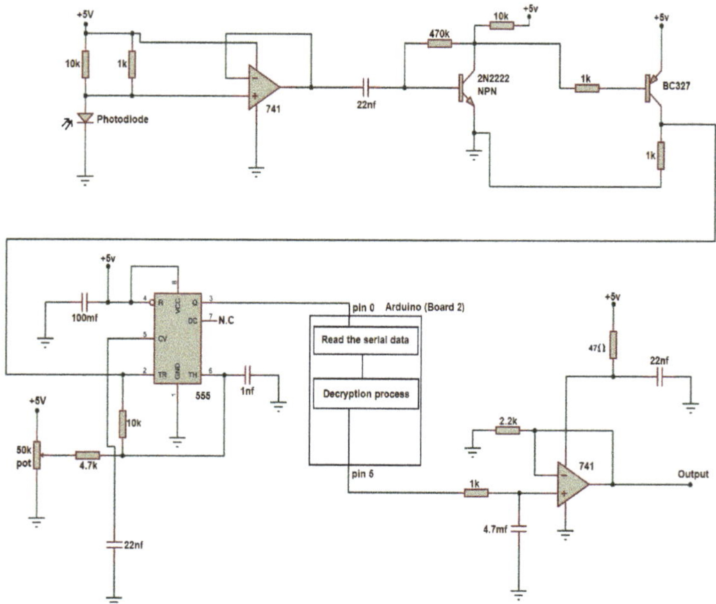

Figure 13. The receiver circuit.

6.4. Decryption Process

The decryption signal is referred to by $M_d(k)$. This output signal is sent to Pin 5, which in turn sends it immediately to delta demodulation, which is designed by an analog electronics via the operating amplifier. Figure 13 shows the delta demodulation composite of a Delta demodulation integrator. An op-amplifier as a low pass filter is used to send the signals and obtain the final $M_r(s)$, which is equal to the original signal M(s).

7. Experimental Implementation

This section investigates the simplicity of the proposed secure communication system in simulation and hardware implementation.

7.1. Simulation Implementation

The simulation results of the proposed secure communication system are presented here by Matlab 2017b programs and implemented in a computer with specification Core i3-2.00 GHz, Intel CPU, and 4 GB RAM. The decimal values of the secret key are selected as $x_0 = 0.9382$, $y_0 = 0.3171$, $\alpha = 4.6516$, $s = 0.5782$, and $z = 149936$. Figure 14a–f depicts the simulation results of the proposed communication system. Empirical correctness of the system can be observed through the retrieved signal $M_r(s)$, which is completely identical to the original signal $M(s)$ that has been encrypted, and then sent through a free-space optical channel.

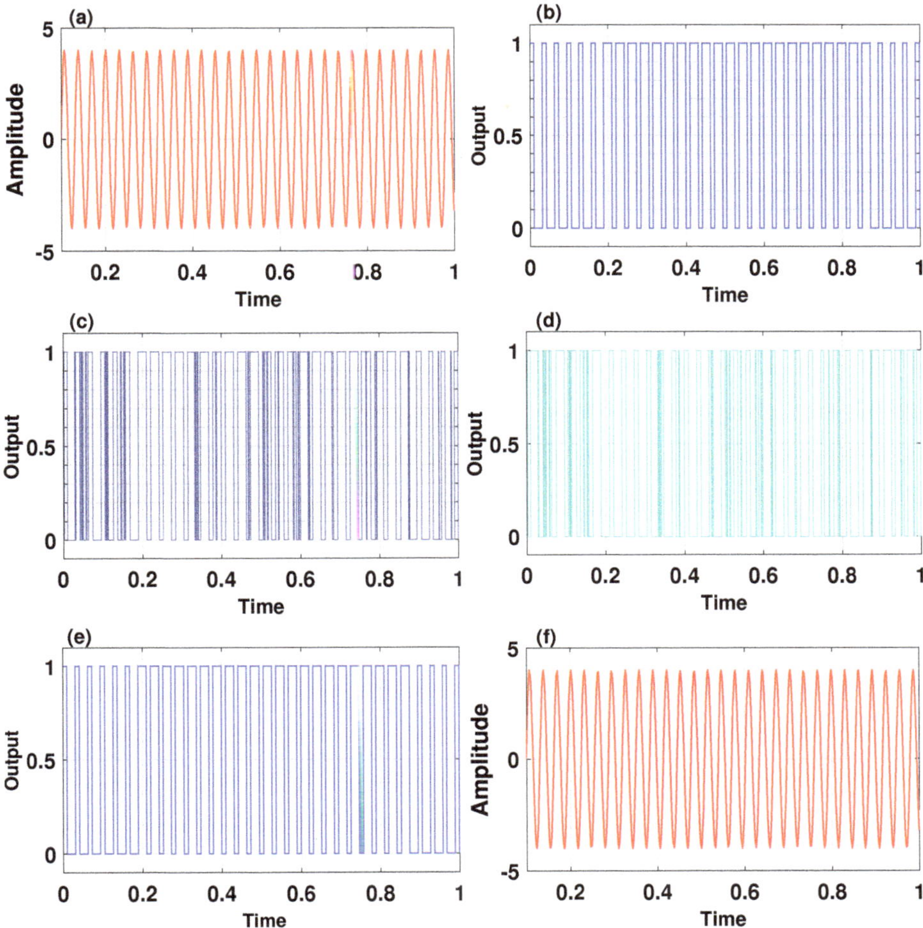

Figure 14. Simulation results: (**a**) the original signal; (**b**) the modulated output $M_b(k)$; (**c**) the encrypted binary data $M_e(k)$; (**d**) the serial binary data; (**e**) the decrypted binary data; (**f**) the retrieved signal.

Key Sensitivity Analysis

The employed key for the encryption scheme is considered highly sensitive when the encrypted message cannot be recovered, as a slight difference in one of the key components would result in an incorrect decrypted ciphertext. Therefore, we hereby investigate the key sensitivity of the proposed secure communication scheme, as shown in Figure 15. In this figure, a signal has been encrypted by k_1, and using the same key, we could recover the signal, as shown in Figure 15a–c. However, when we change the 14th decimal place in the parameters, or initial conditions, or both to obtain three other keys, namely, k_2, k_3, and k_4, respectively. Figure 15d–f demonstrates that these keys fail to recover the original signal, which means that the proposed communication system has a robust and sensitive key.

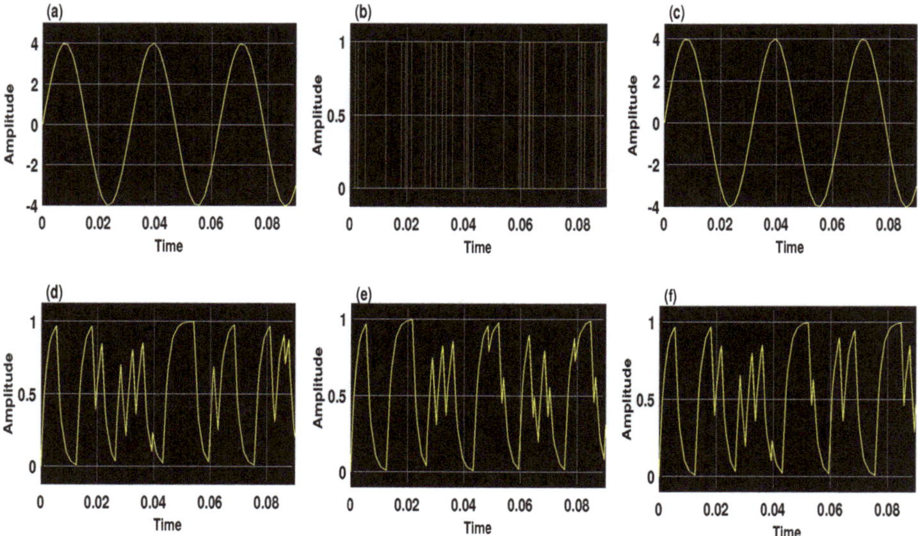

Figure 15. Key sensitivity analysis: (**a**) the original signal; (**b**) the encrypted signal; (**c**) the decrypted with the right key (k_1); (**d**–**f**) the decrypted signal with the wrong keys (k_2, k_3, k_4), in which the change occurred slightly in the parameters and initial conditions, only the parameters, and only the initial condition, respectively.

7.2. Hardware Implementation

The encryption algorithm is applied to a signal between (0 Hz–200 Hz) based on two microcontroller boards of Arduino Uno R3. The properties of these boards are non-expensive, have a simple design, and powerful microcontrollers that depend on the ATmega328 chip. The input and output pins are digital which consists of 14 digits, six of them are used as pulse width modulation (PWM) outputs, the other six are analog inputs. Besides that, the other 2 digits are used as 16 MHz crystal oscillator, the USB connection, and a reset button. These materials can be implemented by utilizing the C++ language. The computer with specification Core i3- 2.00 GHz, Intel CPU, and RAM 4 GB is used to run the software that is designed by the C++/C programing language, which is used to implement the two Arduino boards. These experiments have been implemented in a lab using a 200 MHz digital oscilloscope and a digital function generator. It is shown in Figure 16.

Figure 16. Work environment and laboratory materials.

The experiment was conducted in the laboratory. The results presented in Figure 17a refers to the sent message that appears in blue, while the retrieved message appears in yellow, and Figure 17a refers to the serial binary data.

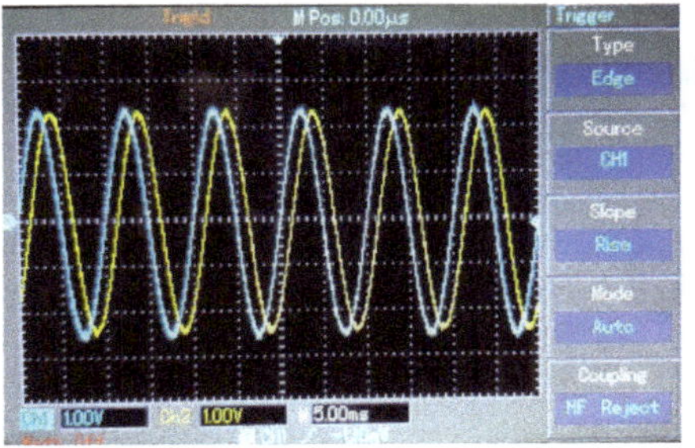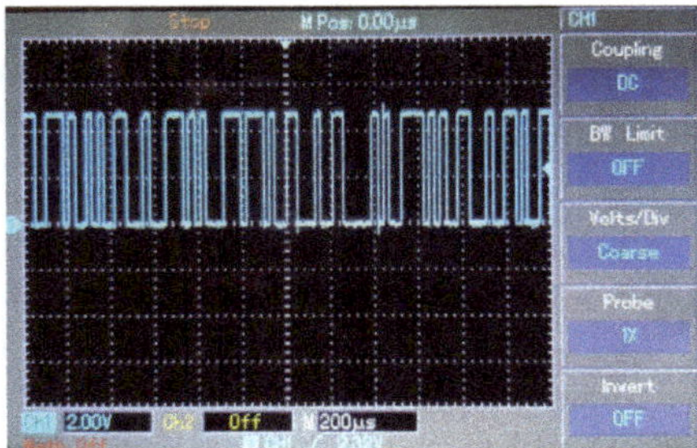

Figure 17. Oscilloscope outputs: (**a**) the sent message appeared in blue, and the reconstructed message appeared in yellow; (**b**) the serial data.

8. Conclusions

In summary, this paper introduces the 2D-ICSM, which is a new hyperchaotic map designed using the 1D infinite collapse model as seed. The fixed points of certain parameters of the 2D-ICSM have been calculated, and then the stability of these points was analyzed by the graphical method. Performance evaluations including Lyapunov exponents, bifurcation diagram, cross-correlation coefficient, phase space diagram, NIST-800-22 randomness test, and Sample Entropy algorithm showed that the 2D-ICSM has a wide hyperchaotic range, high sensitivity, good ergodicity, sufficient level of randomness, and extreme complexity performance. Therefore, the 2D-ICSM could be an ideal source for many chaos-based practical applications. To demonstrate the efficiency of 2D-ICSM, we proposed a secure communication system, which is designed to transmit a message between two points. The input message is modulated using a simple Delta modulator and then encrypted using the 2D-ICSM. In the receiver side, the 2D-ICSM along with Delta demodulation are employed to retrieve the original message. It is crucial to state that the transmitted message by the proposed communication system could be an image, a text, or a sound. Simulation and empirical results have verified the efficiency and simplicity of the proposed secure communication system.

Author Contributions: Conceptualization, N.M.G.A.-S. and D.Y.; methodology, H.N.; software, H.N. and D.Y.; validation, M.A.A. and Z.M.; writing–original draft preparation, N.M.G.A.-S. and H.N.; writing–review and editing, M.R.K.A.; supervision, M.R.K.A.; project administration, M.R.K.A.; funding acquisition, M.R.K.A. All authors have read and agreed to the published version of the manuscript.

Funding: This research received no external funding.

Conflicts of Interest: The authors declare no conflicts of interest.

References

1. Casdagli, M. Nonlinear prediction of chaotic time series. *Phys. Nonlinear Phenom.* **1989**, *35*, 335–356. [CrossRef]
2. Cho, K.; Takaya M. Chaotic cryptography using augmented Lorenz equations aided by quantum key distribution. *IEEE Trans. Circuits Syst. Regul. Pap.* **2014**, *62*, 478–487. [CrossRef]
3. Natiq, H.; Banerjee, S.; He, S.; Said, M.R.M.; Kilicman, A. Designing an M-dimensional nonlinear model for producing hyperchaos. *Chaos Solitons Fractals* **2018**, *114*, 506–515. [CrossRef]
4. Natiq, H.; Banerjee, S.; Ariffin, M.R.K.; Said, M.R.M. Can hyperchaotic maps with high complexity produce multistability? *Chaos Interdiscip. J. Nonlinear Sci.* **2019**, *29*, 011103. [CrossRef]
5. Berry, H.; Daniel, G.P.; Olivier, T. Chaos in computer performance. *Chaos Interdiscip. J. Nonlinear Sci.* **2006**, *16*, 013110. [CrossRef] [PubMed]

6.   Argyris, A.; Syvridis, D.; Larger, L.; Annovazzi-Lodi, V.; Colet, P.; Fischer, I.; Garcia-Ojalvo, J.; Mirasso, C.R.; Pesquera, L.; Shore, K.A. Chaos-based communications at high bit rates using commercial fibre-optic links. *Nature* **2005**, *438*, 343–346. [CrossRef] [PubMed]

7.   Natiq, H.; Said, M.R.M.; Al-Saidi, N.M.; Kilicman, A. Dynamics and complexity of a new 4d chaotic laser system. *Entropy* **2019**, *21*, 34. [CrossRef]

8.   Akgul, A.; Calgan, H.; Koyuncu, I.; Pehlivan, I.; Istanbullu, A. Chaos-based engineering applications with a 3D chaotic system without equilibrium points. *Nonlinear Dyn.* **2016**, *84*, 481–495. [CrossRef]

9.   Farhan, A.K.; Ali, R.S.; Natiq, H.; Al-Saidi, N.M. A new S-box generation algorithm based on multistability behavior of a plasma perturbation model. *IEEE Access* **2019**, *7*, 124914–124924. [CrossRef]

10. Viet-Thanh, P.H.A.M.; Ali, D.S.; Al-Saidi, N.M.; Rajagopal, K.; Alsaadi, F.E.; Jafari, S. A Novel Mega-stable Chaotic Circuit. *Radioengineering* **2020**, *29*, 141.

11. Alvarez, G.; Li, S. Some basic cryptographic requirements for chaos-based cryptosystems. *Int. J. Bifurc. Chaos* **2006**, *16*, 2129–2151. [CrossRef]

12. Hua, Z.; Zhou, Y.; Pun, C.M.; Chen, C.P. 2D Sine Logistic modulation map for image encryption. *Inf. Sci.* **2015**, *297*, 80–94. [CrossRef]

13. Liu, W., Sun, K.; Zhu, C. A fast image encryption algorithm based on chaotic map. *Opt. Lasers Eng.* **2016**, *84*, 26–36. [CrossRef]

14. Hua, Z.; Zhou, Y. Image encryption using 2D Logistic-adjusted-Sine map. *Inf. Sci.* **2016**, *339*, 237–253. [CrossRef]

15. Cao, C.; Sun, K.; Liu, W. A novel bit-level image encryption algorithm based on 2D-LICM hyperchaotic map. *Signal Process.* **2018**, *143*, 122–133. [CrossRef]

16. Natiq, H.; Al-Saidi, N.M.G.; Said, M.R.M.; Kilicman, A. A new hyperchaotic map and its application for image encryption. *Eur. Phys. J. Plus* **2018**, *133*, 1–14. [CrossRef]

17. Moysis, L.; Tutueva, A.; Volos, C.; Butusov, D.; Munoz-Pacheco, J.M.; Nistazakis, H. A Two-Parameter Modified Logistic Map and Its Application to Random Bit Generation. *Symmetry* **2020**, *12*, 829. [CrossRef]

18. Lorenz, E.N. Deterministic nonperiodic flow. *J. Atmos. Sci.* **1963**, *20*, 130–141. [CrossRef]

19. Rössler, O.E. An equation for continuous chaos. *Phys. Lett.* **1976**, *57*, 397–398. [CrossRef]

20. Sprott, J.C. Some simple chaotic flows. *Phys. Rev.* **1994**, *50*, R647. [CrossRef]

21. Chen, G.; Ueta, T. Yet another chaotic attractor. *Int. J. Bifurc. Chaos* **1999**, *9*, 1465–1466. [CrossRef]

22. Chen, A.; Lu, J.; Lü, J.; Yu, S. Generating hyperchaotic Lü attractor via state feedback control. *Phys. Stat. Mech. Appl.* **2006**, *364*, 103–110. [CrossRef]

23. May, R.M. Simple mathematical models with very complicated dynamics. *Nature* **1976**, *261*, 459–467. [CrossRef] [PubMed]

24. Baier, G.; Klein, M. Maximum hyperchaos in generalized Hénon maps. *Phys. Lett.* **1990**, *151*, 281–284. [CrossRef]

25. Natiq, H.; Al-Saidi, M.N.; Said, M.R.M. Complexity and dynamic characteristics of a new discrete-time hyperchaotic model. In Proceedings of the 2017 Second Al-Sadiq International Conference on Multidisciplinary in IT and Communication Science and Applications (AIC-MITCSA), Baghdad, Iraq, 30–31 December 2017.

26. Hussein, W.A.; Al-Saidi, N.M.; Natiq, H. A New 2D Hénon-Logistic Map for Producing Hyperchaotic Behavior. In Proceedings of the 2018 Third Scientific Conference of Electrical Engineering (SCEE), Baghdad, Iraq, 19–20 December 2018.

27. Natiq, H.; Ariffin, M.R.K.; Said, M.R.M.; Banerjee, S. Enhancing the sensitivity of a chaos sensor for internet of things. *Internet Things* **2019**, *7*, 100083. [CrossRef]

28. Chen, C.; Sun, K.; He, S. A class of higher-dimensional hyperchaotic maps. *Eur. Phys. J. Plus* **2019**, *134*, 410. [CrossRef]

29. Hua, Z.; Zhou, B.; Zhou, Y. Sine-transform-based chaotic system with FPGA implementation. *IEEE Trans. Ind. Electron.* **2017**, *65*, 2557–2566. [CrossRef]

30. Zhu, Z.; Leung, H. Identification of linear systems driven by chaotic signals using nonlinear prediction. *IEEE Trans. Circuits Syst. Fundam. Theory Appl.* **2002**, *49*, 170–180.

31. Skrobek, A. Cryptanalysis of chaotic stream cipher. *Phys. Lett.* **2007**, *363*, 84–90. [CrossRef]

32. Natiq, H.; Banerjee, S.; Said, M.R.M. Cosine chaotification technique to enhance chaos and complexity of discrete systems. *Eur. Phys. J. Spec. Top.* **2019**, *228*, 185–194. [CrossRef]

33. Farhan, A.K.; Al-Saidi, N.M.; Maolood, A.T.; Nazarimehr, F.; Hussain, I. Entropy analysis and image encryption application based on a new chaotic system crossing a cylinder. *Entropy* **2019**, *21*, 958. [CrossRef]

34. Liu, C.; Li, K.; Zhao, L.; Liu, F.; Zheng, D.; Liu, C.; Liu, S. Analysis of heart rate variability using fuzzy measure entropy. *Comput. Biol. Med.* **2013**, *43*, 100–108. [CrossRef] [PubMed]

35. Bian, C.; Qin, C.; Ma, Q.D.; Shen, Q. Modified permutation-entropy analysis of heartbeat dynamics. *Phys. Rev.* **2012**, *85*, 021906. [CrossRef]

36. Richman, J.S.; Moorman, J.R. Physiological time-series analysis using approximate entropy and sample entropy. *Am. J. Physiol.-Heart Circ. Physiol.* **2000**, *278*, H2039–H2049. [CrossRef] [PubMed]

37. Hua, Z.; Zhou, B.; Zhou, Y. Sine chaotification model for enhancing chaos and its hardware implementation. *IEEE Trans. Ind. Electron.* **2018**, *66*, 1273–1284. [CrossRef]

38. He, D.; He, C.; Jiang, L.G.; Zhu, H.W.; Hu, G.R. Chaotic characteristics of a one-dimensional iterative map with infinite collapses. *IEEE Trans. Circuits Syst. Fundam. Theory Appl.* **2001**, *48*, 900–906.

39. Gao, Y.; Liu, B. Study on the dynamical behaviors of a two-dimensional discrete system. *Nonlinear Anal. Theory Methods Appl.* **2009**, *70*, 4209–4216. [CrossRef]

40. Rukhin, A.; Soto, J.; Nechvatal, J.; Smid, M.; Barker, E. A Statistical Test Suite for Random and Pseudorandom Number Generators for Cryptographic Applications, Tech. rep., Booz-allen and hamilton inc mclean va 2001. Available online: https://www.nist.gov/publications/statistical-test-suite-random-and-pseudorandom-number-generators-cryptographic (accessed on 2 November 2020).

41. Zapateiro De la Hoz, M.; Acho, L.; Vidal, Y. An experimental realization of a chaos-based secure communication using arduino microcontrollers. *Sci. World J.* **2015**, *2015*, 123080. [CrossRef]

42. Taylor, D.S. *Design of Continuously Variable Slope Delta Modulation Communication Systems.* Motorola Technical Document AN1544. 1996. Available online: http://gamearchive.askey.org/General/Data_Sheets/cvsd_speech_info/an1544_cvsd.pdf (accessed on 2 November 2020).

Publisher's Note: MDPI stays neutral with regard to jurisdictional claims in published maps and institutional affiliations.

Article

A Two-Parameter Modified Logistic Map and Its Application to Random Bit Generation

Lazaros Moysis [1], Aleksandra Tutueva [2], Christos Volos [1,*], Denis Butusov [3], Jesus M. Munoz-Pacheco [4] and Hector Nistazakis [5]

[1] Laboratory of Nonlinear Systems, Circuits & Complexity (LaNSCom), Physics Department, Aristotle University of Thessaloniki, 54124 Thessaloniki, Greece; lmousis@physics.auth.gr

[2] Department of Computer-Aided Design, Saint Petersburg Electrotechnical University "LETI", 5, Professora Popova st., 197376 Saint Petersburg, Russia; avtutueva@etu.ru

[3] Youth Research Institute, Saint-Petersburg Electrotechnical University "LETI", 5, Professora Popova st., 197376 Saint Petersburg, Russia; dnbutusov@etu.ru

[4] Faculty of Electronics Sciences, Autonomous University of Puebla, Puebla 72000, Mexico; jesusm.pacheco@correo.buap.mx

[5] Section of Electronic Physics, Department of Physics, National and Kapodistrian University of Athens, 15784 Athens, Greece; enistaz@phys.uoa.gr

* Correspondence: volos@physics.auth.gr

Received: 20 April 2020; Accepted: 8 May 2020; Published: 18 May 2020

Abstract: This work proposes a modified logistic map based on the system previously proposed by Han in 2019. The constructed map exhibits interesting chaos related phenomena like antimonotonicity, crisis, and coexisting attractors. In addition, the Lyapunov exponent of the map can achieve higher values, so the behavior of the proposed map is overall more complex compared to the original. The map is then successfully applied to the problem of random bit generation using techniques like the comparison between maps, *XOR*, and bit reversal. The proposed algorithm passes all the NIST tests, shows good correlation characteristics, and has a high key space.

Keywords: random bit generation; logistic map; chaos; chaos-based cryptography

---

1. Introduction

Chaos theory has found numerous applications over the last 50 years, including, but not limited to, encryption, engineering, secure communications, robotics, biology, and economics—see, for example, Refs. [1–4] and the references cited therein. Nonlinear systems with chaotic properties are deterministic systems with high sensitivity to small changes in initial conditions and parameters which lead to completely different solution trajectories. This sensitivity, combined with their deterministic nature, makes chaotic systems a perfect basis for designs that require high complexity and increased security.

Due to their abovementioned usability, there is an ongoing demand for constructing novel chaotic systems. Moreover, it is of interest to develop minimal chaotic systems that can provide high performance when implemented on FPGA [5,6]. Most of constructed chaotic maps are modifications of known chaotic systems. These modifications usually follow simple techniques, such as introducing additional nonlinear terms in the system's differential/difference equations, changing an existing term to a higher-order term, or even by adding new variables to make the system hyperchaotic that is, having at least two positive Lyapunov exponents, which can only happen for four-dimensional systems or higher.

The logistic map [7,8] is one of the most well-known one-dimensional discrete time systems with chaotic behavior. Originally considered as a population model, it eventually found numerous applications in encryption, due to its simple and elegant form. This has also lead to many subsequent modifications of the map—see, for example, [9–17].

In this study, we consider a modification of the logistic map considered in [9]. The proposed modification is obtained by combining the map [9] with the conventional logistic map. This is done by multiplying the values of the map in [9] by the values of the logistic map computed from the decimal part of the same map, yielding a more complex behavior. The original map [9] exhibits a symmetric bifurcation diagram and constant chaos with a constant Lyapunov exponent, yet the proposed chaotic map showcases a plethora of chaos related phenomena, like antimonotonicity, crisis, and coexisting attractors, and the symmetric bifurcation diagram is now modified. In addition, the Lyapunov exponent of the new map can reach higher values, so overall the proposed map has a more complex chaotic behavior compared to the original map, and consequently compared to the classic logistic map as well. The emergence of many chaos related phenomena in the proposed system is an indication that the method of combining a given map with the logistic map derived from its decimal part can be generally utilized as a technique to make the behavior of a given system more complex.

Moreover, the proposed map is applied to the problem of pseudo-random bit generation [2,3, 10,11,13,18–36]. The term pseudo comes from the fact that a deterministic system is used to generate the sequence, rather than a random process, which is the case in true random bit generators. If the generator is properly designed though, the resulting sequence will have the characteristics of a random sequence. The method used utilizes various simple techniques like combinations of multiple maps, the comparison between different decimal parts of a number, bit reversal, and the XOR operator. The sequences obtained by the pseudo-random bit generator based on the new map passes all 15 of the NIST statistical tests, shows good correlation and cross-correlation characteristics, and has a satisfactory key space. Thus, it is suitable for encryption related applications.

The rest of the work is structured as follows: In Section 2, the proposed modification of the well-known logistic map is presented and studied. In Section 3, the considered chaotic system is applied to the problem of random bit generation. Finally, Section 4 concludes the work with a discussion on future research topics.

2. The Proposed Map

In [9], the following modified logistic map was proposed:

$$x_i = 2\beta - \frac{x_{i-1}^2}{\beta} \tag{1}$$

The behavior of one-dimensional map Equation (1) depends on a single parameter β. This map exhibits constant chaos for all values of its parameter, with a full mapping of the state values on $x_i \in [-2\beta; 2\beta]$ provided that $x_0 \in [-2\beta; 2\beta]$. Its symmetric bifurcation diagram is shown in Figure 1 and the diagram of its Lyapunov exponent in Figure 2.

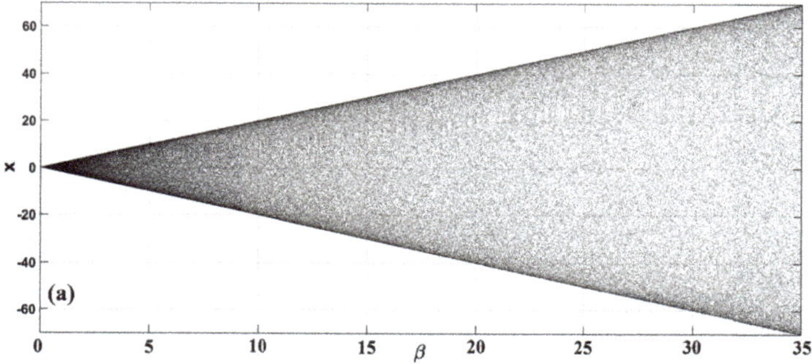

Figure 1. Bifurcation diagram of Equation (1), with respect to parameter β.

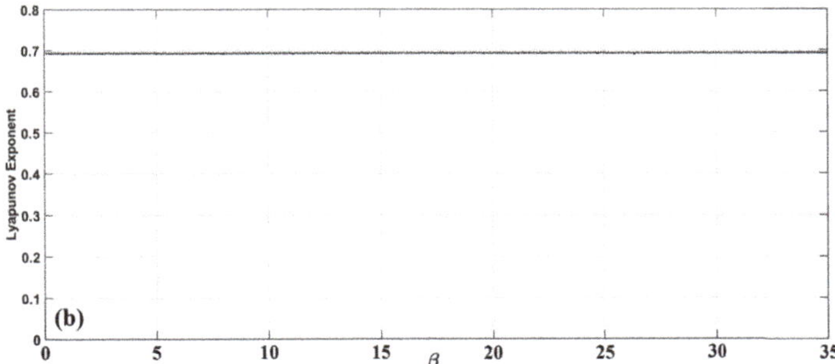

Figure 2. Diagram of the Lyapunov exponent of Equation (1), with respect to parameter β.

Here, a modified version of Equation (1) is proposed, given by

$$x_i \;=\; p_i\left(2\beta - \frac{x_{i-1}^2}{\beta}\right) \tag{2}$$

where $p_i = r \cdot \bmod(x_{i-1}, 1) \cdot (1 - \bmod(x_{i-1}, 1))$. With the above modification, the values of the chaotic map Equation (1) are multiplied by the value $p_i \in [0;1]$ which is actually the classic logistic map with bifurcation parameter r, computed using the decimal part of x_{i-1}, $\bmod(x_{i-1}, 1) \in [0;1]$ instead of x_{i-1}. The mod operator is used here to take the decimal part of x_{i-1}, so that p_i is bounded on the interval $[0;1]$.

The bifurcation diagram of map Equation (2) with respect to parameter β and $r = 4$ is shown in Figure 3. The initial condition in each iteration is chosen as $x_0 = 0.1$. From Figure 3, it can be seen that the system exhibits a similar but more complex behavior compared to Label (1), with small periodic windows appearing. This behavior can be seen more clearly in the zoomed subures. It is observed that the system exhibits crisis phenomena, where it exists abruptly from chaos and reenters it following a period doubling route. What is also interesting is that there are small windows where the phenomenon of antimonotonicity appears. This is when the system enters chaos by following a period doubling route, and then exists from chaos by following a reverse period halving route. This is observed in the subfigures around the value of $\beta = 1.02$. The chaotic oscillation mode is verified by the diagram of the Lyapunov exponent shown in Figure 4. In addition, Figure 5 shows a full plot for the Lyapunov exponent up to $\beta = 150$. From this figure, it can be seen that the Lyapunov exponent slowly increases to reach a value higher than 5, while there are also very small periodic windows appearing.

Similar phenomena can be observed for different values of the parameter r. For example, the bifurcation diagram and the curve of the Lyapunov exponent with respect to β for $r = 3.8$ can be seen in Figures 6 and 7. Again, antimonotonicity appears around the value of $\beta = 1.1$. The system also exits abruptly from chaos and re-enters it through a period doubling route.

In addition to the rich dynamical behavior with respect to parameter β, the proposed map also exhibits chaotic oscillations with respect to parameter r, as seen in Figures 8 and 9 where $\beta = 10$. The system here exhibits crisis phenomena again.

Moreover, as can be seen from Figures 4, 5, 7 and 9, it is important to note that the system can achieve a Lyapunov exponent value that is higher than that of the system in [9] and also the classic logistic map, which both achieve the higher value at around 0.7.

To study the existence of coexisting attractors in the system, its continuation diagram is plotted. The continuation diagram is similar to the bifurcation diagram, with the difference that, in each iteration, the initial value of the chaotic map is taken to be equal to the final value of its previous simulation. The continuation diagram can thus be computed as the bifurcation parameter increases or decreases. Figure 10 shows the bifurcation diagram (black, $x_0 = 0.1$.) of the map with respect to β with $r = 4$, overlapping with its forward (red) and backward (green) continuation diagrams.

This plot reveals coexisting attractors for the system around the value of $\beta = 1.05$. This means that, depending on the initial condition of the system, its steady-state behavior may converge to different attracting regions.

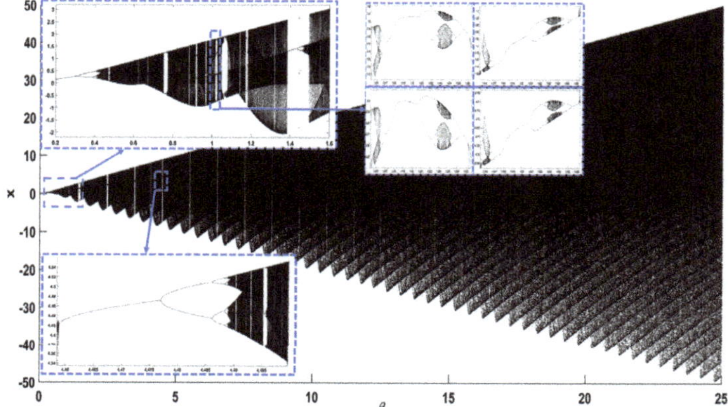

Figure 3. Bifurcation diagram of Equation (2), with respect to parameter β, for $r = 4$.

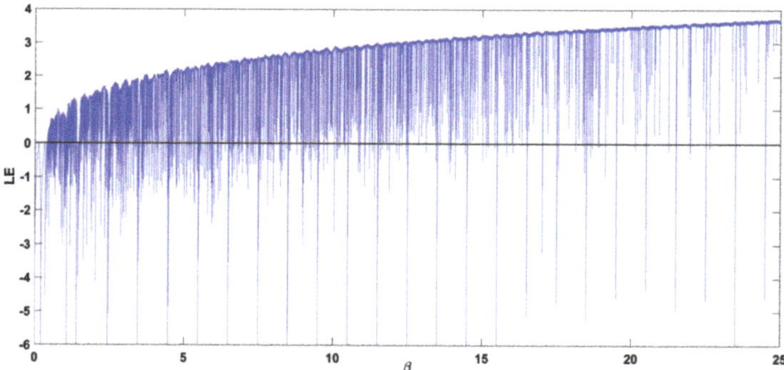

Figure 4. Diagram of the Lyapunov exponent Equation (2), with respect to parameter β, for $r = 4$.

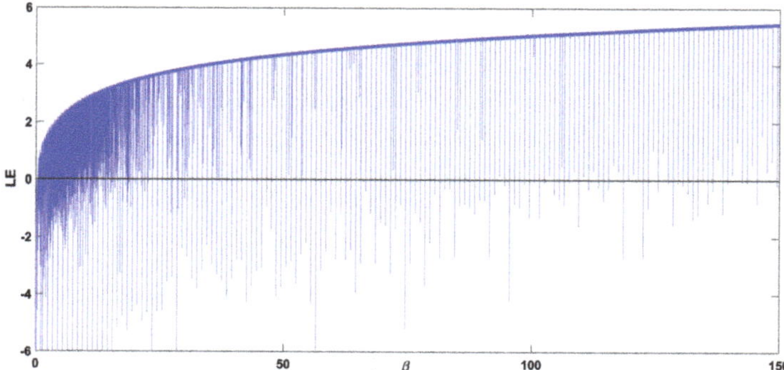

Figure 5. Wider diagram of the Lyapunov exponent Equation (2), with respect to parameter β, for $r = 4$.

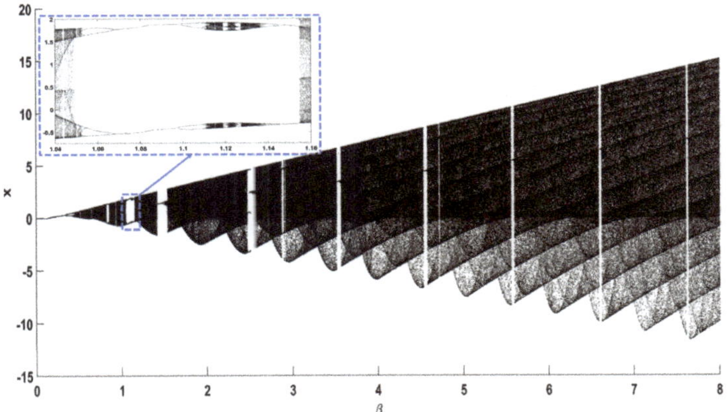

Figure 6. Bifurcation diagram of Equation (2), with respect to parameter β, for $r = 3.8$.

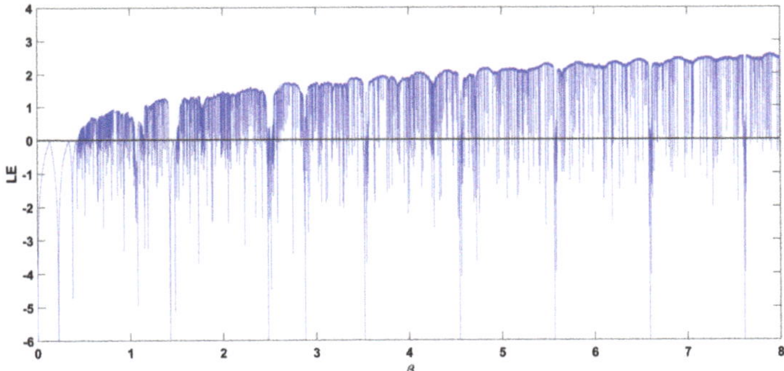

Figure 7. Diagram of the Lyapunov exponent Equation (2), with respect to parameter β, for $r = 3.8$.

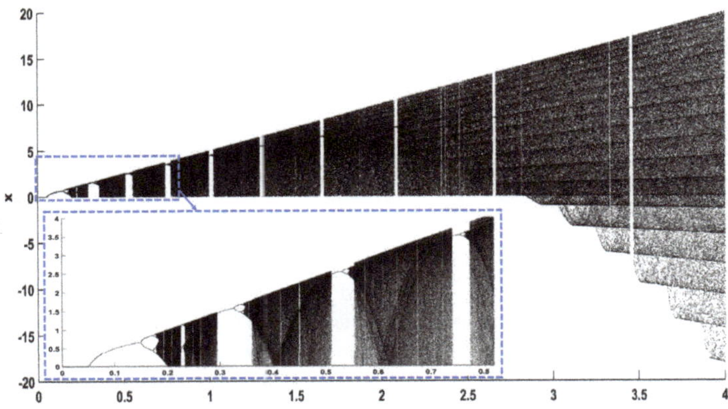

Figure 8. Bifurcation diagram of Equation (2), with respect to parameter r, for $\beta = 10$.

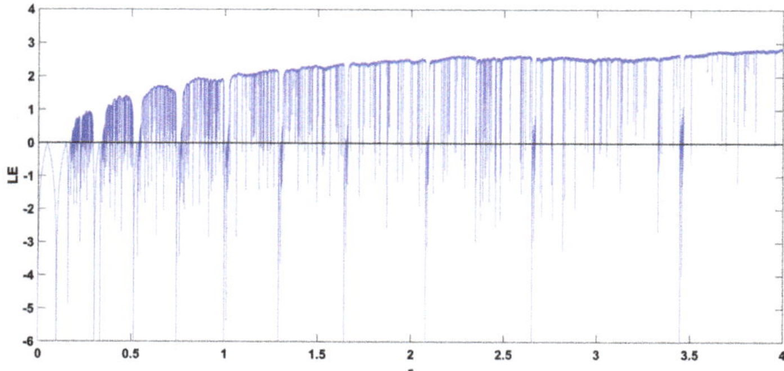

Figure 9. Diagram of the Lyapunov exponent Equation (2), with respect to parameter r, for $\beta = 10$.

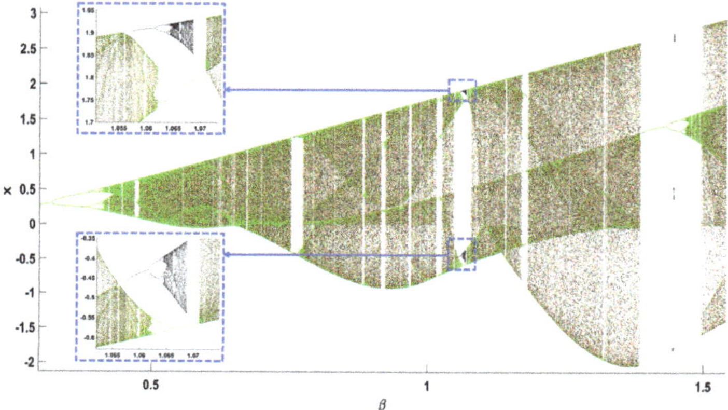

Figure 10. Bifurcation diagram (black), forward continuation diagram (red), and backward continuation diagram (green) for Equation (2), with respect to parameter β, for $r = 4$.

3. Application to Random Bit Generation

To apply the proposed map to pseudo-random bit generation, an algorithm was devised with the aim of having pseudo-random properties, weak correlation, and high key space. The proposed algorithm utilizes the techniques of comparing different decimal parts from different maps, as was performed in [23] for one map, the technique of bit reversal [19,22], performed here in a chaotic way, depending on the values of a logistic map, and also the *XOR* operator, which is commonly used in PRBGs [19–21,31,37].

The algorithm is outlined as follows:

Step 1. First, two modified logistic maps x_0, y_0, one classic logistic map z_0, as well as two bit sequences b_0, d_0 are initialized, and the maps' parameters are chosen.

Step 2. In every iteration, the decimal part of $x_i + y_i$ is compared to the decimal part of $10^6(x_i \cdot y_i)$ and depending on the result a 0 or 1 is produced and saved in b_i. Similarly, the decimal part of $10^6(x_i + y_i)$ is compared to the decimal part of $x_i \cdot y_i$ and depending on the result a 0 or 1 is produced and saved in d_i.

Step 3. For every 10 iterations, the value of the logistic map $z(\frac{i}{10})$ is compared to the decimal part of $x_i + y_i$. Depending on the result, a bit reversal is performed on the last ten digits of b or d.

Step 4. Once the desired bitstream length is reached, the obtained sequence is computed using $XOR(b, d)$.

A full description of the proposed technique is described in Algorithm 1. Note that, when the decimal part is computed in each iteration, its sign is discarded, so a positive value is always returned. The modulo operation is performed using the *rem* command in Matlab.

Algorithm 1 The Proposed Random Bit Generator.

Data: Initialize initial conditions: x_0, y_0, z_0, parameter values: $r_x, r_y, r_z, \beta_x, \beta_y$, Bit subsequences b_0, d_0 and bitstream length: ℓ.

---

  for i=1:ℓ **do**

$$x_i = r_x \cdot \ \mathrm{mod}\ (x_{i-1}, 1)(1 - \ \mathrm{mod}\ (x_{i-1}, 1))\left(2\beta_x - \frac{x_{i-1}^2}{\beta_x}\right)$$

$$y_i = r_y \cdot \ \mathrm{mod}\ (y_{i-1}, 1)(1 - \ \mathrm{mod}\ (y_{i-1}, 1))\left(2\beta_y - \frac{y_{i-1}^2}{\beta_y}\right)$$

    if $\ \mathrm{mod}\ (x_i + y_i, 1) \leq \ \mathrm{mod}\ (10^6(x_i \cdot y_i), 1)$ **then**

      $b_i = 0$

    else

      $b_i = 1$

    end if

    if $\ \mathrm{mod}\ (10^6(x_i + y_i, 1)) \leq \ \mathrm{mod}\ (x_i \cdot y_i, 1)$ **then**

      $d_i = 0$

    else

      $d_i = 1$

    end if

    % Perform bit reversal for every 10 bits

    if $\ \mathrm{mod}\ (i, 10) = 0$ **then**

$$z_{\frac{i}{10}} = r_z z_{\frac{i}{10}-1}\left(1 - z_{\frac{i}{10}-1}\right)$$

      if $z_{\frac{i}{10}} \leq \ \mathrm{mod}\ (x_i + y_i, 1)$ **then**

        $b_{(i-9):i} = b_{i:-1:(i-9)}$

      else

        $d_{(i-9):i} = d_{i:-1:(i-9)}$

      end if

    end if

  end for

  bitstream= $XOR(b, d)$

The proposed technique was tested using the National Institute of Standards and Technology (NIST) statistical test package [38]. The suite consists of 15 tests that are used to test the randomness of a sequence. For each test, a *p*-value is calculated. If the value exceeds a significance value *a*, the test is passed. A set of 50 bit sequences of 10^6 bits each was considered, for parameter values $\beta_x = \beta_y = 40$, $r_x = r_y = r_z = 4$ and arbitrarily chosen initial values. The results are shown in Table 1, where it can be seen that all the tests are passed. For tests that have multiple case runs, only the last *p*-value is printed.

In addition, Figure 11 depicts the autocorrelation and cross-correlation plots for a bit sequence of length 10^5, generated for parameter values $\beta_x = \beta_y = 30$, $r_x = r_y = r_z = 4$. For pseudo-random sequences, the auto-correlation should have a delta like form, and the cross-correlation should be close to zero [3,13], which is verified in Figure 11a. For the cross-correlation, two bit sequences were generated for the same parameter values and initial conditions, with the only difference taken as follows: In (b), the initial condition of the first map was chosen as $x_0 = 0.1$ and $x_0' = x_0 + 10^{-16}$. In (c), the parameter of the first map is chosen as $r_x = 4$ and $r_x' = 4 - 10^{-15}$. In (d), the parameter of the first map is chosen as $\beta_x = 40$ and $\beta_x' = 40 + 10^{-14}$.

As for the key space, the proposed technique utilizes two modified logistic maps and one logistic map, each with different initial conditions and parameters. Thus, in the case of the floating-point data type with *double* precision [39], there are overall eight key parameters, so the upper bound for the key space is $(4 - 3.6)10^{8 \cdot 16} = 2^2 10^{-1} 10^{128} = 2^2 10^{127} \approx 2^2 (10^3)^{42.3} \approx 2^2 2^{423} = 2^{425}$. This is higher than the value of 2^{100} that is required to resist brute force attacks, as reported in [40].

Table 1. NIST statistical test results, with $a = 0.01$.

	If $P \geq \alpha$, the Test Is Successful			
No.	Statistical Test	*p*-Value	Proportion	Result
1	Frequency	0.289667	49/50	success
2	Block Frequency	0.383827	48/50	success
3	Cumulative Sums	0.419021	49/50	success
4	Runs	0.122325	50/50	success
5	Longest Run	0.383827	50/50	success
6	Rank	0.616305	49/50	success
7	FFT	0.191687	49/50	success
8	Non-Overlapping Template	0.991468	49/50	success
9	Overlapping Template	0.739918	50/50	success
10	Universal	0.699313	50/50	success
11	Approximate Entropy	0.534146	50/50	success
12	Random Excursions	0.407091	32/32	success
13	Random Excursions Variant	0.066882	32/32	success
14	Serial	0.171867	50/50	success
15	Linear Complexity	0.911413	48/50	success

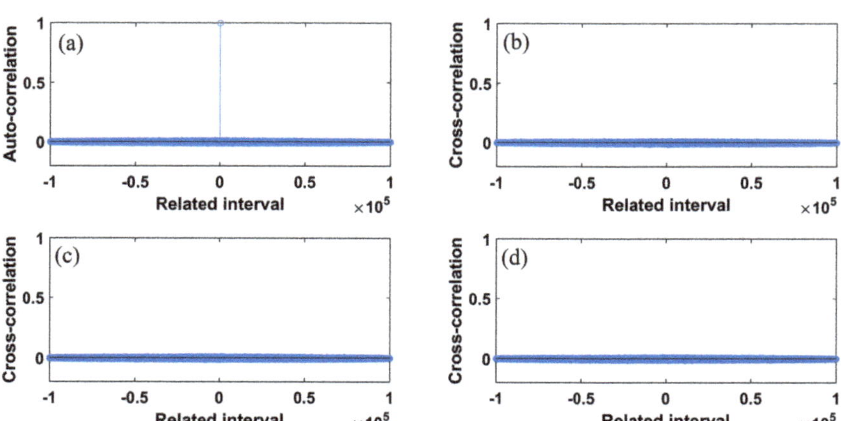

Figure 11. (**a**) auto-correlation; (**b**) cross-correlation, with the initial condition of the first map being chosen as $x_0 = 0.1$ and $x_0' = x_0 + 10^{-16}$; (**c**) cross-correlation, with the parameter of the first map is chosen as $r_x = 4$ and $r_x' = 4 - 10^{-15}$; (**d**) cross-correlation, with the parameter of the first map being chosen as $\beta_x = 40$ and $\beta_x' = 40 + 10^{-14}$.

4. Conclusions

In the present work, the modified version of the logistic map proposed in [9] was presented. The extensive dynamical analysis has shown that the proposed system exhibits phenomena like crisis, antimonotonicity, and coexisting attractors. This technique to increase the complexity of a map can be tested on other systems in the future. The map was then applied to constuct the chaos-based pseudo-random bit generator, utilizing techniques like multiple map comparison, bit reversal, and *XOR*. In addition, the key space of the proposed algorithm is much higher than the indicated threshold of 2^{100}. Future aspects of this work will consider the application of the proposed PRBG to image encryption, the generation of multiple bits per iteration, as well as fractional versions of the map. It is of interest to develop the adaptive chaotic maps with controllable symmetry of a higher order based on the proposed map. It was previously shown that such systems are prospective for stream encryption algorithms [29,41]. Moreover, the obtained results can be applied to other cryptographic problems, including encoding multimedia data, creating watermarks and QR codes, generating checksums using chaotic hash functions, etc.

Author Contributions: C.V., L.M., A.T., D.B., J.M.M.-P. and H.N. contributed equally to this paper. All authors have read and agreed to the published version of the manuscript.

Funding: This research is co-financed by Greece and the European Union (European Social Fund- ESF) through the Operational Programme «Human Resources Development, Education and Lifelong Learning» in the context of the project "Reinforcement of Postdoctoral Researchers—2nd Cycle" (MIS-5033021), implemented by the State Scholarships Foundation (IKY).

Acknowledgments: The authors would like to thank the anonymous reviewers for theis comments.

Conflicts of Interest: The authors declare no conflict of interest.

References

1. Strogatz, S.H. *Nonlinear Dynamics and Chaos: With Applications to Physics, Biology, Chemistry, and Engineering;* CRC Press: Boca Raton, FL, USA, 2018.
2. Volos, C.K.; Kyprianidis, I.M.; Stouboulos, I.N. Image encryption process based on chaotic synchronization phenomena. *Signal Process.* **2013**, *93*, 1328–1340. [CrossRef]
3. Huang, X.; Liu, L.; Li, X.; Yu, M.; Wu, Z. A New Pseudorandom Bit Generator Based on Mixing Three-Dimensional Chen Chaotic System with a Chaotic Tactics. *Complexity* **2019**, 6567198. [CrossRef]
4. Moysis, L.; Petavratzis, E.; Volos, C.; Nistazakis, H.; Stouboulos, I. A chaotic path planning generator based on logistic map and modulo tactics. *Robot. Auton. Syst.* **2020**, *124*, 103377. [CrossRef]
5. Nepomuceno, E.G.; Lima, A.M.; Arias-García, J.; Perc, M.; Repnik, R. Minimal digital chaotic system. *Chaos Solitons Fractals* **2019**, *120*, 62–66. [CrossRef]
6. Wang, H.; Song, B.; Liu, Q.; Pan, J.; Ding, Q. FPGA design and applicable analysis of discrete chaotic maps. *Int. J. Bifurc. Chaos* **2014**, *24*, 1450054. [CrossRef]
7. May, R.M. Simple mathematical models with very complicated dynamics. *Nature* **1976**, *261*, 459–467. [CrossRef] [PubMed]
8. Ausloos, M.; Dirickx, M. *The Logistic Map and the Route to Chaos: From the Beginnings to Modern Applications;* Springer Science & Business Media: Berlin/Heidelberg, Germany, 2006.
9. Han, C. An image encryption algorithm based on modified logistic chaotic map. *Optik* **2019**, *181*, 779–785. [CrossRef]
10. Wang, Y.; Liu, Z.; Ma, J.; He, H. A pseudorandom number generator based on piecewise logistic map. *Nonlinear Dyn.* **2016**, *83*, 2373–2391. [CrossRef]
11. Murillo-Escobar, M.; Cruz-Hernández, C.; Cardoza-Avendaño, L.; Méndez-Ramírez, R. A novel pseudorandom number generator based on pseudorandomly enhanced logistic map. *Nonlinear Dyn.* **2017**, *87*, 407–425. [CrossRef]
12. Radwan, A.G. On some generalized discrete logistic maps. *J. Adv. Res.* **2013**, *4*, 163–171. [CrossRef]
13. Liu, L.; Miao, S.; Hu, H.; Deng, Y. Pseudorandom bit generator based on non-stationary logistic maps. *IET Inf. Secur.* **2016**, *10*, 87–94. [CrossRef]
14. Liu, L.; Miao, S. A new image encryption algorithm based on logistic chaotic map with varying parameter. *SpringerPlus* **2016**, *5*, 289. [CrossRef] [PubMed]
15. Chen, S.L.; Hwang, T.; Lin, W.W. Randomness enhancement using digitalized modified logistic map. *IEEE Trans. Circuits Syst. II Express Briefs* **2010**, *57*, 996–1000.

16. Borujeni, S.E.; Ehsani, M.S. Modified logistic maps for cryptographic application. *Appl. Math.* **2015**, *6*, 773. [CrossRef]

17. Li, S.; Yin, B.; Ding, W.; Zhang, T.; Ma, Y. A nonlinearly modulated logistic map with delay for image encryption. *Electronics* **2018**, *7*, 326. [CrossRef]

18. Irfan, M.; Ali, A.; Khan, M.A.; Ehatisham-ul Haq, M.; Mehmood Shah, S.N.; Saboor, A.; Ahmad, W. Pseudorandom Number Generator (PRNG) Design Using Hyper-Chaotic Modified Robust Logistic Map (HC-MRLM). *Electronics* **2020**, *9*, 104. [CrossRef]

19. Ahmad, M.; Doja, M.; Beg, M.S. A new chaotic map based secure and efficient pseudo-random bit sequence generation. In Proceedings of the International Symposium on Security in Computing and Communication, Bangalore, India, 19–22 September 2018; pp. 543–553.

20. Ge, R.; Yang, G.; Wu, J.; Chen, Y.; Coatrieux, G.; Luo, L. A Novel Chaos-Based Symmetric Image Encryption Using Bit-Pair Level Process. *IEEE Access* **2019**, *7*, 99470–99480. [CrossRef]

21. François, M.; Grosges, T.; Barchiesi, D.; Erra, R. Pseudo-random number generator based on mixing of three chaotic maps. *Commun. Nonlinear Sci. Numer. Simul.* **2014**, *19*, 887–895. [CrossRef]

22. Alawida, M.; Samsudin, A.; Teh, J.S. Enhanced digital chaotic maps based on bit reversal with applications in random bit generators. *Inf. Sci.* **2020**, *512*, 1155–1169. [CrossRef]

23. Wang, X.Y.; Xie, Y.X. A design of pseudo-random bit generator based on single chaotic system. *Int. J. Mod. Phys. C* **2012**, *23*, 1250024. [CrossRef]

24. Patidar, V.; Sud, K.K.; Pareek, N.K. A pseudo random bit generator based on chaotic logistic map and its statistical testing. *Informatica* **2009**, *33*, 441–452.

25. Stojanovski, T.; Kocarev, L. Chaos-based random number generators-part I: analysis [cryptography]. *IEEE Trans. Circuits Syst. Fundam. Theory Appl.* **2001**, *48*, 281–288. [CrossRef]

26. Volos, C.; Kyprianidis, I.; Stouboulos, I. Experimental investigation on coverage performance of a chaotic autonomous mobile robot. *Robot. Auton. Syst.* **2013**, *61*, 1314–1322. [CrossRef]

27. Hamza, R. A novel pseudo random sequence generator for image-cryptographic applications. *J. Inf. Secur. Appl.* **2017**, *35*, 119–127. [CrossRef]

28. Nepomuceno, E.G.; Nardo, L.G.; Arias-Garcia, J.; Butusov, D.N.; Tutueva, A. Image encryption based on the pseudo-orbits from 1D chaotic map. *Chaos Interdiscip. J. Nonlinear Sci.* **2019**, *29*, 061101. [CrossRef]

29. Tutueva, A.V.; Nepomuceno, E.G.; Karimov, A.I.; Andreev, V.S.; Butusov, D.N. Adaptive chaotic maps and their application to pseudo-random numbers generation. *Chaos Solitons Fractals* **2020**, *133*, 109615. [CrossRef]

30. Akgül, A.; Arslan, C.; Arıcıoğlu, B. Design of an Interface for Random Number Generators based on Integer and Fractional Order Chaotic Systems. *Chaos Theory Appl.* **2019**, *1*, 1–18.

31. Khanzadi, H.; Eshghi, M.; Borujeni, S.E. Image encryption using random bit sequence based on chaotic maps. *Arab. J. Sci. Eng.* **2014**, *39*, 1039–1047. [CrossRef]

32. Andrecut, M. Logistic map as a random number generator. *Int. J. Mod. Phys. B* **1998**, *12*, 921–930. [CrossRef]

33. Wang, L.; Cheng, H. Pseudo-Random Number Generator Based on Logistic Chaotic System. *Entropy* **2019**, *21*, 960. [CrossRef]

34. Meranza-Castillón, M.; Murillo-Escobar, M.; López-Gutiérrez, R.; Cruz-Hernández, C. Pseudorandom number generator based on enhanced Hénon map and its implementation. *AEU-Int. J. Electron. Commun.* **2019**, *107*, 239–251. [CrossRef]

35. Persohn, K.; Povinelli, R.J. Analyzing logistic map pseudorandom number generators for periodicity induced by finite precision floating-point representation. *Chaos Solitons Fractals* **2012**, *45*, 238–245. [CrossRef]

36. Phatak, S.; Rao, S.S. Logistic map: A possible random-number generator. *Phys. Rev. E* **1995**, *51*, 3670. [CrossRef] [PubMed]

37. Volos, C.K.; Kyprianidis, I.; Stouboulos, I. Text Encryption Scheme Realized with a Chaotic Pseudo-Random Bit Generator. *J. Eng. Sci. Technol. Rev.* **2013**, *6*, 9–14. [CrossRef]

38. Rukhin, A.; Soto, J.; Nechvatal, J.; Smid, M.; Barker, E. *A Statistical Test Suite for Random and Pseudorandom Number Generators for Cryptographic Applications*; Technical Report; Booz-Allen and Hamilton Inc.: Mclean, VA, USA, 2001.

39. Kahan, W. IEEE standard 754 for binary floating-point arithmetic. *Lect. Notes Status IEEE* **1996**, *754*, 11.

40. Alvarez, G.; Li, S. Some basic cryptographic requirements for chaos-based cryptosystems. *Int. J. Bifurc. Chaos* **2006**, *16*, 2129–2151. [CrossRef]
41. Butusov, D.N.; Karimov, A.I.; Pyko, N.S.; Pyko, S.A.; Bogachev, M.I. Discrete chaotic maps obtained by symmetric integration. *Phys. A Stat. Mech. Its Appl.* **2018**, *509*, 955–970. [CrossRef]

Article

A Simple Chaotic Flow with Hyperbolic Sinusoidal Function and Its Application to Voice Encryption

Saleh Mobayen [1], Christos Volos [2,*], Ünal Çavuşoğlu [3] and Sezgin S. Kaçar [4]

[1] Future Technology Research Center, National Yunlin University of Science and Technology, 123 University Road, Section 3, Douliou, Yunlin 64002, Taiwan; mobayens@yuntech.edu.tw

[2] Laboratory of Nonlinear Systems, Circuits & Complexity (LaNSCom), Department of Physics, Aristotle University of Thessaloniki, GR-54124 Thessaloniki, Greece

[3] Department of Software Engineering, Sakarya University, Sakarya 54050, Turkey; unalc@sakarya.edu.tr

[4] Department of Electrical and Electronics Engineering, Sakarya University of Applied Sciences, Sakarya 54050, Turkey; skacar@subu.edu.tr

* Correspondence: volos@physics.auth.gr

Received: 29 October 2020; Accepted: 8 December 2020; Published: 10 December 2020

Abstract: In this article, a new chaotic system with hyperbolic sinusoidal function is introduced. This chaotic system provides a new category of chaotic flows which gives better perception of chaotic attractors. In the proposed chaotic flow with hyperbolic sinusoidal function, according to the changes of parameters of the system, the self-excited attractor and two forms of hidden attractors are occurred. Dynamic behavior of the offered chaotic flow is studied through eigenvalues, bifurcation diagrams, phase portraits, and spectrum of Lyapunov exponents. Moreover, the existence of double-scroll attractors in real word is considered via the Orcard-PSpice software through an electronic execution of the new chaotic flow and illustrative results between the numerical simulation and Orcard-PSpice outcomes are obtained. Lastly, random number generator (RNG) design is completed with the new chaos. Using the new RNG design, a novel voice encryption algorithm is suggested and voice encryption use and encryption analysis are performed.

Keywords: chaotic flow; hyperbolic sinusoidal function; hidden attractor; voice encryption; symmetry

---

1. Introduction

Chaotic flows are mathematical models originated from the rules of defining chaotic behaviors [1,2]. In the former decades, the chaos theory has been employed in numerous fields such as digital signature [3], secure cryptography [4], pseudorandom number generation [5], secure communication [6], weak signal detection [7], DC-DC boost converter [8], image encryption [9], neurophysiology [10], secure data transmission [11], etc. For the control and synchronization purposes of chaotic systems, several techniques like active control [12], fuzzy control [13], linear matrix inequality (LMI) [14], sampled-data control [15], impulsive adaptive control [16], intermittent control [17] and sliding mode control (SMC) [18] have been introduced.

Recently, Wei (2011) announced a chaotic system with no equilibrium point [19]. Jafari et al. (2013) discovered a set of 17 elementary quadratic chaos systems with no equilibrium points [20]. A chaos system possessing a stable equilibrium point was recently found in [21,22]. It is observed that Shilnikov method [23,24] is not applicable to check chaos behavior in special dynamical systems with no equilibrium point or with stable equilibrium points. Such dynamical systems can be viewed as systems with hidden chaotic attractors in scientific computing [24–26]. Chaotic systems with hidden attractors can result in unexpected disastrous behavior in mechanical systems and electronic circuits.

It is stimulating that chaotic flows containing infinite number of equilibrium points have achieved much consideration in the past decade. Especially, structures with uncountable equilibrium points are categorized as systems with hidden attractors [27,28]. Hidden attractors do not have basins of attraction related to the unstable equilibria. As stated by the recent investigations, hidden attractors are fundamental in engineering usages, for instance, radio-physical oscillator [29], multilevel DC/DC converter [30], electromechanical systems [31] or relay system with hysteresis [32].

In recent years, some new chaotic flows have been planned via cascade chaos, dimension expansion, and physics modelling [1,33]. An extensive body of scientists has been devoted on counteracting degradation and performance improvement of existing chaotic flows. As chaos is broadly employed in nonlinear control, synchronization, and other usages, the design problem of the chaotic flows with complex chaotic behaviors is more attractive [34].

In addition, a widespread application of chaotic systems is that of encryption schemes, voice, text or image. In these schemes, RNGs are the most basic constructs. The fact that the numbers used in encryption have a high randomness and a big impact on the quality of the encryption. In the last few years, some encryption schemes, especially for sound messages, based mainly on discrete chaotic maps, have been presented. In 2016, Sadkhan et al. presented a new speech scrambling system using a hybrid use of different chaotic maps [35]. In 2018, Mobayen et al. proposed the implementation of a sound encryption method based on a novel chaotic system with boomerang-shaped equilibrium [36]. On the same year, Raheema et al. presented an efficient Simulink model, speech scrambling based chaotic maps for encryption of data such as voice, video and text, because it possesses high sensitive to initial values and model external parameters [37].

The objective of this article is to investigate a novel chaotic flow with hyperbolic sinusoidal function. The proposed chaotic flow provides a new category of chaotic systems which helps in more perception of chaotic attractors. In this chaotic flow, because of the variations of the parameters, the self-excited attractor and two forms of hidden attractors (no equilibrium point and line of equilibria) are created. Next, the proposed chaotic system with hidden attractors has been used in the design of an RNG algorithm. Finally, this RNG algorithm is used in a sound encryption scheme.

The rest of this work is organized as follows. In the following section, mathematical form of new chaotic structure is given and different scenarios are proposed. Moreover, some discussions for chaotic flow covering dynamic features such as spectrum of the Lyapunov exponents, bifurcation diagrams, and Poincaré map are proposed. In Section 3, the circuit design of presented chaotic flow is provided and PSpice representation of the chaotic attractors is presented. In Section 4, the engineering application containing RNG algorithm design and voice encryption algorithm is described. As a final point, conclusions are provided in Section 5.

2. Chaotic Flow with a Hyperbolic Sinusoidal Function

In the search for chaos flows with hyperbolic sinusoidal function, we study the form of a three-dimensional chaotic structure as:

$$\begin{aligned}
\dot{x} &= a_1 x + a_2 y + a_3 z + a_4 xy + a_5 xz + a_6 yz + b, \\
\dot{y} &= a_7 xy + a_8 xz + a_9 yz + a_{10} sinh(y) + c, \\
\dot{z} &= a_{11} x + a_{12} y + a_{13} z + a_{14} xy + a_{15} xz + a_{16} yz
\end{aligned} \tag{1}$$

where x, y and z denote the system states; $a_1, \dots, a_{16}$ indicate the coefficients of the terms; b and c are two scalars which define the chaos behavior.

A computer examination is executed investigating millions of combinations of different forms, various initial states and different constant values, looking for dissipative cases for which the largest Lyapunov exponent is bigger than 0.001. The system is in chaos state if the largest Lyapunov exponent is bigger than zero, and the system is in steady period state if the largest Lyapunov exponent is smaller than zero [38,39]. Therefore, in the present work, a three-dimensional chaotic flow is reported which is specified by:

$$\dot{x} = x - ayz + b,$$
$$\dot{y} = xz - \sinh(y) + c, \qquad (2)$$
$$\dot{z} = x$$

where the parameter a_6 in (1) has been denoted as parameter a in system (2).

Next, three different scenarios depending on the values of system's (2) parameters b, c are discussed in details.

2.1. Scenario A: Line of Equilibria

If $b = c = 0$, the chaotic flow (2) will have a line of equilibria, i.e., $E_A = [0, 0, z^*]^T$, where z^* is the equilibrium point value in z axis and T means the transpose of the vector. To analyze the state trajectories in the vicinity of the equilibrium point, the Jacobian matrix is obtained from (2) as:

$$J = \begin{bmatrix} 1 & -az & -ay \\ z & -\cosh(y) & x \\ 1 & 0 & 0 \end{bmatrix}. \qquad (3)$$

For equilibrium point $E_A = [0, 0, z^*]^T$, the Jacobian matrix is defined as

$$J = \begin{bmatrix} 1 & -az^* & 0 \\ z^* & -1 & 0 \\ 1 & 0 & 0 \end{bmatrix}. \qquad (4)$$

Therefore, the eigenvalues of the linearized system are achieved as

$$|\lambda I - J| = \begin{vmatrix} \lambda - 1 & az^* & 0 \\ -z^* & \lambda + 1 & 0 \\ -1 & 0 & \lambda \end{vmatrix} = \lambda\left(\lambda^2 + az^{*2} - 1\right) = 0 \Rightarrow \lambda_1 = 0, \lambda_{2,3} = \pm \sqrt{1 - az^{*2}}. \qquad (5)$$

The equilibrium is a saddle node for $-\frac{1}{\sqrt{a}} < z^* < \frac{1}{\sqrt{a}}$. The equilibrium point is an unstable node for $z^* = \pm\frac{1}{\sqrt{a}}$. For the values $z^* > \frac{1}{\sqrt{a}}$ and $z^* < -\frac{1}{\sqrt{a}}$, since one eigenvalue is zero and two eigenvalues are imaginary, the stability of the equilibrium point cannot be determined by this method; the equilibrium point may be stable, unstable or marginally stable.

If one design parameter is varied and the norm of the state variables vector is plotted for finding the fixed points of the system versus the changing parameter, finally the bifurcation diagram is obtained [40]. In the bifurcation diagrams, the fixed points maybe disappear, appear, or change their stability nature when the design parameter is changed. Those variations may occur even for infinitesimal changes in the parameter. Bifurcation diagram is used for the stability analysis of a dynamical system [41,42]. Moreover, the Lyapunov exponents spectrum makes it possible to qualitatively quantify a local property with respect to the attractor's stability. The positive/negative values of the Lyapunov exponents can be observed as a measure of the averaged exponential divergence/convergence of neighborhood trajectories [43,44]. The bifurcation diagram for y, when the states cut the plane $z = 0$ with $dz/dt < 0$, as well as the spectrum of system's Lyapunov exponents (LE_i, $i = 1, 2, 3$), by varying the value of a to explore the dynamical form of system (2), while keeping the initial states as $[x_0, y_0, z_0] = [2, 0.2, 1]$, are depicted in Figure 1. Therefore, the suggested structure (2) is integrated via the classical Runge-Kutta integration algorithm [45], numerically. For all of the parameters, the simulation calculations are executed via the parameters and variables in extended precision mode. In addition, the spectrum of the Lyapunov exponents are found via the Wolf's algorithm [46].

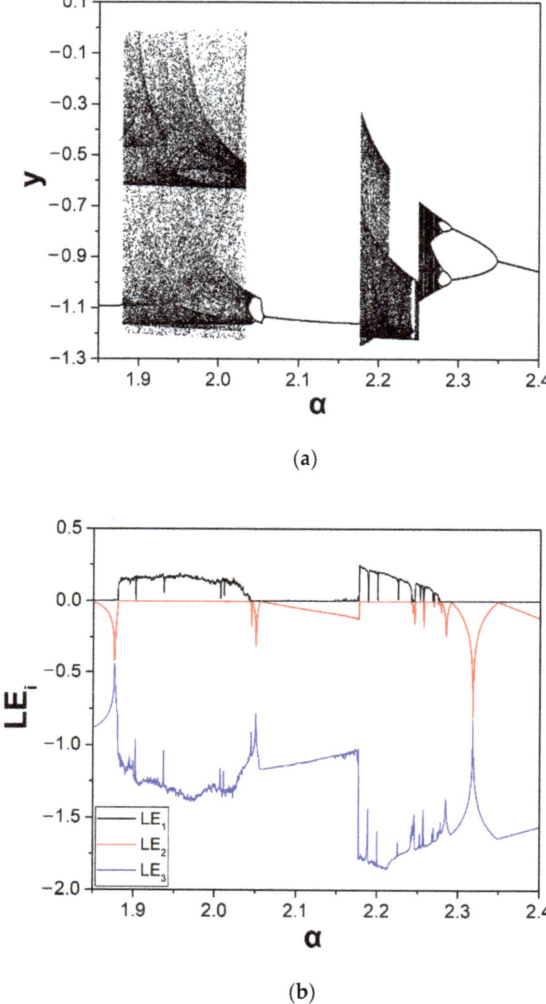

Figure 1. (**a**) Bifurcation diagram, (**b**) Lyapunov exponents spectrum of dynamics (2), when changing *a* from 1.85 to 2.4, and $b = c = 0$.

The dynamics (2) shows a chaotic attractor, for $a = 2$ (Figure 2), and a limit cycle of Period-1, for $a = 2.35$ (Figure 3). The spectrum of Lyanpunov exponents (Figure 1b) approves the dynamic behavior of the system as it has been revealed via bifurcation diagram.

2.2. Scenario B: No Equilibrium Point

If $b \neq 0$, $c = 0$ and by keeping $a = 2$, for obtaining the equilibrium point, we solve $\dot{x} = 0$, $\dot{y} = 0$ and $\dot{z} = 0$, that is

$$\begin{aligned}
x - 2yz + b &= 0, \\
xz - \sinh(y) &= 0, \\
x &= 0.
\end{aligned} \tag{6}$$

Consequently, the chaotic flow has no equilibrium point in this case. Therefore, it belongs to the category of chaotic systems containing hidden attractors.

Taking the bifurcation diagram of y (Figure 4a), along with the Lyapunov exponents spectrum (Figure 4b) by changing b for $0 < b < 0.005$, in order to explore the dynamics (2), for initial conditions $[x_0, y_0, z_0] = [2, 0.2, 1]$, interesting dynamical behavior has been investigated. As it is obtained from bifurcation diagram (Figure 4a), the system passes from a chaotic region, for $b \in [0, 0.075)$, to a periodic one as the parameter b increases.

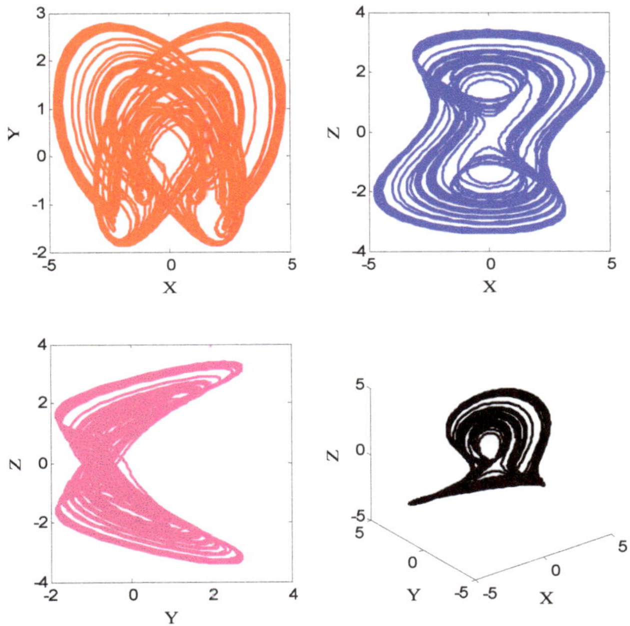

Figure 2. Strange chacs attractor for $a = 2$ and $b = c = 0$ in Scenario A.

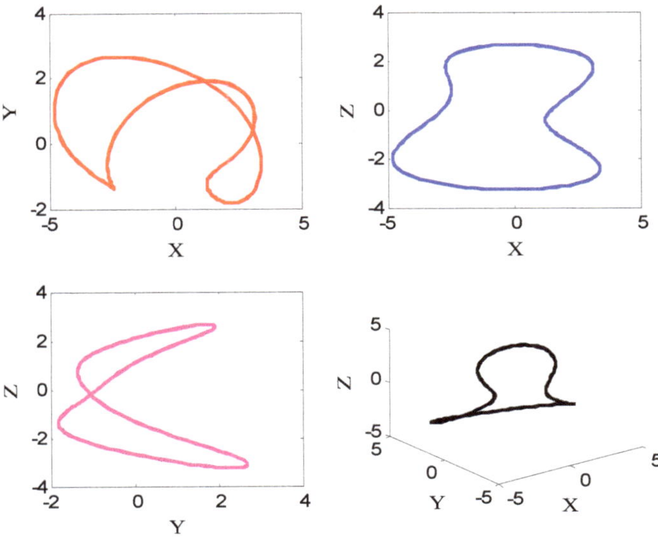

Figure 3. Limit cycle of period-1 for $b = c = 0$ and $a = 2.35$ in Scenario A.

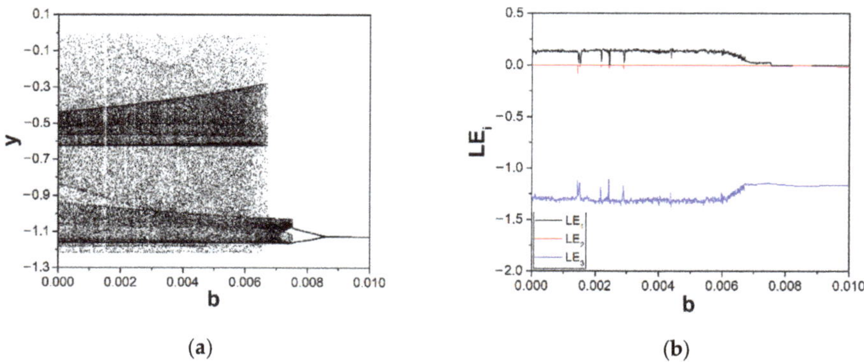

Figure 4. (**a**) Bifurcation diagram, (**b**) spectrum of Lyapunov exponents of (2), when changing b from 0 to 0.01, for $a = 2, c = 0$.

The strange attractors of the system (2) are displayed for $a = 2$, $b = 0.005$ and $c = 0$ in Figure 5. In this case, the Lyapunov exponents are $LE_1 = 0.14107$, $LE_2 = 0$, $LE_3 = -1.33835$, which confirmed the chaotic behavior of the system (2). The Kaplan-York dimension of the chaotic flow is $D_{KY} = 2.1054$. Besides, the Poincaré map in x-y plane presents the folding properties of chaos when $z = 0$ with $dz/dt < 0$ (Figure 6).

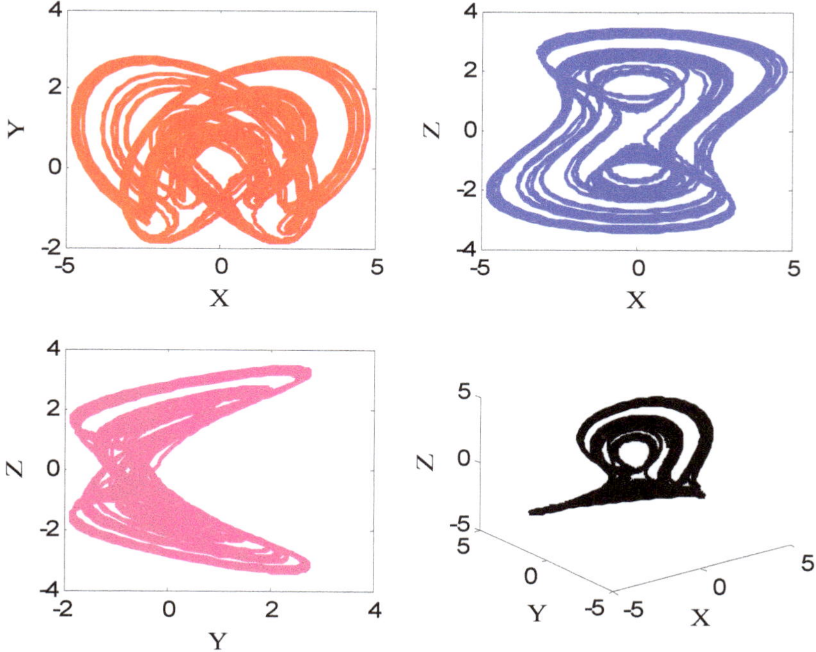

Figure 5. Strange chaotic attractors for $a = 2$, $b = 0.005$ and $c = 0$.

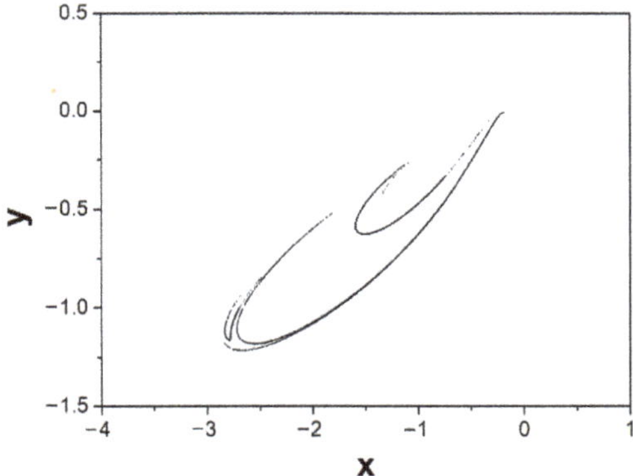

Figure 6. Poincaré map of chaotic system (2) in the x–y plane, for $a = 2$, $b = 0.005$ and $c = 0$.

2.3. Scenario C: Self-Excited Attractor

If $b = 0$, $c \neq 0$ and $a = 2$ this chaotic flow has only one equilibrium $E_C = \left[0, sinh^{-1}(b), 0\right]^T$. For the equilibrium point E_C, the Jacobian matrix is found as:

$$J = \begin{bmatrix} 1 & 0 & -2sinh^{-1}(b) \\ 0 & -cosh(sinh^{-1}(b)) & 0 \\ 1 & 0 & 0 \end{bmatrix}. \tag{7}$$

Then, the eigenvalues of the linearized chaotic flow are obtained as:

$$\begin{aligned} |\lambda I - J| &= \begin{vmatrix} \lambda - 1 & 0 & 2sinh^{-1}(c) \\ 0 & \lambda + cosh(sinh^{-1}(c)) & 0 \\ -1 & 0 & \lambda \end{vmatrix} \\ &= \left(\lambda + cosh(sinh^{-1}(c))\right)\left(\lambda^2 - \lambda + 2sinh^{-1}(c)\right) = 0 \\ \Rightarrow \lambda_1 &= -cons(sinh^{-1}(c)) = -\sqrt{1 + c^2}, \lambda_{2,3} = \frac{1 \pm \sqrt{1 - 8sinh^{-1}(c)}}{2}. \end{aligned} \tag{8}$$

For $c > 0.1253$, the eigenvalues of the chaotic flow are found as $\lambda = -\sqrt{1 + c^2}, \frac{1 \pm i\omega}{2}$, and the equilibrium point is a saddle focus. For $c < 0.1253$, the eigenvalues of the chaotic flow are obtained as $\lambda = -\sqrt{1 + c^2}, \frac{1 \pm \sqrt{\Delta}}{2}$, and the equilibrium point is a saddle node.

Figure 7 depicts the bifurcation diagram of variable y as well as the spectrum of Lyapunov exponents by varying c, for $0 < c < 0.05$, to explore the dynamics (2), for initial states $[x_0, y_0, z_0] = [2, 0.2, 1]$. It is shown from bifurcation diagram (Figure 7a) that the system passes from a chaotic region for $c \in [0, 0.0285)$ to a periodic one as the parameter c increases. The respective spectrum of Lyapunov exponents to parameter c displays the aforementioned system's (2) dynamical behavior for $a = 2$ and $b = 0$.

The strange attractors of (2) for $a = 2$, $b = 0$, and $c = 0.02$ are demonstrated in Figure 8. For these parameter's values, the Lyapunov exponents are $LE_1 = 0.09822$, $LE_2 = 0$, $LE_3 = -1.27669$, which confirmed the chaotic behavior of system (2). The Kaplan-York dimension is $D_{KY} = 2.07699$. Furthermore, the Poincaré maps in x–y plane, when $z = 0$ with $dz/dt < 0$ (Figure 9) presents the folding features of chaos.

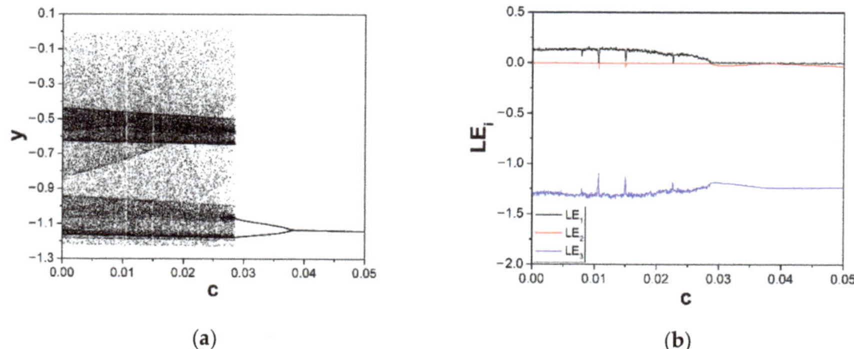

Figure 7. (a) Bifurcation diagram, (b) Lyapunov exponents spectrum of (2) when changing c from 0 to 0.05, for $a = 2$, $b = 0$.

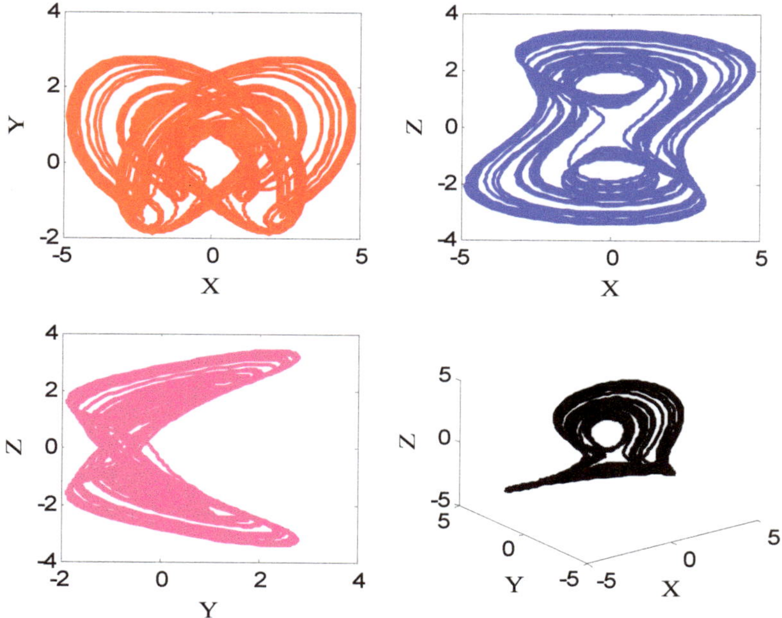

Figure 8. Strange attractors for $a = 2$, $b = 0$ and $c = 0.02$.

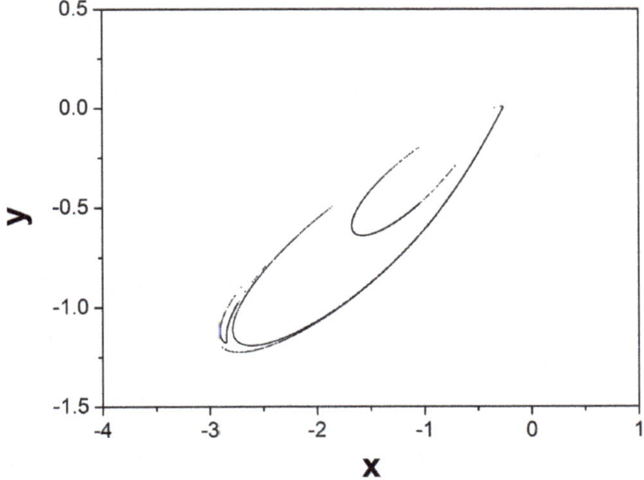

Figure 9. Poincaré map of system (2) in the *x-y* plane, for $a = 2$, $b = 0$ and $c = 0.02$.

3. Circuit Design of the Proposed Chaotic Flow

In recent years, the physical realizations of theoretical chaos forms have been investigated extensively for approving the feasibility and employing them in practical usages [47–50]. Therefore, in this section, a circuit realization with the hyperbolic sinusoidal nonlinearity is presented. For the reason of easiness, the general design methodology is applied according to the operational amplifiers [51,52]. The circuit is designed by using the common electronic components as displayed in Figure 10. There are an inverting amplifier (U4), three integrators (U1–U3), and two analog multipliers (U7, U8) of type AD633. The circuit for simulating the hyperbolic sinusoidal nonlinearity, in the dotted frame, includes three resistors (R_{S1}–R_{S3}), two operational amplifiers (U5, U6) and two diodes (D_1, D_2).

Based on Figure 8, via the Kirchhoff's laws, the circuital equation of the circuit is found as

$$\dot{X} = \frac{1}{RC}\Big[X - \frac{R}{R_1 1V}YZ + V_b\Big],$$
$$\dot{Y} = \frac{1}{RC}\Big[\frac{XZ}{1V} - \frac{2I_S R_{S3} R}{R_4}\sinh\Big(\frac{R_{S2}}{nV_T R_{S1}}\Big) + V_c\Big],$$
$$\dot{Z} = \frac{1}{RC}X \tag{9}$$

where I_S, n and V_T are diode's reverse bias saturation current, the diode's ideality factor, and the thermal voltage, correspondingly. Normalizing the Equation (9) with $\tau = t/RC$, the dimensionless structure can be designated by

$$\dot{X} = X - \frac{R}{R_1 1V}YZ + V_b,$$
$$\dot{Y} = \frac{XZ}{1V} - \frac{2I_S R_{S3} R}{R_4}\sinh\Big(\frac{R_{S2}}{nV_T R_{S1}}\Big) + V_c,$$
$$\dot{Z} = X \tag{10}$$

The variables (X, Y, Z) are equivalent to output voltages of integrators (U1–U3), when the power supply is ±15 V_{DC}. The system (10) corresponds to the suggested system with the hyperbolic sinusoidal nonlinear function (2). The electronic components are selected for $a = 2$, and $b = c = 0$; then we have $R = R_4 = 30$ kΩ, $R_1 = 15$ kΩ, $R_2 = 10$ kΩ, $R_3 = 90$ kΩ, $R_{S1} = 100$ kΩ, $R_{S2} = 50.66$ kΩ, $R_{S3} = 18.65$ MΩ and $C = 10$ nF. The planned circuit of Figure 10 has been executed in Multisim, and some PSpice results are presented in Figure 11. One can obviously confirm the consistency of the simulations (Figure 11) and numerical outcomes (Figure 2).

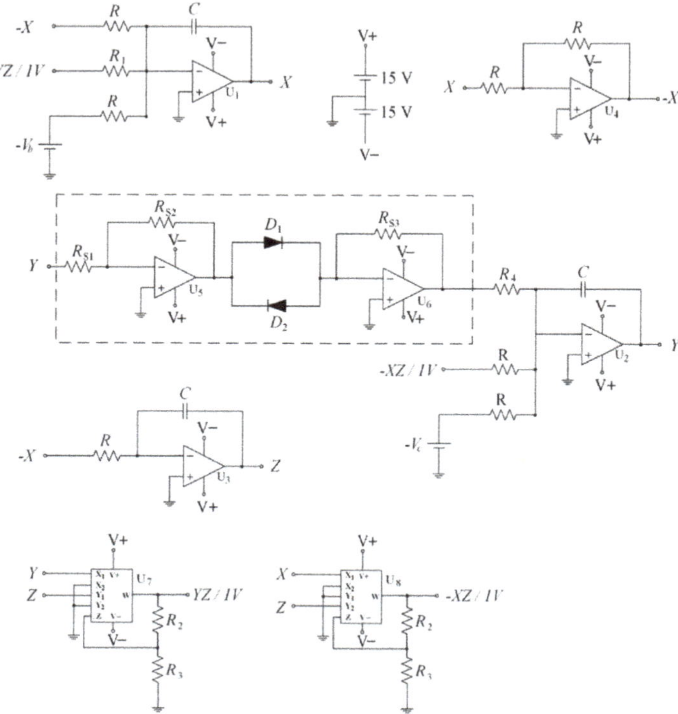

Figure 10. Schematic of circuit simulation for the system with hyperbolic sinusoidal nonlinear function (2).

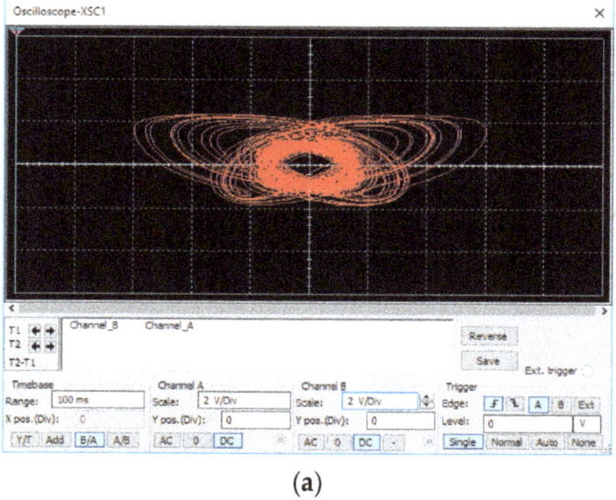

(a)

Figure 11. *Cont.*

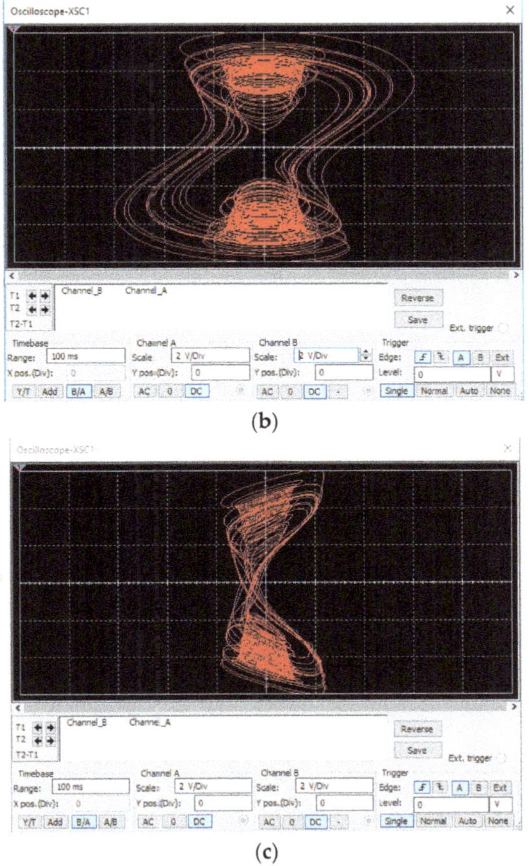

Figure 11. PSpice chaotic attractors of system with hyperbolic sinusoidal nonlinearity in (**a**) *X–Y* plane, (**b**) *X–Z* plane, and (**c**) *Y–Z* plane (*x*: 2 V/Div, *y*: 2 V/Div).

4. Voice Encryption Algorithm Design and Its Analysis

Chaos-based cryptography is one of the topics that has been intensively studied in recent years because of the randomness and rich dynamics of chaotic systems [53–58]. These works usually focus on image encryption. A new RNG algorithm design is performed by using the developed chaotic form and National Institute of Standards and Technology (NIST) 800-22 [59] randomness tests are employed to study the randomness of the obtained random numbers. It is noted that the NIST statistical test suit is used for the evaluation of the advanced encryption standard candidate algorithms. At first, a voice encryption algorithm is proposed using the obtained random number. Then, the voice encryption is executed by using the proposed algorithm and frequency spectrum analysis of the encryption procedure is executed.

4.1. RNG Algorithm Design and NIST 800-22 Test Results

In this subsection, an RNG algorithm design is developed via the newly introduced chaos in order to obtain the random numbers to be employed in the algorithm. The design process of the RNG algorithm is carried out as exposed in Algorithm 1 (see Appendix A). In the design process of the RNG algorithm, firstly the initial states and parameters of the chaotic system are defined. Then, the sampling interval of the system is determined and the chaotic system is considered by using fourth-order Runge-Kutta (RK-4) integration algorithm using this sampling value. As an outcome of the system analysis, float values are found for each cycle from each phase. On the float values obtained from each phase, the step values of the decimal parts after the comma are subjected to the mode 2 operation. As a result, 15 bits are generated from each phase in the each cycle. Further, these obtained values are added to the number sequences for each phase (rngx, rngy, rngz). This process continues for each number sequence until the 1 M. bit is generated for the NIST 800-22 randomness examinations. Because at least 1 M. bit is needed for NIST 800-22 tests. After 1 M bits are generated from each phase, the phases are subjected to Exclusive Or (XOR) processing in binary form and new random number sequences are generated as named rngxy, rngxz, rngyz, rngxyz in the Algorithm 1. The generated values from the x and y phases are subjected to XOR processing to obtain a rngxy random number sequence. Similarly, generated from the phases y and z for rngyz, the x and z phases for rngyz and the x, y, z phases for rngxyz are subjected to XOR processing. Finally, NIST 800-22 tests are employed to all obtained random bit sequences. When random bit sequences are tested singularly, they cannot pass some tests. For this reason, random bit sequences generated are subjected to 2 or 3 XOR operations.

For the safe use of random numbers, they must have an appropriate randomness. The NIST 800-22 tests are a set of internationally accepted and frequently used tests in the literature that define the numbers' randomness via a variety of different tests. The NIST 800-22 test outcomes for the random number sequences originated from the developed RNG algorithm are displayed in Table 1. According to the test outcomes, it is seen that all the random numbers created passed all the examinations.

Table 1. NIST 800-22 NIST Test Results.

NIST Statistical Tests	*p*-Value $(x \oplus y)$	*p*-Value $(x \oplus z)$	*p*-Value $(y \oplus z)$	*p*-Value $(x \oplus y \oplus z)$	Results
Frequency (Monobit) Test	0.32708	0.70840	0.07409	0.83679	Passed
Block-Frequency Test	0.05028	0.44384	0.87530	0.24483	Passed
Cumulative-Sums Test	0.49997	0.79399	0.12548	0.88754	Passed
Runs Test	0.28235	0.05285	0.01530	0.34010	Passed
Longest-Run Test	0.91108	0.88963	0.46730	0.057827	Passed
Binary Matrix Rank Test	0.17994	0.15263	0.55596	0.39136	Passed
Discrete Fourier Transform Test	0.17441	0.92688	0.52063	0.20211	Passed
Overlapping Templates Test	0.63213	0.12006	0.96148	0.29966	Passed
Maurer's Universal Statistical Test	0.59708	0.81350	0.40059	0.48723	Passed
Approximate Entropy Test	0.95048	0.38285	0.27585	0.52635	Passed
Random-Excursions Test ($x = -4$)	0.82604	0.57997	0.40488	0.34822	Passed
Random-Excursions Variant Test ($x = -4$)	0.74935	0.63538	0.19136	0.46211	Passed
Serial Test-1	0.53650	0.89087	0.74965	0.92028	Passed
Serial Test-2	0.13577	0.48589	0.27236	0.75602	Passed
Linear-Complexity Test	0.72956	0.94527	0.31945	0.78612	Passed

4.2. Voice Encryption Algorithm Design and Its Application

A new voice encryption algorithm is developed via RNG algorithm introduced in the previous section, voice encryption usage and its analysis are performed. The block diagram of the encryption process is presented in Figure 12. In the encryption algorithm, after entering initial states and parameters of chaotic system, these values are transmitted to the receiving side as a key for generation of the random number sequences to be employed in the decryption process. In order to realize the bit-based encryption, the values are obtained from the voice file consisting of float values with appropriate sampling step and converted into binary form. With random bit sequences past all NIST 800-22 randomness tests obtained from the RNG, the voice file in the binary form is encrypted. XOR operation is used in encryption process. After the encryption operation, the encrypted binary bit array is changed to the float form to generate the encrypted voice file. After the encrypted voice data is sent to the receiver side in this way, the decryption process is performed by applying the reverse operations in the encryption. Thus, the original voice file is attained on the receiving side.

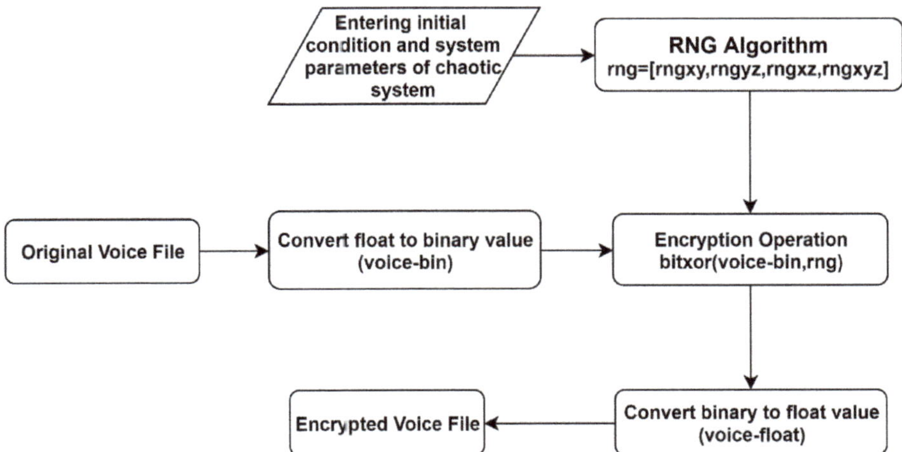

Figure 12. The block diagram of encryption process.

The voice files in encryption procedure are shown in Figure 13. The original, encrypted and decrypted voice file is demonstrated in Figure 13a–c, respectively. When comparing the original and encrypted voice file in Figure 13a,b; it is seen that a very different file is gotten than the original and the encryption process is successful. When Figure 13a,c are examined, it is observed that the decryption procedure is successful. Figure 14 shows the frequency spectrum analysis results of the encryption process. Frequency spectrum analysis is carried out to determine the frequency range of voice files. To determine the success of the encryption process, frequency spectrum analyzes are performed on original and encrypted voice files. The spectrum analysis results of original and encrypted voice files are illustrated in Figure 14a,b. If we compare these two graphs, it seems that the encrypted voice file has a rather wide frequency spectrum range than the original. When these results are evaluated, it shows the success of the encryption process.

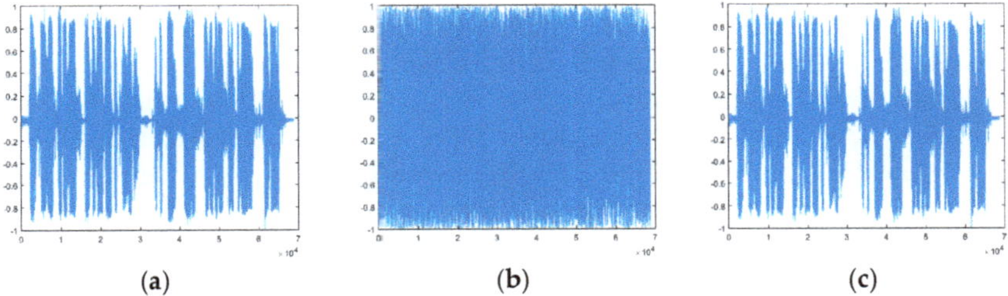

Figure 13. The voice file (**a**) original; (**b**) encrypted; (**c**) decrypted.

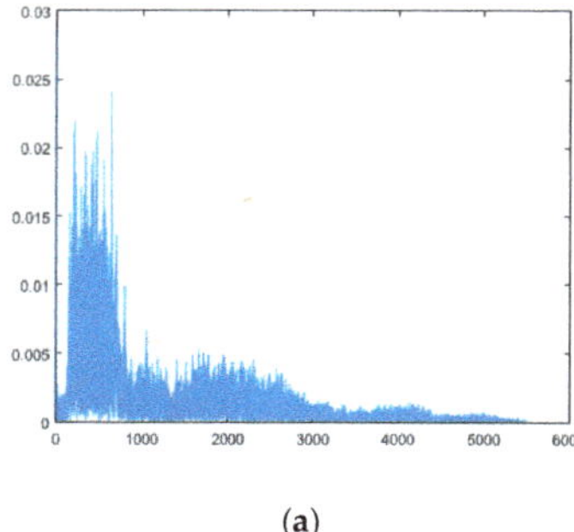 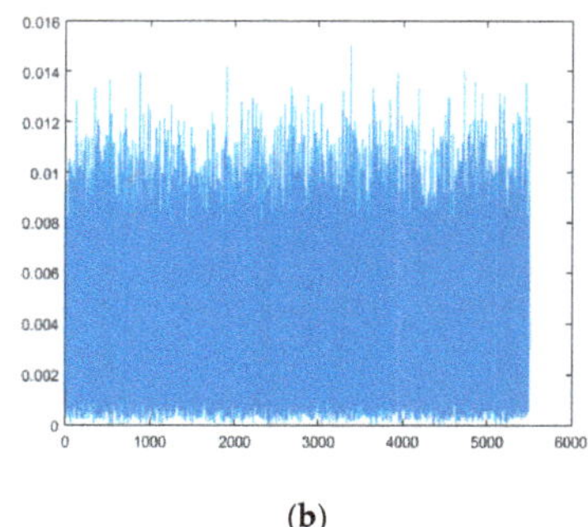

(a)   (b)

Figure 14. The spectrum analysis outcomes of original and encrypted voice files. (**a**) Original, (**b**) Encrypted.

5. Conclusions and Discussion

In this article, a new chaotic system with hyperbolic sinusoidal nonlinearity is designed. The proposed system belongs to a new category of dynamical systems with hidden chaotic flows, which assist in further understanding of chaotic attractors and also to use them in interesting applications like cryptography and secure communication schemes. The feature of hidden chaotic attractors, such as in systems with line of equilibria or in systems with no equilibrium point, makes them more suitable for the aforementioned applications, due to the fact that using systems with hidden attractors adds complexity to the dynamical system, which is used in this kind of applications. Therefore, in this work a voice encryption scheme, which is based on the specific systems was studied. Based on the variations of parameters of the system, this flow presented two classes of hidden attractors (with line of equilibria and no equilibrium point) plus a self-excited attractor, which has been reported to the literature for the first time. Dynamical behavior of the proposed system was explored and its bifurcation diagram and spectrum of Lyapunov exponents were propounded. For the appropriate selection of the parameters, the flow could display periodic oscillations and double-scroll chaos attractors. The system's electronic simulation investigated the confirmation of the double-scroll chaos attractor in real word. Via the proposed chaotic system, a novel RNG design was realized and random number generation was performed. NIST 800-22 randomness examinations were employed to the produced numbers and it was determined that all tests passed. By using RNG design, a novel voice encryption algorithm was established and encryption process was done. Frequency spectrum analysis of the voice encryption procedure was executed. In line with the analysis outcomes, it has been found that the new RNG design produces high random numbers and that the suggested encryption algorithm effectively achieves the encryption process. Therefore, authors aim with this work to attract the interest of the research community in the use of chaotic dynamical systems with hidden attractors in encryption schemes, as the results are proved to be very promising. Finally, as a future plan, the hardware implementation of the specific approach has been planned.

Author Contributions: S.M. was responsible for formal analysis and writing—original draft preparation. C.K.V. and Ü.Ç. were responsible to conceptualization, methodology, and software. S.K. was responsible for software, supervision and editing. All authors have read and agreed to the published version of the manuscript.

Funding: This research received no external funding.

Conflicts of Interest: The authors of this paper declare that they have no conflicts of interest.

Date Availability: The MATLAB files data used to support the findings of this paper are available from the corresponding author upon request.

Appendix A

Algorithm 1 RNG Design Algorithm Pseudo Code

1:    **Start**
2:    rngx=[[] rngy=[[] rngz=[[] rngxy=[[] rngyz=[[] rngxz=[[] rngxyz=[[]
3:    Entering system parameters and initial conditions of chaotic systems
4:    Determination of the appropriate value of (Δh=0.001)
5:    **while** i<=1000000 **do**
6:    Sampling with determination Δh value
7:    Solving the chaotic system using RK4 algorithm
8:    Obtaining time series float values (x,y,z) (each value 15 digit)
9:    **for** k = 0 **to** 14 **do**
10:       rngx[i]=mod(x[k], 2);
11:       rngy[i]=mod(y[k], 2);
12:       rngz[i]=mod(z[k], 2);
13:         i=i+1
14:      **end for**
15:    **end while**
16:    rngxy=bitxor(rngx, rngy);
17:    rngyz=bitxor(rngy, rngz);
18:    rngxz=bitxor(rngx, rngz);
19:    rngxyz = bitxor(rngxy, rngz);
20:    The implementation of NIST tests for each new array (rngxy, rngyz, rngxz, rngxyz)
21:    rng=[rngxy,rngyz,rngxz,rngxyz]
22:    **End**

References

1. Hua, Z.; Zhou, B.; Zhou, Y. Sine-Transform-Based Chaotic System With FPGA Implementation. *IEEE Trans. Ind. Electron.* **2018**, *65*, 2557–2566. [CrossRef]
2. Thoai, V.P.; Kahkeshi, M.S.; Huynh, V.V.; Ouannas, A.; Pham, V.-T. A Nonlinear Five-Term System: Symmetry, Chaos, and Prediction. *Symmetry* **2020**, *12*, 865. [CrossRef]
3. Chain, K.; Kuo, W.-C. A new digital signature scheme based on chaotic maps. *Nonlinear Dyn.* **2013**, *74*, 1003–1012. [CrossRef]
4. Muthukumar, P.; Balasubramaniam, P.; Ratnavelu, K. Sliding mode control design for synchronization of fractional order chaotic systems and its application to a new cryptosystem. *Int. J. Dyn. Control* **2017**, *5*, 115–123. [CrossRef]
5. Deng, Y.; Hu, H.; Liu, L. Feedback control of digital chaotic systems with application to pseudorandom number generator. *Int. J. Mod. Phys. C* **2015**, *26*, 1550022. [CrossRef]
6. Castro-Ramírez, J.; Martínez-Guerra, R.; Cruz-Victoria, J.C. A new reduced-order observer for the synchronization of nonlinear chaotic systems: An application to secure communications. *Chaos Interdiscip. J. Nonlinear Sci.* **2015**, *25*, 103128. [CrossRef]
7. Wang, G.; Chen, D.; Lin, J.; Chen, X. The application of chaotic oscillators to weak signal detection. *IEEE Trans. Ind. Electron.* **1999**, *46*, 440–444. [CrossRef]
8. Sakthivel, R.; Santra, S.; Anthoni, S.M.; Kuppili, V. Synchronisation and anti-synchronisation of chaotic systems with application to DC–DC boost converter. *IET Gener. Transm. Distrib.* **2017**, *11*, 959–967. [CrossRef]
9. Chen, E.; Min, L.; Chen, G. Discrete Chaotic Systems with One-Line Equilibria and Their Application to Image Encryption. *Int. J. Bifurc. Chaos* **2017**, *27*, 1750046. [CrossRef]

10. Glushkov, A.V.; Khetselius, O.; Brusentseva, S.V.; Zaichko, P.A.; Ternovsky, V.B. Studying interaction dynamics of chaotic systems within a non-linear prediction method: Application to neurophysiology. *Adv. Neural Netw. Fuzzy Syst. Artif. Intell.* **2014**, *21*, 69–75.

11. Aguilar-López, R.; Martínez-Guerra, R.; Perez-Pinacho, C.A. Nonlinear observer for synchronization of chaotic systems with application to secure data transmission. *Eur. Phys. J. Spec. Top.* **2014**, *223*, 1541–1548. [CrossRef]

12. Radwan, A.; Moaddy, K.; Salama, K.N.; Momani, S.; Hashim, I. Control and switching synchronization of fractional order chaotic systems using active control technique. *J. Adv. Res.* **2014**, *5*, 125–132. [CrossRef] [PubMed]

13. Boulkroune, A.; Bouzeriba, A.; Hamel, S.; Bouden, T. Adaptive fuzzy control-based projective synchronization of uncertain nonaffine chaotic systems. *Complexity* **2015**, *21*, 180–192. [CrossRef]

14. Mobayen, S. Finite-time stabilization of a class of chaotic systems with matched and unmatched uncertainties: An LMI approach. *Complexity* **2016**, *21*, 14–19. [CrossRef]

15. Ma, D.; Sun, Q.; Li, X. Synchronization of master-slave chaotic system with coupling time-varying delay based on sampled-data control. In Proceedings of the Control and Decision Conference (CCDC), 2015 27th Chinese, Qingdao, China, 23–25 May 2015; pp. 6545–6550.

16. Xiong, W.; Huang, J. Finite-time control and synchronization for memristor-based chaotic system via impulsive adaptive strategy. *Adv. Differ. Equ.* **2016**, *2016*, 101. [CrossRef]

17. Song, Q.; Huang, T. Stabilization and synchronization of chaotic systems with mixed time-varying delays via intermittent control with non-fixed both control period and control width. *Neurocomputing* **2015**, *154*, 61–69. [CrossRef]

18. Mobayen, S.; Baleanu, D.; Tchier, F. Second-order fast terminal sliding mode control design based on LMI for a class of non-linear uncertain systems and its application to chaotic systems. *J. Vib. Control* **2017**, *23*, 2912–2925. [CrossRef]

19. Wei, Z. Dynamical behaviors of a chaotic system with no equilibria. *Phys. Lett. A* **2011**, *376*, 102–108. [CrossRef]

20. Jafari, S.; Sprott, J.C.; Golpayegani, S. Elementary quadratic chaotic flows with no equilibria. *Phys. Lett. A* **2013**, *377*, 699–702. [CrossRef]

21. Wang, X.; Chen, G. A chaotic system with only one stable equilibrium. *Commun. Nonlinear Sci. Numer. Simul.* **2012**, *17*, 1264–1272. [CrossRef]

22. Molaie, M.; Jafari, S.; Sprott, J.C.; Golpayegani, S. Coexisting hidden attractors in a 4-D simplified Lorenz system. *Int. J. Bifurc. Chaos* **2013**, *23*, 1350188. [CrossRef]

23. Shilnikov, L.P. A case of the existence of a denumerable set of periodic motions. *Sov. Math.* **1965**, *24*, 163–166.

24. Leonov, G.A.; Kuznetsov, N.V.; Kuznetsova, O.A.; Seledzhi, S.M.; Vagaitsev, V.I. Hidden oscillations in dynamical systems. *Trans. Syst. Contr.* **2011**, *6*, 54–67.

25. Leonov, G.A.; Kuznetsov, N.V.; Vagaitsev, V.I. Hidden attractor in smooth chua systems. *Phys. D Nonlinear Phenom.* **2012**, *241*, 1482–1486. [CrossRef]

26. Leonov, G.A.; Kuznetsov, N.V. Hidden attractors in dynamical systems. From hidden oscillations in Hilbert Kolmogorov, Aizerman, and Kalman problems to hidden chaotic attractor in Chua circuits. *Int. J. Bifurc. Chaos* **2013**, *23*, 1330002. [CrossRef]

27. Wang, Z.; Volos, C.; Kingni, S.T.; Azar, A.T.; Pham, V.-T. Four-wing attractors in a novel chaotic system with hyperbolic sine nonlinearity. *Opt. Int. J. Light Electron Opt.* **2017**, *131*, 1071–1078. [CrossRef]

28. Pham, V.-T.; Volos, C.; Kingni, S.T.; Kapitaniak, T.; Jafari, S. Bistable Hidden Attractors in a Novel Chaotic System with Hyperbolic Sine Equilibrium. *Circuitssyst. Signal Process.* **2017**, *37*, 1028–1043. [CrossRef]

29. Kuznetsov, A.; Kuznetsov, S.; Mosekilde, E.; Stankevich, N. Co-existing hidden attractors in a radio-physical oscillator system. *J. Phys. A Math.* **2015**, *48*, 125101. [CrossRef]

30. Zhusubaliyev, Z.T.; Mosekilde, E. Multistability and hidden attractors in a multilevel DC/DC converter. *Math. Comput. Simul.* **2015**, *109*, 32–45. [CrossRef]

31. Kiseleva, M.A.; Kuznetsov, N.V.; Leonov, G.A. Hidden attractors in electromechanical systems with and without equilibria. *IFAC Pap.* **2016**, *49*, 51–55. [CrossRef]

32. Zhusubaliyev, Z.T.; Mosekilde, E.; Rubanov, V.G.; Nabokov, R.A. Multistability and hidden attractors in a relay system with hysteresis. *Phys. D Nonlinear Phenom.* **2015**, *306*, 6–15. [CrossRef]

33. Yu, M.; Sun, K.; Liu, W.; He, S. A hyperchaotic map with grid sinusoidal cavity. *Chaossolitons Fractals* **2018**, *106*, 107–117. [CrossRef]

34. Zhang, X.; Li, C.; Lei, T.; Liu, Z.; Tao, C. A symmetric controllable hyperchaotic hidden attractor. *Symmetry* **2020**, *12*, 550. [CrossRef]

35. Sadkhan, S.B.; Ali, H. A proposed speech scrambling based on hybrid chaotic key generators. In Proceedings of the 2016 Al-Sadeq IEEE International Conference on Multidisciplinary in IT and Communication Science and Applications (AIC-MITCSA), Al-Najaf, Iraq, 9–10 May 2016; pp. 1–6.

36. Mobayen, S.; Vaidyanathan, S.; Sambas, A.; Kacar, S.; Çavuşoğlu, Ü. A novel chaotic system with boomerang-shaped equilibrium, its circuit implementation and application to sound encryption. *Iran. J. Sci. Technol. Trans. Electr. Eng.* **2019**, *43*, 1–12.

37. Raheema, A.M.; Sadkhan, S.B.; Sattar, S.M.A. Design and implementation of speech encryption based on hybrid chaotic maps. In Proceedings of the 2018 IEEE International Conference on Engineering Technology and Their Applications (IICETA), Al-Najaf, Iraq, 8–9 May 2018; pp. 112–117.

38. Nosrati, K.; Volos, C. Bifurcation Analysis and Chaotic Behaviors of Fractional-Order Singular Biological Systems. In *Nonlinear Dynamical Systems with Self-Excited and Hidden Attractors*; Springer: Berlin/Heidelberg, Germany, 2018; pp. 3–44.

39. Wu, F.; Ma, J. The chaos dynamic of multiproduct Cournot duopoly game with managerial delegation. *Discret. Dyn. Nat. Soc.* **2014**, *2014*. [CrossRef]

40. Barrera, J.; Flores, J.J.; Fuerte-Esquivel, C. Generating complete bifurcation diagrams using a dynamic environment particle swarm optimization algorithm. *J. Artif. Evol. Appl.* **2007**, *2008*. [CrossRef]

41. Ouannas, A.; Khennaoui, A.A.; Wang, X.; Pham, V.-T.; Boulaaras, S.; Momani, S. Bifurcation and chaos in the fractional form of Hénon-Lozi type map. *Eur. Phys. J. Spec. Top.* **2020**, *229*, 2261–2273. [CrossRef]

42. Zhu, X.; Du, W.-S. New chaotic systems with two closed curve equilibrium passing the same point: Chaotic behavior, bifurcations, and synchronization. *Symmetry* **2019**, *11*, 951. [CrossRef]

43. Awrejcewicz, J.; Krysko, A.V.; Erofeev, N.P.; Dobriyan, V.; Barulina, M.A.; Krysko, V.A. Quantifying chaos by various computational methods. Part 1: Simple systems. *Entropy* **2018**, *20*, 175. [CrossRef]

44. Kong, G.; Zhang, Y.; Khalaf, A.J.M.; Panahi, S.; Hussain, I. Parameter estimation in a new chaotic memristive system using ions motion optimization. *Eur. Phys. J. Spec. Top.* **2019**, *228*, 2133–2145. [CrossRef]

45. Huang, W.; Kamenski, L.; Lang, J. Conditioning of implicit Runge–Kutta integration for finite element approximation of linear diffusion equations on anisotropic meshes. *J. Comput. Appl. Math.* **2019**. [CrossRef]

46. Wolf, A.; Swift, J.B.; Swinney, H.L.; Vastano, J.A. Determining Lyapunov exponents from a time series. *Phys. D Nonlinear Phenom.* **1985**, *16*, 285–317. [CrossRef]

47. Bouali, S.; Buscarino, A.; Fortuna, L.; Frasca, M.; Gambuzza, L. Emulating complex business cycles by using an electronic analogue. *Nonlinear Anal. Real World Appl.* **2012**, *13*, 2459–2465. [CrossRef]

48. Banerjee, T.; Biswas, D. Theory and experiment of a first-order chaotic delay dynamical system. *Int. J. Bifurc. Chaos* **2013**, *23*, 1330020. [CrossRef]

49. Zhou, W.-J.; Wang, Z.-P.; Wu, M.-W.; Zheng, W.-H.; Weng, J.-F. Dynamics analysis and circuit implementation of a new three-dimensional chaotic system. *Opt. Int. J. Light Electron Opt.* **2015**, *126*, 765–768. [CrossRef]

50. Lai, Q.; Wang, L. Chaos, bifurcation, coexisting attractors and circuit design of a three-dimensional continuous autonomous system. *Opt. Int. J. Light Electron Opt.* **2016**, *127*, 5400–5406. [CrossRef]

51. Gokyildirim, A.; Uyaroglu, Y.; Pehlivan, I. A novel chaotic attractor and its weak signal detection application. *Opt. Int. J. Light Electron Opt.* **2016**, *127*, 7889–7895. [CrossRef]

52. Hajipour, A.; Tavakoli, H. Analysis and circuit simulation of a novel nonlinear fractional incommensurate order financial system. *Opt. Int. J. Light Electron Opt.* **2016**, *127*, 10643–10652. [CrossRef]

53. Wang, Y.; Wong, K.-W.; Liao, X.; Chen, G. A new chaos-based fast image encryption algorithm. *Appl. Soft Comput.* **2011**, *11*, 514–522. [CrossRef]

54. Çavuşoğlu, Ü.; Kaçar, S.; Pehlivan, I.; Zengin, A. Secure image encryption algorithm design using a novel chaos based S-Box. *Chaossolitons Fractals* **2017**, *95*, 92–101. [CrossRef]

55. Bakhache, B.; Ghazal, J.M.; El Assad, S. Improvement of the security of zigbee by a new chaotic algorithm. *IEEE Syst. J.* **2014**, *8*, 1024–1033. [CrossRef]

56. Khan, M. A novel image encryption scheme based on multiple chaotic S-boxes. *Nonlinear Dyn.* **2015**, *82*, 527–533. [CrossRef]

57. Çavuşoğlu, Ü.; Zengin, A.; Pehlivan, I.; Kaçar, S. A novel approach for strong S-Box generation algorithm design based on chaotic scaled Zhongtang system. *Nonlinear Dyn.* **2017**, *87*, 1081–1094. [CrossRef]

58. Hua, Z.; Zhou, Y.; Pun, C.-M.; Chen, C.P. 2D Sine Logistic modulation map for image encryption. *Inf. Sci.* **2015**, *297*, 80–94. [CrossRef]
59. Rukhin, A.; Soto, J.; Nechvatal, J.; Smid, M.; Barker, E. *A Statistical Test Suite for Random and Pseudorandom Number Generators for Cryptographic Applications*; Booz-Allen and Hamilton Inc Mclean Va: McLean, VA, USA, 2001.

Publisher's Note: MDPI stays neutral with regard to jurisdictional claims in published maps and institutional affiliations.

Article

Symmetric Key Encryption Based on Rotation-Translation Equation

Borislav Stoyanov * and Gyurhan Nedzhibov

Department of Computer Informatics, Faculty of Mathematics and Informatics, Konstantin Preslavski University of Shumen, 9712 Shumen, Bulgaria; g.nedzhibov@shu.bg
* Correspondence: borislav.stoyanov@shu.bg

Received: 12 November 2019; Accepted: 25 December 2019; Published: 2 January 2020

Abstract: In this paper, an improved encryption algorithm based on numerical methods and rotation–translation equation is proposed. We develop the new encryption-decryption algorithm by using the concept of symmetric key instead of public key. Symmetric key algorithms use the same key for both encryption and decryption. Most symmetric key encryption algorithms use either block ciphers or stream ciphers. Our goal in this work is to improve an existing encryption algorithm by using a faster convergent iterative method, providing secure convergence of the corresponding numerical scheme, and improved security by a using rotation–translation formula.

Keywords: nonlinear equations; iterative methods; rotation–translation formula; symmetric encryption

1. Introduction

Cryptography is a practice and study of techniques of hidden data transfer so that only the intended receivers can extract and read the data [1]. It is the study of mathematical methods related to different aspects of informational security such as data origin, entity authentication, data integrity and confidentiality. The source data, which is to be protected by cryptography, is called plaintext. The procedure of transforming plaintext into an unreadable form termed ciphertext is called encryption. Decryption is the reverse process, recovering the plaintext back from a ciphertext. A cryptographic system is a set of algorithms, seeded by key that encrypt given messages into ciphertext and recover them back into input data. The scheme for a secret key encryption is first proposed by Shannon [2].

There are two categories of key-based cryptographic algorithms: *symmetric key* (secret key) cryptography and *public key* (asymmetric key) cryptography. In the first category, a sender and recipient share a private key known only to both of them. The same key is used for encryption and decryption. The most commonly used symmetric algorithms are AES (Advanced Encryption Standard) [3], Cha Cha [4], Blowfish [5], and IDEA (International Data Encryption Algorithm) [6]. By contrast, for asymmetric key cryptography, two keys are used: the first one is made publicly available to senders for encrypting plaintext while the second key is kept secret and is used by the receivers for decrypting the ciphertext. The most ordinarily exploited asymmetric schemes are the Rivest–Shamir–Adleman (RSA) cryptosystem [7] and ECC (Elliptic-curve cryptography) [8]. Symmetric encryption schemes are usually faster than public key counterparts and thus are preferred for encrypting big data.

In symmetric key cryptography, either *stream ciphers* or *block ciphers* can be used. An example of stream cipher is the Vigenere Cipher. These types of ciphers encrypt the letters or digits (typically bytes) of a message one at a time, while block ciphers take a number of bits and encrypt them as a single unit. Until now, many symmetric data encryption algorithms have been proposed. Some of them use classical schemes for text encryption. In [9], an extension of a public key cryptographic scheme to support a private key cryptographic scheme which is a mix of AES and ECC is presented.

Plain text encryption based on AES, Blowfish, and SALSA20 is designed and experimentally evaluated in [10].

Some of them use chaotic equations for text encryption. In Reference [11], a novel scheme for digital image encryption based on a mix of chaos theory and DNA calculation is presented. In [12], a chaos-based pseudorandom generation scheme based on a six-dimensional chaotic system is proposed. A text encryption architecture is given. Novel symmetric data encryption algorithms based on logistic chaotic formula are presented in [13–15]. A chaotic logistic map filtered with binary function is proposed to text encryption scheme in Reference [16]. In [17], a chaos-based encryption technique based on logistic, pinchers, and sine-circle maps is proposed. An algorithm of chaotic data encryption system by using private characteristic of electrocardiogram (ECG) signal and logistic map is designed in [18]. In [19,20], the chaotic behaviour of a Chua system is used in novel text encryption scheme designs. A novel pseudorandom bit generation scheme based on rotation equations is proposed in [21]. The technique has good statistical properties measured by test packages. A novel encryption method based on modified pulsed-coupled spiking neurons circuit is presented in Reference [22]. In [23], a modified quadratic map for numeric sensor data encryption is proposed.

2. Symmetric Key Encryption Algorithms Based on Numerical Methods

One of the first published works that consider *symmetric key encryption algorithm based on numerical methods* is by Ghosh in [24] (see also [25,26]), where it is shown that any nonlinear function with one variable $f(z)$ can be defined as a key. The encryption process then is defined as finding the solution of the equation

$$f(z) - c_i = 0, \tag{1}$$

where c_i represents the numerical code of the *i*th symbol in the plaintext (e.g., the ASCII code). The function $f(z)$ must be chosen in such a way that the corresponding formula (1) has at least one real root for any *i*. Then, the set of roots $\{z_i^*\}$ represents the ciphertext. On the receiver side, each entry z_i^* is decoded by substituting it into $f(z)$ giving rise to the plaintext character $c_i = f(z_i^*)$ (the value $f(z_i^*)$ must be appropriately rounded to recover c_i). In [24], as a key function $f(z)$, the authors use a cubic polynomial and, for the numerical solution of equations $f(z) - c_i = 0$, they use the Newton's iterative method. We have to mention that, in solving nonlinear Equation (1), we can use different iterative methods. Analogous to this algorithm, an example of a public key cryptosystem based on numerical methods is considered in [27].

It is important to say that the main weaknesses of such algorithms can be summarized in the following:

1. Lack of rules on how to choose the function f and suitable iterative method so that the convergence of the process is always guaranteed.
2. Vulnerability to attack because in these types of algorithms the same letter is encoded with the same real number of each occurrence in the plaintext.

Our aim in the present work is to develop a new algorithm that solves the disadvantages mentioned above. In order to achieve this, the new scheme will be based on employing numerical iterative methods and rotation–translation formula.

3. On Numerical Methods and Rotation–Translation Equation

3.1. On Numerical Methods for Solving Nonlinear Equations

Although the Newton's iterative method

$$z_{k+1} = z_k - \frac{f(z_k)}{f'(z_k)}, \quad k = 0, 1, 2, \dots \tag{2}$$

is one of the most popular and commonly used methods, *numerical analysis* offers many iterative methods that can be used in the stage of solution of Equation (1). In general to calculate the roots of nonlinear equations (of the type (1)), we have to use approximate (iterative) methods. When studying an iterative method, two of the most important aspects to consider are:

- the convergence speed of the iteration,
- an interval of convergence and the rules for choosing the initial approximations.

Most of the known iterative algorithms for solving nonlinear equations are only locally convergent, i.e., before using such a method, we need to locate the unknown root at a sufficiently small interval. Even if the root sought is located at the appropriate interval, if we do not choose the initial approximation in a proper way, the process may not be convergent. Usually, iterative methods of this type require the following convergence conditions:

- need to have an interval $[a, b]$ containing a single root of f, and
- the derivatives f' and f'' must not have zeros in the interval $[a, b]$.

Then, the corresponding iterative process converges to the sought root for an initial approximation z_0 which is the end of the interval $[a, b]$, where $f(z_0)f''(z_0) > 0$ (or $f(z_0)f''(z_0) < 0$).

For some examples of more computationally efficient and higher order iterative methods, we refer the reader to [28].

In the encryption algorithm that we will introduce later, we will use the following iterative function

$$z_{k+1} = z_k - \frac{h(z_k)}{2} \left(\frac{3f'(u_k) + f'(z_k)}{3f'(u_k) - f'(z_k)} \right), \quad k = 0, 1, 2, \dots \tag{3}$$

where $h(z_k) = \frac{f(z_k)}{f'(z_k)}$ and $u_k = z_k - \frac{2}{3}h(z_k)$. This iterative algorithm is explored by Jarrat in [29], and it is known as *Jarrat's method* (see also [30]).

The reason we prefer iteration method (3) over method (2) is its faster convergence. The order of convergence of Jarrat's method is four, while the one of Newton's method is only two (see [30]). In addition, method (3) has higher computational efficiency, although at each step of the iteration one value of f and two values of f' are calculated (while in the Newton's method, one value of f and one value of f' are calculated). Thus, if the function f is a polynomial, then calculating the value of the function f is always more complex than calculating its derivative f'.

3.2. Base of Rotation–Translation Equation

In order to avoid the vulnerability to statistical attack, we include additional randomness by using the following space contraction formula based on rotation–translation equation of the form [31]

$$\begin{aligned}
x_{k+1} &= a + b(x_k \cos \theta_k - y_k \sin \theta_k), \\
y_{k+1} &= b(x_k \sin \theta_k + y_k \cos \theta_k),
\end{aligned} \tag{4}$$

where the angle of rotation is

$$\theta_k = c + \frac{d}{x_k^2 + y_k^2}. \tag{5}$$

The translation value is $a = 6$, the space contraction value is $b = 0.8 < 1$, and rotation values are $c = a/2$ and $d = a$. The rotation–translation Equation (4) with initial conditions $x_0 = 0.233$, $y_0 = -0.67$ is presented in Figure 1.

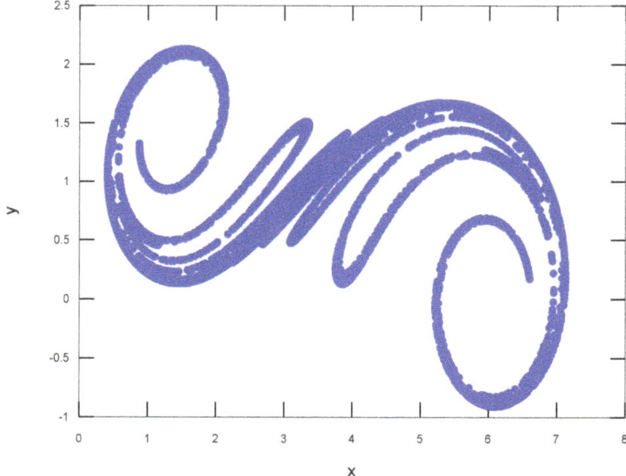

Figure 1. Space contraction.

4. Proposed Encryption Algorithm Based on Numerical Method and Rotation–Translation Equation

Here, we describe an encryption algorithm based on space contraction and numerical method. Any nonlinear function or polynomial f with one variable can be defined as part of a key.

We consider plaintext P with byte length of L. The initial values x_0 and y_0 from Equation (4), and an initial iteration number M_0, are determined. The rotation–translation formula, Equation (4), is iterated for M_0 times.

The proposed algorithm based on numerical method and space contraction is given below:

1. Read the symbols from the plaintext data and get the ASCII values of the different symbols;
2. Construct a system of L nonlinear equations by subtracting the ASCII values from the function f and equate with zero;
3. Solve individually the nonlinear equations and put the results α_i into an array B;
4. The loop of Equation (4) continues, and as an output, two real numbers x_i and y_i are generated. We take the sum of x_i and y_i to produce the real number $d_i = x_i + y_i$, which is put into an array R.
5. Return to Step 4 until a stream of real numbers R with length L is reached.
6. We get the sum of the two arrays B and R to produce E, the output array of real numbers.

Remark 1. *In Step 2 of Encryption algorithm, it is desirable that the function or polynomial f is such that each equation of the type $f(z) - c_i = 0$ has at least one real root, and it is easy to determine the initial approximation, which guarantees the convergence of the iterative process.*

4.1. Approaches for Choosing a Nonlinear Function

In the following, we consider two example functions that are suitable for selecting in the above algorithm.

4.1.1. Nonlinear Function

Let $f(z)$ has the following form

$$f(z) = e^z - z^2 - p, \tag{6}$$

where p is a real parameter such that $p \geq 1$. For its first derivative,

$$f'(z) = e^z - 2z,$$

we conclude that $f'(z) > 0$ for all $z \in \mathbb{R}$, i.e., the function $f(z)$ is monotonically increasing for all $z \in \mathbb{R}$. This and the limits

$$\lim_{z \to -\infty} f(z) = -\infty \quad \text{and} \quad \lim_{z \to \infty} f(z) = \infty$$

show that the function $f(z)$ has only one real root. From the second derivative of f

$$f''(z) = e^z - 2,$$

and because $f''(z) > 0$ for $\forall z > \ln 2$, it follows that the function $f(z)$ is convex for $z \in (\ln 2, \infty)$. Therefore, all these properties are also valid for the functions

$$g_i(z) = f(z) - c_i,$$

where c_i is an integer value (the corresponding ASCII code). Then, for any i, the function $g_i(z)$ is monotonous, convex, and has a real root in the interval $(\ln 2, \infty)$. Indeed, it can be shown that each one function $g_i(z)$ has a real root in the finite interval $(\ln 2, 6)$.

4.1.2. Polynomial Function

We consider a fifth degree monic polynomial having the following form:

$$f(z) = z^5 - z^4 + z^3 - pz^2 + qz - (p + 2q), \tag{7}$$

where p and q are real parameters such that $p, q \in [1, 10]$. From the fundamental theorem of algebra, it follows that $f(z)$ has at least one real root. Using the Descartes' rules of sign, we can prove that $f(z)$ has no negative real root, hence it has at least one real positive root (see Appendix A). Examining the first two derivatives of f, it can be shown that the function $f(z)$ is monotonically increasing, convex and has a real root in the interval $\left(\frac{9}{8}, \infty\right)$, for any $p, q \in [1, 10]$. By using the bounding theorems (see Appendix A), it can be shown that $f(z)$ has a real root in the finite interval $\left(\frac{9}{8}, 2|p + 2q|^{1/5}\right)$.

4.2. An Example of Encryption

In order to demonstrate the proposed algorithm, we will use the following example:
The text to be encrypted: "Shumen university".
As a key function, we use the polynomial

$$f(z) = z^5 - z^4 + z^3 - z^2 + z - 3,$$

which is obtained by Equation (7) in the case of $p = q = 1$. During the encryption process (Step 3 of the Algorithm), we have to solve in series nonlinear equations of the type

$$f(z) - c_i = 0, \tag{8}$$

where c_i represents the ASCII code of the i-th character in the text, i.e., $c_i \in [1, 255]$. From the analysis of the polynomial (7) and using the bounding theorems for the roots of polynomials, we deduce that Equation (8) has a real root in the interval $\left(\frac{9}{8}, 6\right)$ for any $c_i \in [1, 255]$. Moreover, this interval is such that the iterative process (3) is convergent to the solution for any initial approximation $z_0 \in \left(\frac{9}{8}, 6\right)$. For this reason, we use the same initial approximation for each Equation (8) obtained during the encryption process, namely the middle point of the interval: $z_0 = \frac{6 + 9/8}{2} \approx 3.56$.

We solve all the equations by iterative function (3) and by using the following stopping criteria

- $|f(z_k)| \leq \epsilon$, and
- $|z_k - z_{k+1}| \leq \epsilon$,

where $\epsilon = 10^{-15}$.

As a result, for all the equations, the stopping criteria are reached after three iterations. For comparison, if we use the Newton iterative method (2) instead of the Jarrats' method for solving the corresponding equations, with the same initial approximation, we get six iterations for each equation, see Table 1.

Table 1. Number of iterations for Jarrats'method (JM) and Newton method (NM), and generated arrays.

Letter (Char)	ASCII Code	NM Iterations	JM Iterations	Array B Reached Root (α_i)	Array R d_i	Array E $e_i = \alpha_i + d_i$
S	83	6	3	2.596938615169214	1.13761418319195	3.73455279836116
h	104	6	3	2.707594514758099	3.83246813052273	6.54006264528082
u	117	6	3	2.767550880788345	2.58986907946370	5.35741996025204
m	109	6	3	2.731316748315844	4.91511042783787	7.64642717615371
e	101	6	3	2.692927857503279	2.09200715087053	4.78493500837380
n	110	6	3	2.735958159508397	7.52948634851868	10.2654445080271
	94	6	3	2.657327240630354	0.10782288092075	2.76515012155110
U	85	6	3	2.608365856583876	5.70292778455835	8.31129364114222
n	110	6	3	2.735958159508397	1.30796668243219	4.04392484194058
i	105	6	3	2.712409561369016	7.38162948307289	10.0940390444419
v	118	6	3	2.771941812496392	0.13478345799064	2.90672527048704
e	101	6	3	2.692927857503279	5.23813805178474	7.93106590928801
r	114	6	3	2.754198397484480	1.66157719938621	4.41577559687069
s	115	6	3	2.758679632039476	7.40857522965636	10.1672548616958
i	105	6	3	2.712409561369016	0.16954303210037	2.88195259346938
t	116	6	3	2.763130305077092	6.32301454738480	9.08614485246189
y	121	6	3	2.784941120909602	0.81874887373720	3.60368999464681

The output array of real numbers E is in the last column of Table 1, and this is the encrypted text that the recipient receives.

4.3. Brute-Force Attack Analysis

The set of all initial values constitutes the key size. The key size of the novel encryption algorithm has the following initial key values x_0, y_0, M_0 and at least three real coefficients a_i of the polynomial f (for monic polynomial f of degree $n \geq 3$). The two seeds x_0 and y_0 are constructed by randomly choosing two floating-point values that belonging to the intervals $[0.5, 7]$ and $[-0.8, 2]$, respectively. The novel encryption algorithm does not propose weak keys. As stated in the IEEE Standard for floating-point arithmetic [32], the computational precision of the 64-bit floating point variable is about $10^{-15} \approx 2^{49}$. The key size of the novel encryption is $(2^{49})^5 + 2^{32} > 2^{248}$, which is sufficient enough to defeat brute-force attack [33]. The key space is comparable to state-of-the-art chaos-based encryption algorithms; for example, [10,13,16].

4.4. Statistical Test Analysis of the Proposed Encryption

In an attempt to evaluate randomness of the improved encryption algorithm, we used NIST [34], ENT [35], and PractRand [36] statistical test applications. The output numbers e_i from array E are converted to bytes as follows: $s_i = mod(abs(integer(e_i \times 10^{15}))), 256)$, where $integer(e)$ calculates the integer part of e, truncating the value at the decimal point, $abs(e)$ calculates the absolute value of e, and $mod(e, w)$ calculates the reminder after division. The bytes s_i are produced. Using the improved encryption, 10^3 sequences of 125,000 bytes are produced.

The NIST suite software (version sts-2.1.2) includes 15 statistical tests: monobit, block frequency, cumulative sums forward and reverse, runs, longest run of ones, rank, Fourier, non-overlapping templates, overlapping templates, universal, approximate entropy, serial one and two, linear complexity, random excursion, and random excursion variant.

The output results of the first 13 tests are in Table 2. The minimum hit rate for each statistical test with the excluding of the random excursion variant test is approximately 980 for a sample size of 1000 byte stings. The minimum hit rate for the random excursion variant test is approximately 600 for a sample size of 614 byte strings. The random excursion test outputs 8 p-values which are tabulated in Table 3. The random excursion variant test calculates 18 randomness probability numbers: p-values, and they are in Table 4.

The improved encryption algorithm passed successfully all the NIST tests.

Table 2. NIST test suite results.

NIST Test	p-Value	Success Rate
Monobit	0.556460	992/1000
Block frequency	0.010093	981/1000
Cumulative sums forward	0.399442	993/1000
Cumulative sums reverse	0.299736	993/1000
Runs	0.605916	986/1000
Longest run of ones	0.605916	988/1000
Rank	0.830808	988/1000
Fourier	0.200115	980/1000
Non overlapping templates	0.498222	990/1000
Overlapping templates	0.859637	992/1000
Universal	0.653773	988/1000
Approximate entropy	0.693142	988/1000
Serial one	0.894918	990/1000
Serial two	0.282626	986/1000
Linear complexity	0.051942	995/1000

Table 3. NIST Random excursion test results.

State	p-Value	Success Rate
-4	0.696617	610/614
-3	0.746463	606/614
-2	0.211467	610/614
-1	0.501472	606/614
$+1$	0.933509	607/614
$+2$	0.584363	605/614
$+3$	0.873629	610/614
$+4$	0.672912	608/614

Table 4. NIST Random excursion variant test results.

State	p-Value	Success Rate
-9	0.283657	608/614
-8	0.444875	607/614
-7	0.699986	609/614
-6	0.775401	607/614
-5	0.876173	610/614
-4	0.921867	607/614
-3	0.135745	607/614
-2	0.036332	610/614
-1	0.574229	612/614
$+1$	0.345203	609/614
$+2$	0.366645	607/614
$+3$	0.517714	610/614
$+4$	0.024235	612/614
$+5$	0.990938	612/614

Table 4. *Cont.*

State	p-Value	Success Rate
+6	0.447934	610/614
+7	0.232430	609/614
+8	0.193732	611/614
+9	0.659297	611/614

The ENT application includes six tests to bit or byte sequences. We tested a stream of 125,000,000 bytes (1,000,000,000 bits) of the improved encryption and tabulated the output results in Table 5. The novel encryption passed successfully all the ENT tests.

Table 5. ENT test results.

ENT Test	Input of Bits	Input of Bytes
Entropy	1.000000	7.999999
Optimum compression	Reduce size by 0%	Reduce size by 0%
χ^2 square	0.16, exceed 68.56 %	242.28, exceed 70.66%
Arithmetic mean value	0.5000	127.5055
Monte Carlo for π	3.141226994 (error 0.01%)	3.141226994 (error 0.01%)
Serial correlation	-0.000002	0.000180

The third suite is PractRand. We tested our improved encryption algorithm for strings up to 1 GB (bytes) in length, passing all statistical tests successfully as shown in Table 6.

Table 6. PractRand test results.

Test Name	Raw	Processed	Evaluation
BCFN(2,13):!	R = +0.0	"pass"	normal
BCFN(2+0,13−0)	R = −0.7	$p = 0.608$	normal
BCFN(2 + 1,13 − 0)	R = +2.3	$p = 0.172$	normal
BCFN(2 + 2,13 − 1)	R = −0.1	$p = 0.504$	normal
BCFN(2 + 3,13 − 1)	R = −2.2	$p = 0.812$	normal
BCFN(2 + 4,13 − 2)	R = −4.4	$p = 0.968$	normal
BCFN(2 + 5,13 − 3)	R = −1.1	$p = 0.669$	normal
BCFN(2 + 6,13 − 3)	R = −4.1	$p = 0.960$	normal
BCFN(2 + 7,13 − 4)	R = +4.8	$p = 0.032$	normal
BCFN(2 + 8,13 − 5)	R = +3.3	$p = 0.093$	normal
BCFN(2 + 9,13 − 5)	R = −0.3	$p = 0.524$	normal
BCFN(2 + 10,13 − 6)	R = −4.3	$p = 0.981$	normal
BCFN(2 + 11,13 − 6)	R = −0.9	$p = 0.614$	normal
BCFN(2 + 12,13 − 7)	R = +1.6	$p = 0.219$	normal
BCFN(2 + 13,13 − 8)	R = −2.7	$p = 0.914$	normal
DC6-9x1Bytes-1	R = −1.0	$p = 0.795$	normal
Gap-16:!	R = +0.0	"pass"	normal
Gap-16:A	R = +0.0	$p = 0.614$	normal
Gap-16:B	R = −3.2	$p = 0.987$	normal
(Low1/8)BCFN(2,13):!	R = +0.0	"pass"	normal
(Low1/8)BCFN(2+0,13 − 1)	R = −1.7	$p = 0.754$	normal
(Low1/8)BCFN(2+1,13 − 2)	R = +1.0	$p = 0.336$	normal
(Low1/8)BCFN(2+2,13 − 3)	R = +1.7	$p = 0.243$	normal
(Low1/8)BCFN(2+3,13 − 3)	R = −0.7	$p = 0.605$	normal
(Low1/8)BCFN(2+4,13 − 4)	R = +2.7	$p = 0.138$	normal
(Low1/8)BCFN(2+5,13 − 5)	R = −0.3	$p = 0.528$	normal
(Low1/8)BCFN(2+6,13 − 5)	R = −0.9	$p = 0.626$	normal

Table 6. Cont.

Test Name	Raw	Processed	Evaluation
(Low1/8)BCFN(2+7,13 − 6)	R = −2.3	$p = 0.838$	normal
(Low1/8)BCFN(2+8,13 − 6)	R = −2.4	$p = 0.853$	normal
(Low1/8)BCFN(2+9,13 − 7)	R = +1.6	$p = 0.223$	normal
(Low1/8)BCFN(2+10,13 − 8)	R = +3.1	$p = 0.096$	normal
(Low1/8)DC6-9x1Bytes-1	R = −0.5	$p = 0.730$	normal
(Low1/8)Gap-16:!	R = +0.0	"pass"	normal
(Low1/8)Gap-16:A	R = −0.1	$p = 0.675$	normal
(Low1/8)Gap-16:B	R = −1.7	$p = 0.888$	normal

The different statistical tests clearly show the high quality of the proposed algorithm. Table 7 summarizes some of the computed values of our proposed scheme with other algorithms. The performance test of the novel scheme is based on the average response time with data size of 1 MB. The execution is done on mobile Dell Inspiron computer i7-3630QM (2.4 GHz, 8GB RAM).

Table 7. Comparison of our improved symmetric key encryption with other algorithms.

Algorithm	Key Size	Correlation	Entropy	Arithmetic Mean	Performance Evaluation
Proposed	2^{248}	−0.000002	7.999999	127.5055	0.105
[14] Murillo-Escobar	2^{128}	−0.002100	7.994500	-	-
[21] Stoyanov 2015	2^{100}	0.000001	7.999998	127.4982	0.19
[37,38] AES-128	2^{128}	−0.002100	7.954880	127.5281	0.12

Based on the good test outputs, we can infer that the novel text encryption based on numerical method and rotation–translation formula has satisfying statistical characteristics and provides a reasonable level of security.

5. Conclusions

We have presented an improved encryption algorithm based on numerical method and rotation–translation formula. The new method uses a faster convergent iterative algorithm and adds additional randomness by using the space contraction equation. Two exemplary ways of constructing nonlinear functions or polynomials with corresponding properties are described. In the examples considered, we demonstrate how to determine the interval containing the desired root and in which the iterative method is guaranteed to be convergent. Our security analysis shows that the improved encryption scheme can be successfully used for information security.

Author Contributions: B.S. and G.N. wrote and edited the manuscript. All authors have read and agreed to the published version of the manuscript.

Funding: The paper is partially supported by the National Scientific Program "Information and Communication Technologies for a Single Digital Market in Science, Education and Security (ICTinSES)", financed by the Ministry of Education and Science, Bulgaria for Borislav Stoyanov.

Acknowledgments: The authors are grateful to the reviewer's valuable comments that significantly improved the scientific paper.

Conflicts of Interest: The authors declare no conflict of interest.

Appendix A

Appendix A.1. Real Roots Counting of Polynomials

Consider a monic polynomial of degree n

$$f(x) = x^n + a_{n-1}x^{n-1} + \ldots + a_1 x + a_0.$$

From the fundamental theorem of algebra, it follows that f has n real or complex roots, counting multiplicities. If the coefficients $a_0, a_1, \ldots, a_{n-1}$ are all real, then the complex roots occur in conjugate pairs.

Using the following Descartes' rules of sign, we can count the number of real positive zeros of f.

Descartes' rules

Let p be the number of variations in the sign of the coefficients $a_n, a_{n-1}, \ldots, a_0$ (where $a_n = 1$ and the zero coefficients are ignored). Let m be the number of real positive zeros of f. Then,

- $m \leq p$;
- $p - m$ is an even integer.

A negative zero of $f(x)$, if exists, is a positive zero of $f(-x)$.

Appendix A.2. Bounds of Real Roots of Polynomials

The first result in the theory of the location of polynomial zeros is due to Gauss, which is improved by Cauchy in [39], where he proves the following theorem.

Theorem A1 (Cauchy). *Let*

$$f(x) = a_n x^n + a_{n-1} x^{n-1} + \ldots + a_1 x + a_0$$

be a polynomial with complex coefficients, where $n \geq 1$ and $a_n \neq 0$. Then, all the zeros of $f(x)$ lie inside the circle of radius

$$R = 1 + \max_{0 \leq k \leq n-1} \left| \frac{a_k}{a_n} \right|$$

about the origin.

Another bound given by Lagrange is:

Let

$$f(x) = a_n x^n + a_{n-1} x^{n-1} + \ldots + a_1 x + a_0$$

be a polynomial with complex coefficients, where $n \geq 1$ and $a_n \neq 0$. Then, all the zeros of $f(x)$ lie inside the circle of radius

$$R = 2 \max \left(\left| \frac{a_{n-1}}{a_n} \right|, \left| \frac{a_{n-2}}{a_n} \right|^{1/2}, \ldots, \left| \frac{a_0}{a_n} \right|^{1/n} \right)$$

about the origin.

The next theorem is about bounding positive real roots of polynomials with real coefficients due to Cauchy.

Theorem A2 (Cauchy). *Let*

$$f(x) = a_n x^n + a_{n-1} x^{n-1} + \ldots + a_1 x + a_0$$

be a polynomial with real coefficients, where $n \geq 1$ and $a_n > 0$ and which has $s > 0$ strictly negative coefficients. Then, every positive real root of $f(x)$ is no larger than r:

$$R = \max \left(\left| s \frac{a_{n-1}}{a_n} \right|^{1/i} : 1 \leq i \leq n \text{ and } a_{n-i} < 0 \right).$$

More recent and sharper results are obtained by Joyal, Labelle, and Rahman [40] by proving.

Theorem A3. *If* $M = \max_{0 \leq i < n-1} |a_i|$*, then all the zeros of the monic polynomial*

$$f(x) = x^n + a_{n-1}x^{n-1} + \ldots + a_1 x + a_0$$

are contained in the disc

$$|x| \leq \frac{1}{2}\left(1 + |a_{n-1}| + \sqrt{(1 - |a_{n-1}|)^2 + 4M}\right).$$

References

1. Stallings, W. *Cryptography and Network Security: Principles and Practice*; Pearson: Upper Saddle River, NJ, USA, 2017.
2. Shannon, C.E. Communication theory of secrecy systems. *Bell Syst. Tech. J.* **1949**, *28*, 656–715. [CrossRef]
3. Daemen, J.; Rijmen, V. The Rijndael block cipher: AES proposal. In Proceedings of the First, Candidate Conference (AeS1), Rome, Italy, 22–23 March 1999; pp. 343–348.
4. Bernstein, D.J. ChaCha, a variant of Salsa20. In Proceedings of the Workshop Record of SASC, Lausanne, Switzerland, 13–14 February 2008; Volume 8, pp. 3–5.
5. Schneier, B. Description of a new variable-length key, 64-bit block cipher (Blowfish). In *Fast Software Encryption*; Anderson, R., Ed.; Springer: Berlin/Heidelberg, Germany, 1994; pp. 191–204.
6. Lai, X.; Massey, J.L. A Proposal for a New Block Encryption Standard. In *Advances in Cryptology—EUROCRYPT '90*; Damgård, I.B., Ed.; Springer: Berlin/Heidelberg, Germany, 1991; pp. 389–404.
7. Rivest, R.L.; Shamir, A.; Adleman, L.M. Cryptographic Communications System and Method. U.S. Patent 4,405,829, 20 September 1983.
8. Koblitz, N. Elliptic curve cryptosystems. *Math. Comput.* **1987**, *48*, 203–209. [CrossRef]
9. Mathur, N.; Bansode, R. AES Based Text Encryption Using 12 Rounds with Dynamic Key Selection. *Procedia Comput. Sci.* **2016**, *79*, 1036–1043. [CrossRef]
10. Panda, M.; Nag, A. Plain Text Encryption Using AES, DES and SALSA20 by Java Based Bouncy Castle API on Windows and Linux. In Proceedings of the 2015 Second International Conference on Advances in Computing and Communication Engineering, Rohtak, India, 1–2 May 2015; pp. 541–548. [CrossRef]
11. Babaei, M. A novel text and image encryption method based on chaos theory and DNA computing. *Natural Comput.* **2013**, *12*, 101–107. [CrossRef]
12. Min, L.; Lan, X.; Hao, L.; Yang, X. A 6 Dimensional Chaotic Generalized Synchronization System and Design of Pseudorandom Number Generator with Application in Image Encryption. In Proceedings of the 2014 Tenth International Conference on Computational Intelligence and Security, Yunnan, China, 15–16 November 2014; pp. 356–362. [CrossRef]
13. Murillo-Escobar, M.; Abundiz-Pérez, F.; Cruz-Hernández, C.; López-Gutiérrez, R. A novel symmetric text encryption algorithm based on logistic map. In Proceedings of the International Conference on Communications, Signal Processing and Computers, Guilin, China, 5–8 August 2014; Volume 32, pp. 49–53.
14. Murillo-Escobar, M.; Cruz-Hernández, C.; Cardoza-Avendaño, L.; Méndez-Ramírez, R. A novel pseudorandom number generator based on pseudorandomly enhanced logistic map. *Nonlinear Dyn.* **2017**, *87*, 407–425. [CrossRef]
15. Volos, C.K.; Kyprianidis, I.; Stouboulos, I. Text Encryption Scheme Realized with a Chaotic Pseudo-Random Bit Generator. *J. Eng. Sci. Technol. Rev.* **2013**, *6*, 9–14. [CrossRef]
16. Wang, X.Y.; Gu, S.X. New chaotic encryption algorithm based on chaotic sequence and plain text. *IET Inf. Secur.* **2014**, *8*, 213–216. [CrossRef]
17. Akgül, A.; Kaçar, S.; Arıcıoğlu, B.; Pehlivan, I. Text encryption by using one-dimensional chaos generators and nonlinear equations. In Proceedings of the 2013 IEEE 8th International Conference on Electrical and Electronics Engineering (ELECO), Bursa, Turkey, 28–30 November 2013; pp. 320–323.
18. Chen, C.; Lin, C. Text encryption using ECG signals with chaotic Logistic map. In Proceedings of the 2010 5th IEEE Conference on Industrial Electronics and Applications, Taichung, Taiwan, 15–17 June 2010; pp. 1741–1746. [CrossRef]
19. Volos, C.K.; Andreatos, A.S. Secure text encryption based on hardware chaotic noise generator. *J. Appl. Math. Bioinform.* **2015**, *5*, 15–35.

20. Giakoumis, A.; Volos, C.K.; Munoz-Pacheco, J.M.; del Carmen Gomez-Pavon, L.; Stouboulos, I.N.; Kyprianidis, I.M. Text encryption device based on a chaotic random bit generator. In Proceedings of the 2018 IEEE 9th Latin American Symposium on Circuits Systems (LASCAS), Puerto Vallarta, Mexico, 25–28 February 2018; pp. 1–5. [CrossRef]

21. Stoyanov, B.; Kordov, K. Image Encryption Using Chebyshev Map and Rotation Equation. *Entropy* **2015**, *17*, 2117–2139. [CrossRef]

22. Ge, R.; Yang, G.; Wu, J.; Chen, Y.; Coatrieux, G.; Luo, L. A Novel Chaos-Based Symmetric Image Encryption Using Bit-Pair Level Process. *IEEE Access* **2019**, *7*, 99470–99480, [CrossRef]

23. Nesa, N.; Ghosh, T.; Banerjee, I. Design of a chaos-based encryption scheme for sensor data using a novel logarithmic chaotic map. *J. Inf. Secur. Appl.* **2019**, *47*, 320 – 328. [CrossRef]

24. Ghosh, A.; Saha, A. A Numerical Method Based Encryption Algorithm With Steganography. *Comput. Sci. Inf. Technol.* **2013**, *3*, 149–157. [CrossRef]

25. Othman, H.; Hassoun, Y.; Owayjan, M. Entropy model for symmetric key cryptography algorithms based on numerical methods. In Proceedings of the 2015 International Conference on Applied Research in Computer Science and Engineering (ICAR), Beirut, Lebanon, 8–9 October 2015; pp. 1–2. [CrossRef]

26. Hassoun, Y.; Othman, H. Symmetric Key Cryptography Algorithms Based on Numerical Methods. In Proceedings of the NumAn 2014 Conference, Crete, Greece, 22–27 June 2014; pp. 151–155.

27. AL-Siaq, I.R. Public Key Cryptosystem Based on Numerical Methods. *Glob. J. Pure Appl. Math.* **2017**, *13*, 3105–3112.

28. Traub, J.F. *Iterative Methods for the Solution of Equations*; Prentice-Hall Series in Automatic Computation; Prentice-Hall: Englewood Cliffs, NJ, USA, 1982.

29. Jarrat, P. Some fourth order multipoint iterative methods for solving equations. *Math. Comput.* **1966**, *20*, 434–437. [CrossRef]

30. Nedzhibov, G.H.; Hasanov, V.I.; Petkov, M.G. On some families of multi-point iterative methods for solving nonlinear equations. *Numer. Algorithms* **2006**, *42*, 127–136. [CrossRef]

31. Skiadas, C.H.; Skiadas, C. *Chaotic Modelling and Simulation: Analysis of Chaotic Models, Attractors and Forms*; Chapman and Hall/CRC: London, UK, 2008.

32. *IEEE Standard for Floating-Point Arithmetic*; IEEE Std 754-2008; IEEE Computer Society: NY, USA, 2008; pp. 1–70. [CrossRef]

33. Alvarez, G.; Li, S. Some basic cryptographic requirements for chaos-based cryptosystems. *Int. J. Bifurc. Chaos* **2006**, *16*, 2129–2151. [CrossRef]

34. Rukhin, A.; Soto, J.; Nechvatal, J.; Smid, M.; Barker, E.; Leigh, S.; Levenson, M.; Vangel, M.; Banks, D.; Heckert, A.; et al. A Statistical Test Suite for Random and Pseudorandom Number Generators for Cryptographic Application. In *NIST Special Publication 800-22: Revision 1a, Lawrence E. Bassham III*; NIST: Gaithersburg, MD, USA, 2010

35. Walker, J. *ENT: A Pseudorandom Number Sequence Test Program*; Fourmilab: Switzerland, 2008.

36. Doty-Humphrey, C. PractRand: C++ Library of Pseudo-Random Number Generators And Statistical Tests for RNGs. 2014. Available online: http://pracrand.sourceforge.net/ (accessed on 17 December 2019).

37. Abubaker, S.; Wu, K. DAFA—A Lightweight DES Augmented Finite Automaton Cryptosystem. In *Security and Privacy in Communication Networks*; Keromytis, A.D., Di Pietro, R., Eds.; Springer: Berlin/Heidelberg, Germany, 2013; pp. 1–18.

38. Mushtaq, M.F.; Jamel, S.; Disina, A.H.; Pindar, Z.A.; Shakir, N.S.A.; Deris, M.M. A Survey on the cryptographic encryption algorithms. *Int. J. Adv. Comput. Sci. Appl.* **2017**, *8*, 333–344.

39. Cauchy, A. *Exercises de Mathematiques*; IV Annee de Bure Freres: Paris, France, 1829.

40. Joyal, G.L.; Rahman, Q.I. On the Location of Zeros of Polynomials. *Can. Math. Bull.* **1967**, *10*, 53–63. [CrossRef]

Article

Image Encryption Algorithm Based on Tent Delay-Sine Cascade with Logistic Map

Guidong Zhang, Weikang Ding and Lian Li *

School of Information Science and Engineering, Lanzhou University, Lanzhou 730000, China; zhanggd@lzu.edu.cn (G.Z.); dingwk15@lzu.edu.cn (W.D.)
* Correspondence: lil@lzu.edu.cn

Received: 4 February 2020; Accepted: 19 February 2020; Published: 1 March 2020

Abstract: We propose a new chaotic map combined with delay and cascade, called tent delay-sine cascade with logistic map (TDSCL). Compared with the original one-dimensional simple map, the proposed map has increased initial value sensitivity and internal randomness and a larger chaotic parameter interval. The chaotic sequence generated by TDSCL has pseudo-randomness and is suitable for image encryption. Based on this chaotic map, we propose an image encryption algorithm with a symmetric structure, which can achieve confusion and diffusion at the same time. Simulation results show that after encryption using the proposed algorithm, the entropy of the cipher is extremely close to the ideal value of eight, and the correlation coefficients between the pixels are lower than 0.01, thus the algorithm can resist statistical attacks. Moreover, the number of pixel change rate (NPCR) and the unified average changing intensity (UACI) of the proposed algorithm are very close to the ideal value, which indicates that it can efficiently resist chosen-plain text attack.

Keywords: chaotic map; image encryption; simultaneous confusion and diffusion

---

1. Introduction

With the development of information technologies, data security has aroused wide public concern. As an important data format, images occupy a large proportion of network data. Their secure transmission plays a vital role in personal and military privacy. In recent years, many chaotic image encryption algorithms have been proposed [1–10] due to the excellent properties of chaotic maps, such as initial value sensitivity and intrinsic randomness.

Researchers have improved the single chaotic map or combined multiple chaotic maps to improve chaotic properties, producing larger secret key spaces and more random chaotic sequences. Pak et al. [3] proposed a structure to modify two same chaotic maps to produce better performance than a single map [11–14]. Li et al. [15] improved the logistic map using linear delay. Zhou et al. [6] proved that cascading chaotic maps can increase the Lyapunov exponent, and many chaotic maps can be generated with the cascade model. Hua et al. [4] combined the logistic map and sine map to generate a two-dimensional (2D) map. This paper proposes a new framework that combines the cascade model and delay. This framework rationally integrates three chaotic maps to overcome the performance flaws of one-dimensional (1D) chaotic maps [16]. The experimental results showed that the chaotic maps produced by this model have initial value sensitivity and a large parameter interval.

Based on the proposed chaotic map, we constructed a new image encryption algorithm. Generally, image encryption algorithms can be divided into two steps: confusion and diffusion. Confusion involves randomly changing the position of pixels. The two commonly used confusion algorithms are: performing row and column confusion on a image, and reshaping a two-dimensional image into a vector, and then performing position confusion on it [17–19]. The basic diffusion methods are based on an XOR operation or mod operation after addition [20,21].

In this paper, a new simultaneous confusion and diffusion algorithm is proposed, which is applied in the vertical and horizontal directions based on an XOR operation. After analysis, the algorithm can resist chosen-plain text attacks and statistical attacks.

The rest of the paper is organized as follows. In Section 2, the proposed chaotic map is introduced. Section 3 shows the details of the image encryption algorithm. The proposed algorithm is analyzed and compared with other works in Section 4. Section 5 concludes the work.

2. Chaotic Map

This section proposes a new chaotic map with delay and cascade using tent, sine, and logistic maps, which we have named tent delay-sine cascade with logistic map (TDSCL). Through the combination of these three kinds of maps, we verified that this new chaotic map has excellent chaotic complexity using the following analysis and comparison.

2.1. The Structure of Chaotic Maps

First, this section reviews three chaotic maps including tent map, sine map, and logistic map. Based on these three chaotic maps, the TDSCL map is generated. The tent map is defined mathematically as [22]:

$$x_{n+1} = T_\lambda(x_n) = \begin{cases} 2\lambda x_n & for\ x_n < 0.5 \\ 2\lambda(1-x_n) & for\ x_n \geq 0.5 \end{cases} \tag{1}$$

where λ is the control parameter with the range of $[0, 1]$.

The structure of the sine map is defined as [23]:

$$x_{n+1} = S_\alpha(x_n) = \alpha sin(\pi x_n) \tag{2}$$

where α is the control parameter with a range of $[0, 1]$, and the map is chaotic with $\alpha \in (0.87, 1)$. For all $n \geq 1$, x_n is bounded within $[0, 1]$. The diagrams of bifurcation are shown in Figure 1b.

The logistic map is a simple 1D chaotic map. As a discrete chaotic map, Figure 1c shows its bifurcation, with outputs in the range of $[0, 1]$ and an initial input value in $[0, 1]$. The structure of the logistic map is defined by [24]:

$$x_{n+1} = L_\mu(x_n) = 4\mu x_n(1-x_n) \tag{3}$$

where μ is the control parameter in the range of $[0, 1]$.

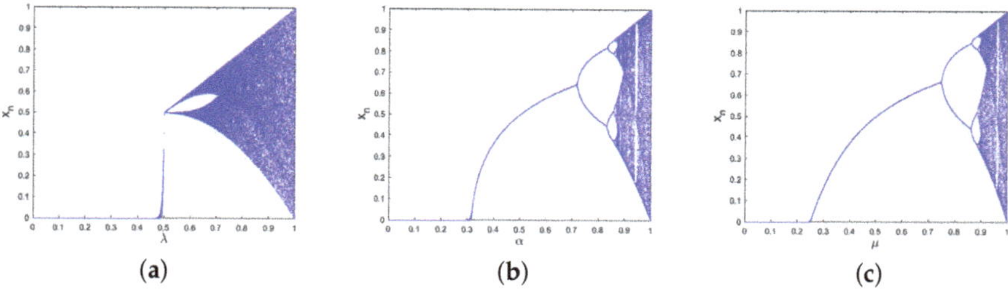

Figure 1. The bifurcation diagram for the (**a**) tent map, (**b**) sine map, and (**c**) logistic map.

The above three chaotic maps all have flaws, producing no chaotic behavior in some ranges of parameters. Specifically, these three maps only show chaotic characteristics at the rightmost part of the parameter variation range, and the chaotic interval may be discontinuous. To overcome these flaws, we designed a novel chaotic map structure, which is shown in Figure 2. As shown in Figure 2,

$T(x)$ represents the tent map with a delay item input, and the sine map is indicated by $S(x)$. Then, the outputs of $T(x)$ and $S(x)$ are added as the input of $f(x)$. The function $f(x)$ is taken as e^x in this paper, and cascaded with $L(x)$, thereby obtaining the output result of the chaotic map.

$$x_{n+1} = \mu f \circ F(x_n)(1 - f \circ F(x_n)) mod1 \tag{4}$$

$$F(x_n) = \begin{cases} 2x_{n-1} + sin(\pi x_n) & x_n < 0.5 \\ 2(1 - x_{n-1}) + sin(\pi x_n) & x_n \geq 0.5 \end{cases} \tag{5}$$

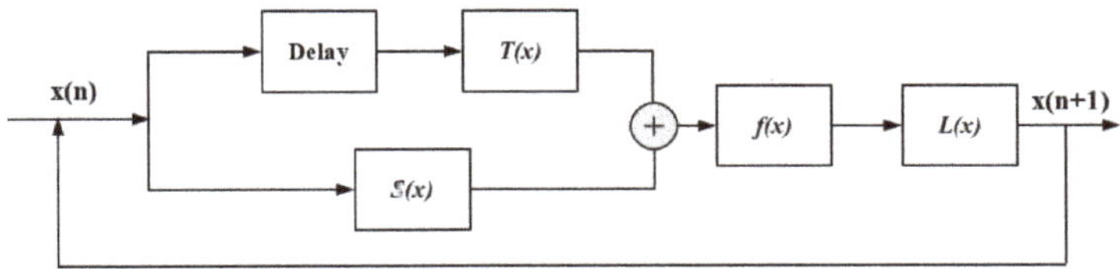

Figure 2. The structure of the tent delay-sine cascade with logistic map (TDSCL).

Here, the control parameters for the tent map and the sine map are set to 1, and the parameter μ for the logistic map is used as the control parameter for this new map. Equations (4) and (5) show the mathematical formulae. The circle symbol in Equation (4) represents the composition of two functions. Compared to the 1D delay and linearly coupled logistic chaotic map (DLCL) [15] and a two-dimensional logistic-modulated sine-coupling logistic chaotic map (LSMCL) [1], the structure of TDSCL produces better chaotic performance. In the following section, we use the trajectory, Lyapunov exponent, and permutation entropy (PE) to analyze the characteristics of chaotic maps.

2.2. Chaotic Performance of TDSCL

2.2.1. Chaotic Trajectory

For a chaotic system, the trajectory on the phase plane can show the randomness of outputs [25]. The larger the space occupied by the trajectory, the better the random outputs of the chaotic systems. Figure 3 shows the trajectories of TDSCL, DLCL, and LSMCL. The trajectory of TDSCL can fill the entire phase space compared to DLCL and LSMCL. This indicates that the sequence generated by the TDSCL chaotic map has better randomness and ergodicity.

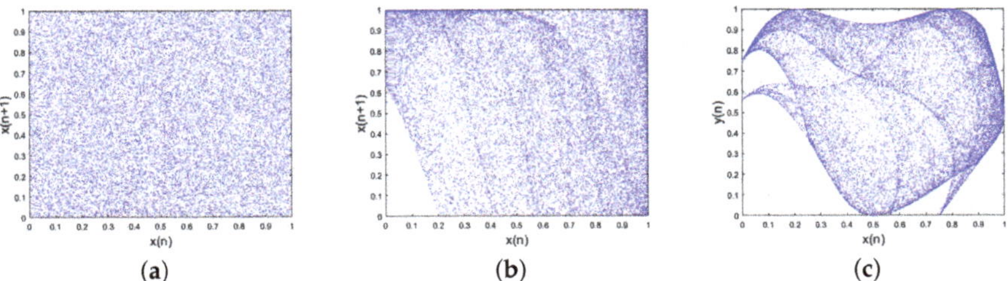

Figure 3. Trajectories for (**a**) TDSCL with $\mu = 1$, (**b**) delay and linearly coupled logistic chaotic map (DLCL) with $\mu = 1$, and (**c**) logistic-modulated sine-coupling logistic chaotic map (LSMCL) with $\theta = 0.75$.

2.2.2. Lyapunov Exponent

One of the most important features of a chaotic system is a strong sensitivity to initial values. The Lyapunov exponent (LE) [26] provides a quantitative description of the initial state sensitivity of a chaotic system. A maximum Lyapunov exponent of the chaotic map greater than 0 indicates that the system is in a chaotic state. For a two-dimensional chaotic system, if the system's two Lyapunov exponents are greater than 0, then the system is in a hyperchaotic state.

In Figure 4a–c, the Lyapunov exponents of TDSCL, DLCL, and LSMCL are calculated. From these diagrams, TDSCL displays hyperchaotic behavior when approximately $\mu \in (0.05, 1]$. When $\mu = 1$, the maximum Lyapunov exponent of TDSCL is close to 9.2. Therefore, compared with the other two maps, TDSCL not only has a larger chaotic state interval, but also a larger Lyapunov exponent in a large continuous interval. Compared with DLCL and LSMCL, TDSCL is more sensitive to small changes in the initial value of the system and has better unpredictability.

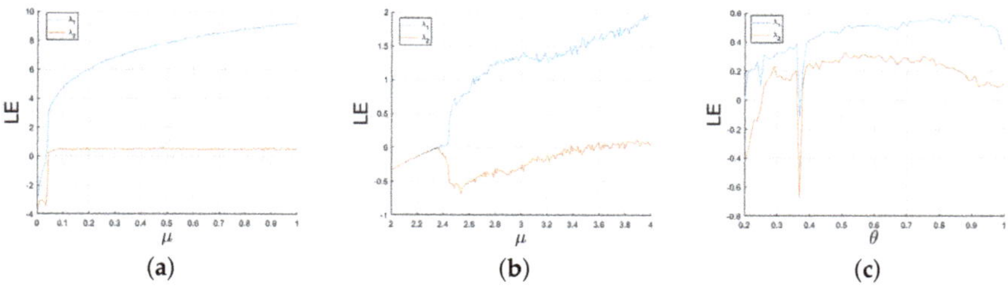

Figure 4. Lyapunov exponent: (**a**) TDSCL, (**b**) DLCL, and (**c**) LSCML.

2.2.3. Permutation Entropy

The permutation entropy can be used to measure the complexity of chaotic sequences [27]. For a given chaotic system, an entropy of the generated chaotic sequence close to 1 indicates that the chaotic system has unpredictability. As shown in Figure 5, the PE of DLCL is close to 1, only when μ in the interval of [0.7, 1], and the permutation entropy of LSMCL is always less than 0.8. The permutation entropy value of TDSCL is very close to 1 when $\mu \in [0.1, 1]$. This indicates that the chaotic sequences generated by TDSCL have more complex dynamic behaviour.

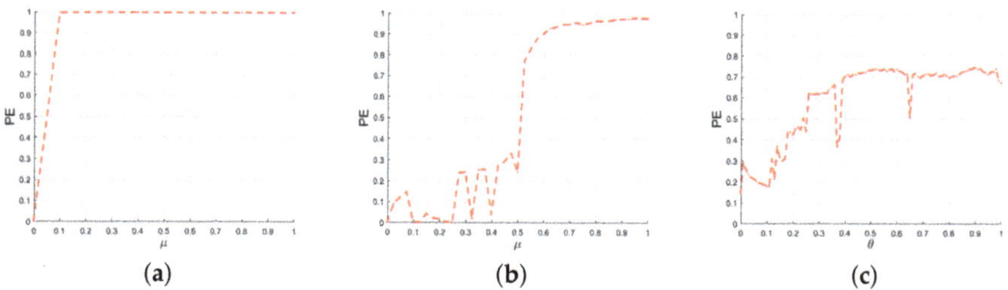

Figure 5. Permutation entropy for (**a**) TDSCL, (**b**) DLCL, and (**c**) LSCML.

3. Image Encryption Algorithm

In the proposed algorithm, the secret key consists of 16 parameters $\{x_1(1), x_1(2), n_1, u_1, x_2(1), x_2(2), n_2, u_2, x_3(1), x_3(2), n_3, u_3, x_4(1), x_4(2), n_4, \text{and } u_4\}$, where $x_i(1), x_i(2)$ are the first two values of the chaotic sequence, u_i is the control parameter of the chaotic map, and n_i is related to the length of the generated chaotic sequence. As shown in Figure 6, the first step in the encryption algorithm obtaining chaotic sequences is based on the key. Then, confusion and diffusion are performed simultaneously. The details of the algorithm are introduced below.

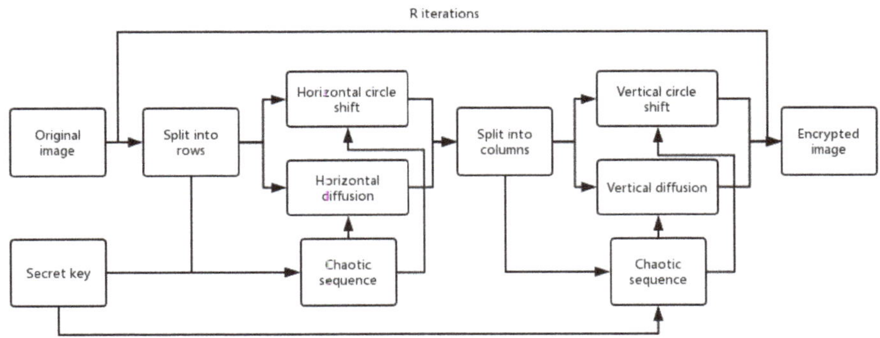

Figure 6. The image encryption architecture.

3.1. Simultaneous Horizontal Confusion and Diffusion

- Step 1. Generate diffusion matrix S_1.

Iterate Equation (4) $n_1 + M \times N$ times with initial value $x_1(1), x_1(2)$ and control parameter u_1. M, N are the height and width of the image I that is being processed, respectively. Then, the diffusion matrix S_1 is obtained by the generated chaotic sequence x_1 using Equation (6).

$$S_1(i,j) = \left\lfloor x_1(n_1 + (i-1) \times M + j) \times 10^6 \right\rfloor mod256. \tag{6}$$

where $i = 1, 2, ..., M$ and $j = 1, 2, ..., N$. S_1 is the matrix of M by N, each value of which is derived from the chaotic sequence x_1. With the given parameter u_1, the generated chaotic sequence x_1 has considerable randomness.

- Step 2. Set $i = 1$.

- Step 3. Obtain begin index b_1^i and circle shift the first row of the image $I(1,:)$ right by t_1^i pixels

Obtain $x_2(1), x_2(2), n_2, u_2$ from the secret key and and calculate the initial value as well as iteration time of chaotic map by adjusting them with the pixel value of image I according to Equation (7).

$$\begin{cases} x_2^i(1) = (x_2(1) + I(r_i, 1)/255)mod1, \\ x_2^i(2) = x_2(2), \\ n_2^i = n_2 + I(r_i, N) \end{cases} \tag{7}$$

where:

$$r_i = \begin{cases} M, & i = 1 \\ i - 1, & else. \end{cases} \tag{8}$$

Then, using initial value $x_2^i(1), x_2^i(2)$ and parameter u_2, iterate Equation (4) $n_2^i + 2$ times. Obtain b_1^i and t_1^i according to Equation (9).

$$\begin{cases} b_1^i = \left\lfloor x_2^i(n_2^i + 1) \times 10^6 \right\rfloor modN \\ t_1^i = \left\lfloor x_2^i(n_2^i + 2) \times 10^6 \right\rfloor modN. \end{cases} \tag{9}$$

- Step 4. Horizontal diffusion.

The horizontal diffusion operation is performed based on the XOR operation. The operation process is as follows:

$$
\begin{aligned}
&for\ j = 1:N \\
&\quad if\ j = 1 \\
&\qquad I(i,c_j) = I(i,c_j) \oplus S_1(i,j) \\
&\quad else \\
&\qquad I(i,c_j) = I(i,c_{j-1}) \oplus I(i,c_j) \oplus S_1(i,j) \\
&\quad end\ if \\
&end\ for
\end{aligned}
$$

where

$$
c_j = \begin{cases} b_1^i, & j = 1 \\ N, & (b_1^i + j - 1) = N \\ (b_1^i + j - 1) mod N, & else. \end{cases} \tag{10}
$$

- Step 5. Circle shift $I(i,:)$ horizontally by t_1^i pixels.

- Step 6. Let $i = i + 1$ and repeat steps 3 to 5 until all rows have been processed.

3.2. Simultaneous Vertical Confusion and Diffusion

The simultaneous operation of vertical confusion and diffusion is similar to the process introduced in the subsection above.

- Step 1. Generate diffusion matrix S_2.

Iterate the formula in Equation (4) $n_3 + M \times N$ times with initial value $x_3(1), x_3(2)$ and control parameter u_3. Then, the diffusion matrix S_3 is obtained according to:

$$
S_2(k,l) = \left\lfloor x_3(n_3 + (k-1) \times M + l) \times 10^6 \right\rfloor mod256. \tag{11}
$$

where $k = 1,2,..,M$ and $l = 1,2,...,N$.

- Step 2. Set $l = 1$.

- Step 3. Generate index b_2^l and circle shift the first column of the image $I(:,1)$ by t_2^l pixels.

Obtain $x_4(1), x_4(2), n_4, u_4$ from the secret key and adjust them with the pixel value of image I according to Equation (12).

$$
\begin{cases} x_4^l(1) = (x_4(1) + I(1,p_l)/255) mod1, \\ x_4^l(2) = x_4(2), \\ n_4^l = n_4 + I(M,p_l) \end{cases} \tag{12}
$$

where:

$$
p_l = \begin{cases} N, & l = 1 \\ l - 1, & else. \end{cases} \tag{13}
$$

Then, using the initial value $x_4^l(1), x_4^l(2)$ and parameter u_4, iterate Equation (4) $n_4^l + 2$ times. Obtain b_2^l and t_2^l according to Equation (14).

$$\begin{cases} b_2^l = \left\lfloor x_4^l(n_4^l + 1) \times 10^6 \right\rfloor mod\, N \\ t_2^l = \left\lfloor x_4^l(n_4^l + 2) \times 10^6 \right\rfloor mod\, N. \end{cases} \tag{14}$$

- Step 4. Vertical diffusion.

The vertical diffusion process is as follows:

$$\begin{aligned} &for\, k = 1 : M \\ &\quad if\, k = 1 \\ &\qquad I(q_k, l) = I(q_k, l) \oplus S_2(k, j) \\ &\quad else \\ &\qquad I(q_k, l) = I(q_{k-1}, l) \oplus I(q_k, l) \oplus S_2(k, l) \\ &\quad end\, if \\ &end\, for \end{aligned}$$

where:

$$q_k = \begin{cases} b_2^l, & k = 1 \\ M, & (b_2^l + k - 1) = M \\ (b_2^l + k - 1) mod\, M, & else. \end{cases} \tag{15}$$

- Step 5. Circle shift $I(:, l)$ vertically by t_2^l pixels.

- Step 6. Let $l = l + 1$, and repeat steps 3 to 5 until all columns have been processed.

4. Experiment Results and Analysis

4.1. Simulation Results

To verify the feasibility of the encryption algorithm proposed in this paper, some pictures were used for testing. Figure 7 shows the original images in the first column, the encrypted images in the second column, and the decrypted images in the last column.

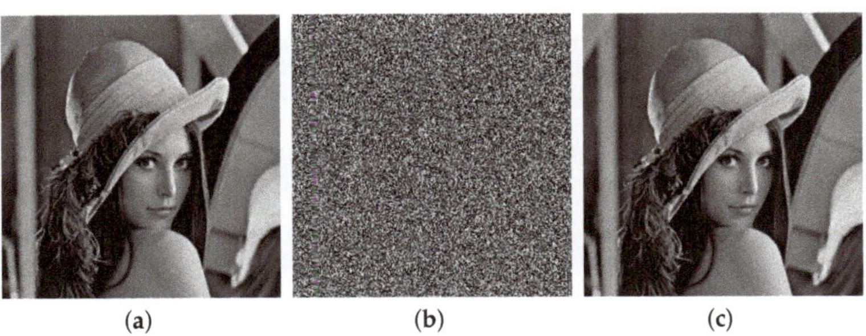

(a)   (b)   (c)

Figure 7. *Cont.*

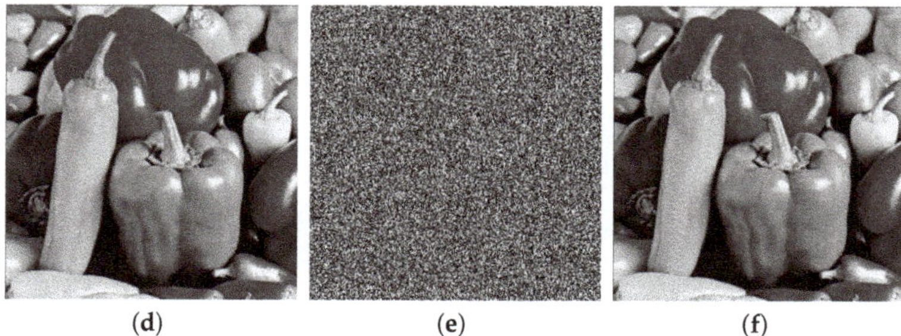

(d)   **(e)**   **(f)**

Figure 7. Simulation results of the proposed image encryption algorithm: (**a,d**) original images, (**b,e**) encrypted images, and (**c,f**) decrypted images.

4.2. Secret Key Space

In the proposed algorithm, the secret key contains 16 parameters. The parameters of the chaotic map are double precise, and the parameters related to the number of iterations are in the range of 0 to 1000. Thus, the secret key space can reach $0.81 \times 10^{192} > 2^{637}$, which is large enough to resist statistical attacks.

4.3. Statistical Analysis

4.3.1. Correlation Coefficient Analysis

In plain images, the correlation between adjacent pixels is fairly strong, and the correlation between adjacent pixels can be used by the attacker to obtain some useful information. Therefore, after image encryption, the correlation between adjacent pixels of the encrypted image is closer to 0, indicating that the pixel distribution is random. We selected 4000 pairs of adjacent pixels in plain images and encrypted images, and then calculated the correlation coefficient of two horizontal, vertical and diagonal adjacent pixels using Equation (16):

$$C_{xy} = \frac{E\{[x - E(x)][y - E(y)]\}}{\sqrt{D(x)}\sqrt{D(y)}} \tag{16}$$

where $E(x)$ and $D(x)$ represent the expectation and variance of variable x, respectively. Table 1 shows the experimental results of the tested images by performing the encryption in two rounds. The correlation coefficient of three directions is close to 0 after the encryption.

Figure 8 shows the correlation of the Lena image and its cipher image. The adjacent pixel pairs of the plain image in all directions are densely on the line of $y = x$, and the adjacent pixel pairs of the cipher image in all directions are evenly distributed in the rectangular area.

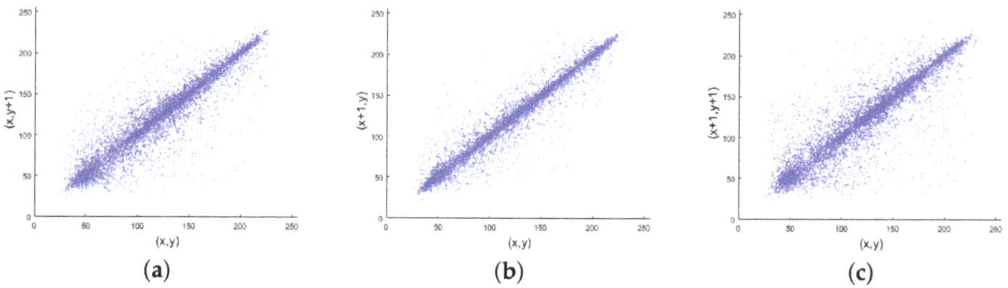

(a)   **(b)**   **(c)**

Figure 8. *Cont.*

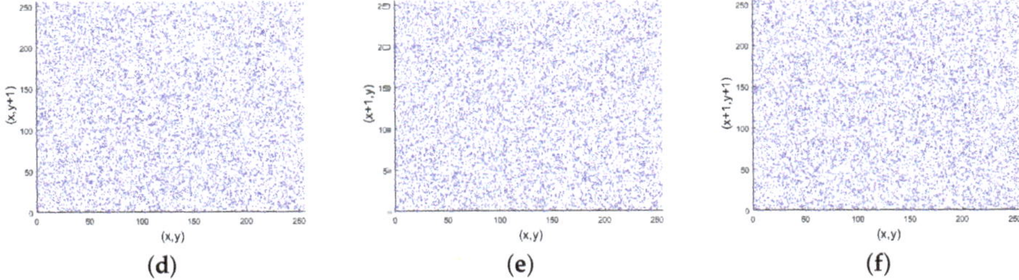

Figure 8. Adjacent pixels correlation analysis: the correlation between two horizontal, vertical, and diagonal pixels in (**a–c**) a plain image and (**d–f**) an encrypted image.

Table 1. The correlation coefficient in three directions for tested images.

Color Image		Horizontal	Vertical	Diagonal
4.2.01.tiff	original	0.9723	0.9843	0.9602
	encrypted	0.0001	0.0013	0.0040
4.2.02.tiff	original	0.9347	0.9413	0.8860
	encrypted	−0.0032	−0.0044	0.0011
4.2.03.tiff	original	0.8736	0.8261	0.7843
	encrypted	0.0075	−0.0012	−0.0014
4.2.04.tiff	original	0.9456	0.9727	0.9213
	encrypted	−0.0040	0.0042	0.0063
4.2.05.tiff	original	0.9364	0.9302	0.8819
	encrypted	0.0007	0.0022	−0.0007
4.2.06.tiff	original	0.9581	0.9564	0.9282
	encrypted	0.0049	−0.0002	−0.0029
4.2.07.tiff	original	0.9634	0.9704	0.9363
	encrypted	−0.0043	−0.0004	−0.0008

4.3.2. Histogram Analysis

An image histogram can reflect the frequency distribution of pixel values in an image [15]. In this experiment, Figure 9 shows the histograms of the plain and cipher images. The histogram of the cipher image has a balanced pixels distribution. This indicates that it is difficult for the attackers to obtain valid statistical information from the encrypted image. As the pixel values of the encrypted image have no obvious regularity, the attacker cannot obtain the original image through brute force analysis of the cipher. Therefore, the encryption system proposed in this paper has the ability to resist statistical attacks.

4.4. Key Sensitivity Test

Key sensitivity can be tested by the number of pixel change rate (NPCR) and the unified average changing intensity (UACI) [28]. In this test, we calculated the NPCR and UACI of two encrypted images based on changing a small value, set to 10^{-15} for keys. The mathematical formulas for calculating NPCR and UACI are defined as [29]:

$$\begin{cases} NPCR = \sum_{j=1}^{M} \sum_{j=1}^{N} \frac{D(i,j)}{M \times N} \times 100\%, \\ UACI = \sum_{j=1}^{M} \sum_{j=1}^{N} \frac{|C(i,j) - C'(i,j)|}{255 \times M \times N} \times 100\%, \end{cases} \tag{17}$$

$$D(i,j) = \begin{cases} 0, & if \ C(i,j) = C'(i,j) \\ 1, & if \ otherwise \end{cases} \tag{18}$$

where $C(i,j)$ and $C'(i,j)$ are the cipher image generated by the original key and the changed key in the key sensitivity test, respectively. The ideal values of NPCR and UACI are 99.6094% and 33.4635% for an 8-bit grey scale image, respectively [1]. Table 2 lists the simulation results. The NPCR and UACI of our proposed algorithm are very close to the expected value. The analysis results showed that the algorithm can resist chosen-plain text attacks.

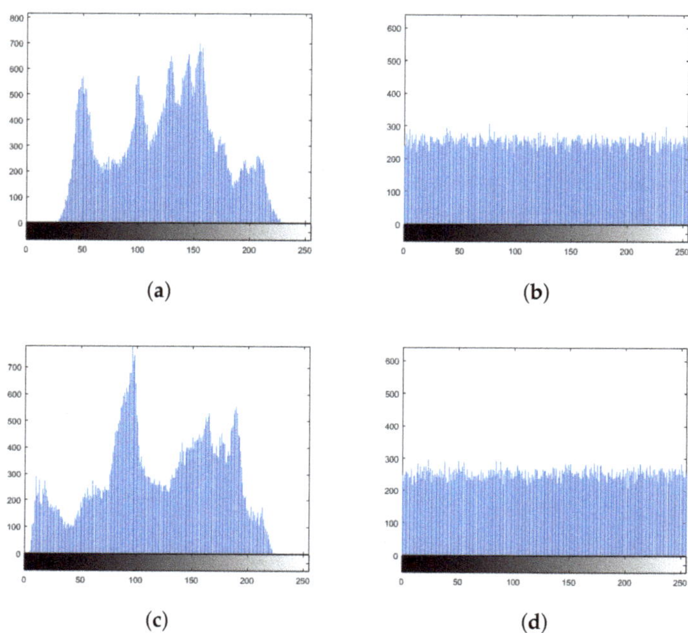

Figure 9. Histograms of (**a**,**b**) Lena and the encrypted image and (**c**,**d**) Pepper and the encrypted image.

Table 2. The number of pixel change rate (NPCR) and the unified average changing intensity (UACI) of different images for key sensitivity.

Image	NPCR	UACI
4.2.01.tiff	0.9959	0.3354
4.2.02.tiff	0.9960	0.3340
4.2.03.tiff	0.9958	0.3345
4.2.04.tiff	0.9958	0.3349
4.2.05.tiff	0.9962	0.3354
4.2.06.tiff	0.9962	0.3357
4.2.07.tiff	0.9960	0.3356

4.5. Resistance Against Chosen-plain Text Attack

To resist chosen-plain text attacks, an encryption system must have strong plaintext sensitivity. Similarly, plaintext sensitivity can use the same key to encrypt two distinct plaintext images, and calculate the NPCR and UACI values of the two images. We calculated the average NPCR and UACI for obtaining two cipher images 200 times, by performing two rounds of encryption in Table 3 .

Table 3. The NPCR and UACI of chosen-plain text analysis.

Image	NPCR	UACI
4.2.01.tiff	0.9961	0.3348
4.2.02.tiff	0.9961	0.3345
4.2.03.tiff	0.9961	0.3346
4.2.04.tiff	0.9961	0.3352
4.2.05.tiff	0.9961	0.3348
4.2.06.tiff	0.9961	0.3350
4.2.07.tiff	0.9961	0.3351

4.6. Information Entropy

Information entropy reflects the uncertainty of an image [30]. The larger the entropy, the greater the uncertainty. The entropy of an image was calculated according to Equation (19)

$$H = -\sum_{i=0}^{L} p(i)\log_2 p(i) \tag{19}$$

where L is the number of pixel grey levels, and $p(i)$ is the probability that the grey value i appears. From Table 4, the information entropy of the encrypted images approaches the ideal value of eight, which indicates that the encrypted images have considerable uncertainty.

Table 4. The information entropy of different images.

Image	Entropy
4.2.01.tiff	7.9969
4.2.02.tiff	7.9973
4.2.03.tiff	7.9973
4.2.04.tiff	7.9972
4.2.05.tiff	7.9971
4.2.06.tiff	7.9973
4.2.07.tiff	7.9971

4.7. Comparison with Other Methods

Table 5 compares the correlation coefficient, the ability against chosen-plain text attacks, and information entropy between the proposed algorithm and others' using 4.2.05.tiff. Our algorithm performs one and two rounds of encryption, and the results are listed in Table 5. The correlation coefficient of our algorithm is closer to 0, which indicates that the encrypted image has less visible information using our algorithm. For NPCR and UACI, our proposed algorithm is closer to the ideal values compared with the other three algorithms. This algorithm has a good ability to resist chosen-plain text attacks.

Table 5. Comparison of the proposed method and other methods.

Paper	Correlation			NPCR	UACI	Entropy
	Horizontal	Vertical	Diagonal			
Paper [31]	0.0062	0.0074	0.0009	0.9942	0.3352	7.9974
Paper [32]	0.0054	0.0089	0.0021	0.9965	0.3351	7.9970
Paper [2]	0.0028	0.0041	0.0010	0.9962	0.3363	7.9970
Proposed with one iteration	0.0001	−0.0007	−0.0025	0.9961	0.3344	7.9971
Proposed with two iteration	0.0007	−0.0022	−0.0007	0.9961	0.3346	7.9977

Moreover, the entropy of the encrypted image using the proposed algorithm with two iterations is larger than others. We show that the entropy of the encrypted image with two iterations is larger than that of one iteration.

4.8. Encryption Efficiency Analysis

In this paper, the simulation is performed on Inter(R) Core(TM) i7-6700K CPU @ 4.00 GHz with 16.0 GB in MATLAB R2019b. The average encryption time of a 256×256 image is 0.425 s, and the decrypted time is 0.452 s. To analyze the proposed encryption algorithm, the encryption throughput (ET) and number of cycles [33] are calculated by:

$$ET = \frac{Image_{size}(byte)}{Encryption_{time}(second)}, \tag{20}$$

$$Number\ of\ cycles\ per\ byte = \frac{CPU_{speed}(Hertz)}{ET(byte)}. \tag{21}$$

The ET of the proposed algorithm is 0.1471 MBps (million byte per second) and the algorithm needs 25,932.68 cpu cycles to finish one-byte operation.

5. Conclusions

In this paper, we constructed a new chaotic map, named TDSCL, which combines the delay tent map with the sine map, which is then cascaded with the logistic map. Compared with the DLCL and LSMCL methods, the simulation results indicated with the chaotic map that our proposed method has a larger Lyapunov exponent and permutation entropy, which demonstrates that it has a better initial value sensitivity and randomness. In addition, we proposed an image encryption algorithm with simultaneous confusion and diffusion in the vertical and horizontal directions. We analyzed the algorithm in terms of the key space, key sensitivity, ability against chosen-plain text attacks, and information entropy. The simulation results showed that this algorithm can resist statistical attacks and chosen-plain text attacks.

Author Contributions: Conceptualization, G.Z.; methodology, G.Z.; software, W.D.; validation, G.Z.; formal analysis, G.Z.; investigation, G.Z.; resources, G.Z.; writing—original draft preparation, G.Z.; writing—review and editing, W.D., L.L.; visualization, G.Z.; supervision, G.Z.; project administration, G.Z.; funding acquisition, G.Z. All authors have read and agreed to the published version of the manuscript.

Funding: The authors would like to thank the National Key R&D Program of China (Grant Nos. 2018YFB1003205 and 2017YFE0111900) and the 2019 Gansu Key Talent project for their support.

Conflicts of Interest: The authors declare no conflict of interest.

References

1. Zhu, H.; Zhao, Y.; Song, Y. 2D Logistic-Modulated-Sine-Coupling-Logistic Chaotic Map for Image Encryption. *IEEE Access* **2019**, *7*, 14081–14098. [CrossRef]
2. Wu, J.; Liao, X.; Yang, B. Image encryption using 2D Hénon-Sine map and DNA approach. *Signal Process.* **2018**, *153*, 11–23. [CrossRef]
3. Pak, C.; Huang, L. A new color image encryption using combination of the 1D chaotic map. *Signal Process.* **2017**, *138*, 129–137. [CrossRef]
4. Hua, Z.; Jin, F.; Xu, B.; Huang, H. 2D Logistic-Sine-coupling map for image encryption. *Signal Process.* **2018**, *149*, 148–161. [CrossRef]
5. Li, B.; Liao, X.; Jiang, Y. A novel image encryption scheme based on logistic map and dynatomic modular curve. *Multimed. Tools Appl.* **2018**, *77*, 8911–8938. [CrossRef]
6. Zhou, Y.; Hua, Z.; Pun, C.M.; Chen, C.P. Cascade chaotic system with applications. *IEEE Trans. Cybern.* **2014**, *45*, 2001–2012. [CrossRef] [PubMed]

7. Xie, J.; Yang, C.; Xie, Q.; Tian, L. An encryption algorithm based on transformed logistic map. In Proceedings of the 2009 International Conference on Networks Security, Wireless Communications and Trusted Computing, Wuhan, China, 25–26 April 2009; Volume 2, pp. 111–114.

8. Wu, X.; Zhu, B.; Hu, Y.; Ran, Y. A novel colour image encryption scheme using rectangular transform-enhanced chaotic tent maps. *IEEE Access* **2017**, *5*, 6429–6436. [CrossRef]

9. Cai, S.; Huang, L.; Chen, X.; Xiong, X. A Symmetric Plaintext-Related Color Image Encryption System Based on Bit Permutation. *Entropy* **2018**, *20*, 282. [CrossRef]

10. Zhu, C.; Wang, G.; Sun, K. Improved cryptanalysis and enhancements of an image encryption scheme using combined 1D chaotic maps. *Entropy* **2018**, *20*, 843. [CrossRef]

11. Pak, C.; An, K.; Jang, P.; Kim, J.; Kim, S. A novel bit-level color image encryption using improved 1D chaotic map. *Multimed. Tools Appl.* **2019**, *78*, 12027–12042. [CrossRef]

12. Wang, X.; Qin, X.; Liu, C. Color image encryption algorithm based on customized globally coupled map lattices. *Multimed. Tools Appl.* **2019**, *78*, 6191–6209. [CrossRef]

13. Wang, H.; Xiao, D.; Chen, X.; Huang, H. Cryptanalysis and enhancements of image encryption using combination of the 1D chaotic map. *Signal Process.* **2018**, *144*, 444–452. [CrossRef]

14. Wang, X.; Zhu, X.; Zhang, Y. An image encryption algorithm based on Josephus traversing and mixed chaotic map. *IEEE Access* **2018**, *6*, 23733–23746. [CrossRef]

15. Li, S.; Ding, W.; Yin, B.; Zhang, T.; Ma, Y. A novel delay linear coupling logistics map model for color image encryption. *Entropy* **2018**, *20*, 463. [CrossRef]

16. Nkandeu, Y.P.K.; Tiedeu, A. An image encryption algorithm based on substitution technique and chaos mixing. *Multimed. Tools Appl.* **2019**, *78*, 10013–10034. [CrossRef]

17. Hua, Z.; Xu, B.; Jin, F.; Huang, H. Image encryption using josephus problem and filtering diffusion. *IEEE Access* **2019**, *7*, 8660–8674. [CrossRef]

18. Huang, L.; Cai, S.; Xiong, X.; Xiao, M. On symmetric color image encryption system with permutation-diffusion simultaneous operation. *Opt. Lasers Eng.* **2019**, *115*, 7–20. [CrossRef]

19. Ur Rehman, A.; Liao, X. A novel robust dual diffusion/confusion encryption technique for color image based on Chaos, DNA and SHA-2. *Multimed. Tools Appl.* **2019**, *78*, 2105–2133. [CrossRef]

20. Li, S.; Yin, B.; Ding, W.; Zhang, T.; Ma, Y. A Nonlinearly Modulated Logistic Map with Delay for Image Encryption. *Electronics* **2018**, *7*, 326. [CrossRef]

21. Li, M.; Lu, D.; Xiang, Y.; Zhang, Y.; Ren, H. Cryptanalysis and improvement in a chaotic image cipher using two-round permutation and diffusion. *Nonlinear Dyn.* **2019**, *96*, 31–47. [CrossRef]

22. Shan, L.; Qiang, H.; Li, J.; Wang, Z.Q. Chaotic optimization algorithm based on Tent map. *Control Decis.* **2005**, *20*, 179–182.

23. Li, C.; Lin, D.; Lü, J.; Hao, F. Cryptanalyzing an image encryption algorithm based on autoblocking and electrocardiography. *IEEE MultiMedia* **2018**, *25*, 46–56. [CrossRef]

24. Pareek, N.K.; Patidar, V.; Sud, K.K. Image encryption using chaotic logistic map. *Image Vis. Comput.* **2006**, *24*, 926–934. [CrossRef]

25. Paar, V.; Buljan, H. Bursts in the chaotic trajectory lifetimes preceding controlled periodic motion. *Phys. Rev. E Stat. Phys. Plasmas Fluids Related Interdisciplinary Top.* **2000**, *62*, 4869–4872. [CrossRef]

26. Amigo, J.M.; Kocarev, L.; Szczepanski, J. Discrete Lyapunov Exponent and Resistance to Differential Cryptanalysis. *IEEE Trans. Circuits Syst. II Express Briefs* **2007**, *54*, 882–886. [CrossRef]

27. Bandt, C.; Pompe, B. Permutation entropy: A natural complexity measure for time series. *Phys. Rev. Lett.* **2002**, *88*, 174102. [CrossRef]

28. Ahmad, J.; Khan, M.A.; Hwang, S.O.; Khan, J.S. A compression sensing and noise-tolerant image encryption scheme based on chaotic maps and orthogonal matrices. *Neural Comput. Appl.* **2017**, *28*, 953–967. [CrossRef]

29. Wu, Y.; Noonan, J.P.; Agaian, S. NPCR and UACI randomness tests for image encryption. *Cyber J. Multidisciplinary J. Sci. Technol. J. Sel. Areas Telecommun. JSAT* **2011**, *1*, 31–38.

30. Wu, Y.; Zhou, Y.; Saveriades, G.; Agaian, S.; Noonan, J.P.; Natarajan, P. Local Shannon entropy measure with statistical tests for image randomness. *Inf. Sci.* **2013**, *222*, 323–342. [CrossRef]

31. Enayatifar, R.; Abdullah, A.H.; Isnin, I.F.; Altameem, A.; Lee, M. Image encryption using a synchronous permutation-diffusion technique. *Opt. Lasers Eng.* **2017**, *90*, 146–154. [CrossRef]

32. Niyat, A.Y.; Moattar, M.H.; Torshiz, M.N. Color image encryption based on hybrid hyper-chaotic system and cellular automata. *Opt. Lasers Eng.* **2017**, *90*, 225–237. [CrossRef]

33. Farajallah, M. Chaos-Based Crypto and Joint Crypto-Compression Systems for Images and Videos. Ph.D. Thesis, Universite de Nantes, Nantes, France, 2015.

Article

A Novel Method for Performance Improvement of Chaos-Based Substitution Boxes

Fırat Artuğer [1] and Fatih Özkaynak [2],*

[1] Department of Computer Engineering, Faculty of Engineering, Munzur University, Tunceli 62000, Turkey; firatartuger@munzur.edu.tr

[2] Department of Software Engineering, Faculty of Technology, Fırat University, Elazığ 23119, Turkey

* Correspondence: ozkaynak@firat.edu.tr; Tel.: +90-424-237-0000

Received: 22 February 2020; Accepted: 16 March 2020; Published: 5 April 2020

Abstract: Symmetry plays an important role in nonlinear system theory. In particular, it offers several methods by which to understand and model the chaotic behavior of mathematical, physical and biological systems. This study examines chaotic behavior in the field of information security. A novel method is proposed to improve the performance of chaos-based substitution box structures. Substitution box structures have a special role in block cipher algorithms, since they are the only nonlinear components in substitution permutation network architectures. However, the substitution box structures used in modern block encryption algorithms contain various vulnerabilities to side-channel attacks. Recent studies have shown that chaos-based designs can offer a variety of opportunities to prevent side-channel attacks. However, the problem of chaos-based designs is that substitution box performance criteria are worse than designs based on mathematical transformation. In this study, a postprocessing algorithm is proposed to improve the performance of chaos-based designs. The analysis results show that the proposed method can improve the performance criteria. The importance of these results is that chaos-based designs may offer opportunities for other practical applications in addition to the prevention of side-channel attacks.

Keywords: chaos; cryptography; substitution box; postprocessing

---

1. Introduction

Developments in Industry 4.0, the Internet of Things (IoT) and artificial intelligence have changed our lives significantly. Although these changes make our lives easier in many ways, guaranteeing the security of the huge quantities information called big data is a serious problem. Strong cryptographic protocols are needed to address this problem. However, cryptology is a complex discipline. It is not enough to demonstrate that only certain security requirements are met. New methods and countermeasures should be constantly researched as new attack techniques are developed [1,2]. Application attacks are an important cryptanalysis technique that threatens existing encryption protocols [3]. One of the attack techniques, called side-channel analysis, is based on the principle of obtaining the secret key of the algorithm with the help of measurements such as sound, heat, light and power consumption after the encryption protocol is implemented on hardware such as a computer, mobile phones or FPGA cards.

Recent studies have shown that chaos-based encryption protocols may be more resistant to side-channel attacks than encryption protocols based on mathematical techniques. In the analysis carried out in [4], first, a side-channel analysis of the AES block encryption algorithm was performed. In the second stage of the analysis, a side-channel analysis of the AES block encryption algorithm was performed using chaotic substitution box (s-box) structures instead of the s-box structure based on mathematical methods proposed by Nyberg [5,6]. The second design is more resistant to side-channel attacks than the standard AES algorithm. In other words, chaos-based s-box structures are more resistant to side-channel attacks than the AES s-box structure, which has the best-known s-box design criteria. However, when a literature review was undertaken, it was shown that even chaos-based designs with the best s-box performance criteria were worse than the Nyberg s-box structure. For example, for nonlinearity measurements, which play an important role in confusion and diffusion requirements, the best achievable value in chaos-based designs is 106.75, while in the Nyberg s-box structure, that value is 112, which is the upper bound value that can be reached [7].

It is therefore possible for chaos-based designs to be more resistant to side-channel attacks than mathematical designs. However, the poor performance criteria for these designs are an important problem. This study seeks to address this problem. Various studies have been published showing that the performance criteria can be improved with the help of optimization algorithms. However, in these approaches, there is another design problem, i.e., the additional processing cost of optimization algorithms. In this study, it has been shown that s-box performance criteria can be improved by applying various postprocessing techniques to chaos-based s-box designs. The practical applicability of the proposed method, its simple structure, and the speed of producing results have been evaluated as the advantages of the proposed method. This also raised a new research question regarding how s-box structures with better performance criteria can be obtained by using different postprocessing techniques in the future.

The rest of the study is organized as follows. In Section 2, the general design principle of chaos-based s-box structures and the basic milestones related to the literature are explained. In Section 3, the details of the proposed postprocessing technique are presented to improve the s-box performance criteria. In Section 4, the success of the proposed method is tested by providing various analysis results. The obtained results are interpreted and a road map for future studies is presented in Section 5.

2. Chaos-Based S-Box Structures

Chaos theory offers researchers various opportunities in many areas of science [8]. The rich dynamics that it contains have always made chaotic systems a popular research area. In addition to its use in modeling and control areas, its random behavior has led cryptography experts to focus on this field [9]. The basic idea behind this interest is that confusion and diffusion requirements can be met with the principle of sensitive dependence on initial conditions and control parameters. Confusion and diffusion requirements are two important properties of encryption protocols. These requirements were identified by Claude Shannon in 1945. "Confusion makes it difficult to find the key from the ciphertext and if a single bit in a key is changed, most or all the bits in the ciphertext will be affected. Diffusion means that if we change a single bit of the plaintext, then (statistically) half of the bits in the ciphertext should change". It has been suggested that these requirements can be met using chaotic systems, since chaotic outputs are extremely sensitive to changes in initial conditions and control parameters, and have a nonlinear characteristic. Researchers have used chaotic systems as an entropy source in cryptographic designs. They used the initial condition and control parameters of chaotic systems as the secret key of cryptographic protocols. It has been suggested that different entropy sources can be produced by using different initial conditions and control parameters, as they will produce different outputs with small changes that may occur in the initial conditions and control parameters. Many cryptographic protocols such as image encryption algorithms [10,11], key generators [12,13] and s-box designs [14] have been proposed using this design idea, as visualized in Figure 1.

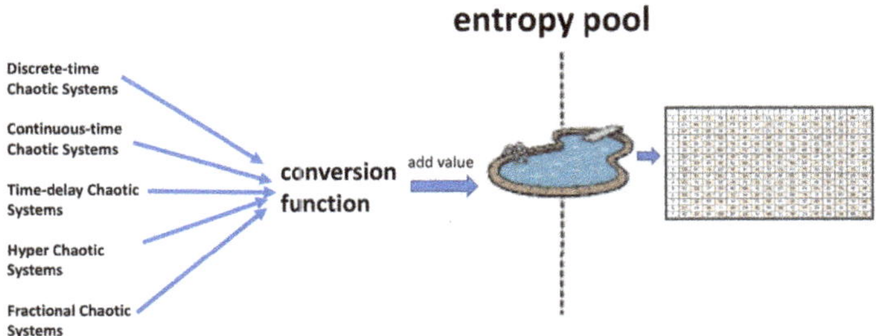

Figure 1. General design approach for chaos-based cryptographic protocol designs.

Although this design approach has been widely studied, the security analysis of some proposals has not been done according to certain criteria, which has caused various security problems. Chaos-based s-box designs stand out as a design class that is not affected by these problems, because the requirements for s-box performance analysis are almost standardized [15,16]. Bijective, nonlinearity, bit independence criterion (BIC), strict avalanche criterion (SAC) and input/output XOR distribution criteria are the standard measurements used in analysis processes of s-boxes. A nonlinearity criterion can be associated with the confusion criterion, which is one of the general characteristics of encryption algorithms; the ideal value for this criterion is 112, and the ideal value for the strict avalanche criterion is 0.5. This value indicates the difficulty of making statistical inferences. Values smaller or greater than 0.5 increase the success of statistical analysis. BIC measurement is related to nonlinearity and SAC measurements through the relationship between input and output bits. Input/output XOR distribution is related to differential cryptanalysis. To show its resistance against differential attacks, the maximum value that can be calculated. The expected value is 4; larger values indicate that differential attacks can be more successful [14–16].

In the simplest terms, s-box structures have the mathematical model given in Equation (1). In other words, it is a bijective function that converts values in a certain range to values in another range. The AES s-box structure is a nonlinear function that maps 256 values between 0 and 255 to 256 values between 0 and 255. Therefore, in the literature, attempts have been made to obtain different s-box structures by converting the chaotic system outputs to 256 different values. Many different s-box structures have been generated by changing the initial conditions and control parameters. Also, different chaotic system classes or different conversion algorithms have been used to improve the s-box performance criteria.

$$S: \begin{matrix} F_2^n \\ (x_1, \ldots, x_n) \end{matrix} \rightarrow \begin{matrix} F_2^m \\ (y_1, \ldots, y_m) \end{matrix} \tag{1}$$

When design studies are classified in terms of chaotic system types, there are two general classes: discrete and continuous-time chaotic systems. Discrete-time systems are among the preferred systems for researchers in the design process [17–21]. The main reason for this is that the systems can produce very fast results due to their simple mathematical models. The biggest advantage of continuous-time systems is that they have more complex mathematical models than discrete-time systems [22–26]. It is thought that this complexity will positively affect the quality of the entropy source. To use this advantage of continuous-time systems most effectively, special chaotic systems such as hyperchaotic [27,28], time-delay [29,30] and fractional-order systems [31,32] have also been used in the design process.

Another remarkable element of the general design architecture visualized in Figure 1 is the conversion function. The purpose of this function is to convert chaotic system outputs into an entropy source. In the literature, two conversion functions are most common. The first is the threshold value function. As stated in Equation (2), the chaotic system outputs are converted to 0 or 1 values by comparing them with a threshold value. Choosing the appropriate threshold value is a critical design problem. It has been shown that successful results can be obtained if 0.5 is selected as the threshold value in many sources [33,34]. The other conversion function is the mode function. It has been shown in various studies that the mode function has various advantages, since it is a one-way function which guarantees various statistical properties [35–37]. Due to these advantages, in the proposed method, the mode function has been used to transform the chaotic entropy source into s-box structures.

$$f_{threshold}(x) : \begin{cases} 0 & x \le 0.5 \\ 1 & x > 0.5 \end{cases} \tag{2}$$

3. Detail of Proposed Method

Block encryption algorithms are ineffective in the encryption of digital images. One of the most important reasons for this problem is the high correlation between the pixel values of an image [38]. Usually, images are represented by a matrix with a size of $m \times n$. The values of m and n indicate the values of the row and column, respectively. One of the proposed approaches to solving the correlation problem is to reposition the matrix cells using the zigzag transformation method, as shown in Figure 2. In this study, we propose the use of the zigzag reading approach as a postprocessing technique.

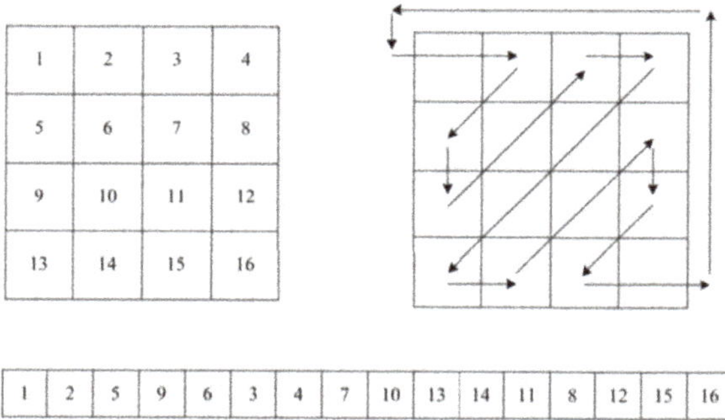

Figure 2. General structure of zigzag transformation approach.

Since AES-like s-box designs comprise a matrix with a size of 16×16, the zigzag transformation approach can be easily performed. The flowchart of the proposed method is given in Figure 3. The operation of the algorithm is given step by step below. Also, the pseudo code is expressed in Table 1 for the logistic map.

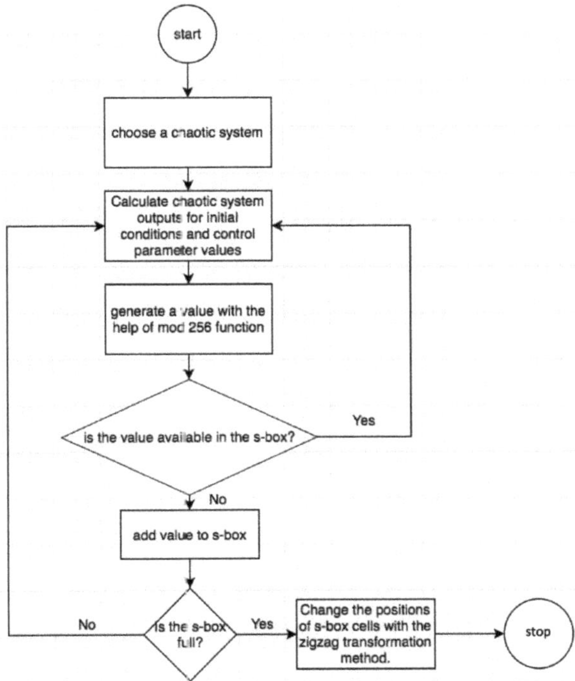

Figure 3. Flowchart of the proposed method.

Step 1. A discrete or continuous time chaotic system is chosen.

Step 2. The initial condition and control parameter values in which the chaotic system can exhibit rich random features are determined.

Step 3. State variable(s) of the chaotic system are calculated. Preferably, the first 1000 values can be ignored to eliminate the effects of transient response.

Step 4. The status variable value, which is the fractional value, is converted to a decimal value between 0–255 by applying mod 256.

Step 5. If the decimal value is not included in the s-box, it is added, otherwise a new state variable value is calculated, which continues until the table is full.

Step 6. The positions of s-box cells are shuffled using zigzag transformation.

Table 1. The pseudo code of chaotic s-box generation.

```
ChaoticSboxGenerate()
begin

sbox=[0:255]
for(k=0;k<256;I++)
sbox[k]=-1
end for

xOld= Random_Selection [0,1]

for(i=0;I<1000;I++)
xNew=4*xOld*(1-xOld)
xOld=xNex
end for

j=0;
while (j<sbox.lenght)
value=(xNex*100000000)%256
if(!contain(sbox,value))
sbox[j]=value
j++;
end if
xNew=r*xOld*(1-xOld)
xOld=xNex
end while

return ZigZagTransform(sbox)
end

contain(array, value)
begin
for(int i=0;i<array.length;i++)
if(array[i]==value)
return true
end if
end for
return false
end
```

---

4. Performance Analysis of Proposed Method

The study is based on a general s-box generator algorithm to examine the effect of the proposed postprocessing technique on the s-box performance. A flowchart of the s-box generator algorithm is given in Figure 3. The details of this algorithm and the program prepared for the Windows operating system can be accessed from [7,14]. Researchers can generate s-box structures using the original method, and verify their performance improvements for new s-box structures modified using the postprocessing technique through the program in [14].

The effect of the proposed method on the performance criteria was analyzed in this section. As explained, there are five basic criteria for s-box performance analysis. The bijective criterion is guaranteed by the proposed method. Therefore, this criterion is not included in the analysis tables. Two main categories can be used to classify chaotic systems. These categories are discrete and continuous-time chaotic systems. Discrete-time systems are first-order difference equations. Continuous-time chaotic systems are at least third-order differential equations [8]. An analysis of six different chaotic systems was carried out using three different chaotic systems for both chaotic system classes. Twenty-five different s-box structures were generated for each chaotic system class. Logistic map, sine map, and circle map are used as discrete-time chaotic systems. Performance comparisons for original and improvement s-box structures are given in Tables 2–4 respectively. Similarly, performance comparisons for the original and improved s-box structures generated for each of the continuous-time Lorenz, Labyrinth Rene Thomas system, and Chua systems are given in Tables 5–7, respectively. To show the success of the proposed method, care was taken to ensure that the average nonlinearity property of all the original s-box structures used in the analysis was less than 103. Performance improvement was observed in all the s-box structures given in the analysis tables.

Table 2. Performance comparisons for original and improved s-boxes based on a logistic map.

Name.	Performance Criteria for Original S-box					Performance Criteria for Improved S-box				
	Average Nonlinearity	SAC	BIC-Non.	BIC-SAC	XOR	Average Nonlinearity	SAC	BIC-Non.	BIC-SAC	XOR
L.map_1	100.75	0.4971	102.71	0.4992	12	105	0.5046	103.64	0.5009	10
L.map_2	102.5	0.5051	104.86	0.5012	12	103	0.5056	102.93	0.5004	12
L.map_3	102.75	0.502	103.21	0.5022	12	104.5	0.5027	102.93	0.4983	12
L.map_4	103.5	0.4985	104.29	0.4981	10	104.75	0.5049	103.71	0.4979	10
L.map_5	101.75	0.4998	103.21	0.4996	10	104.5	0.4983	103.64	0.5011	12
L.map_6	103.25	0.4976	103.64	0.4968	10	103.75	0.4973	103.29	0.5013	12
L.map_7	102	0.5051	103.07	0.5017	12	104.25	0.491	103.64	0.501	12
L.map_8	101.25	0.5056	103.29	0.503	12	103.75	0.4934	103.86	0.4962	12
L.map_9	103.75	0.5059	102.64	0.4997	10	104.5	0.4907	103.86	0.4978	10
L.map_10	103	0.5015	104.71	0.5023	12	104.5	0.498	103.5	0.5018	10
L.map_11	103.5	0.5012	103.36	0.4999	10	104	0.4998	104.14	0.5018	12
L.map_12	103.25	0.5049	103.64	0.4948	10	103.5	0.5	102.36	0.4978	12
L.map_13	102.25	0.5042	103.64	0.503	12	103.25	0.5022	104.07	0.4963	10
L.map_14	102	0.512	103.36	0.4969	12	103	0.4971	102.86	0.5007	12
L.map_15	102.75	0.5007	103.86	0.5001	10	103.25	0.5088	104	0.5005	12
L.map_16	101	0.5039	103.07	0.4976	10	103.5	0.5005	102.86	0.4974	12
L.map_17	102.5	0.5134	102.86	0.5	10	103.5	0.5056	104.29	0.4978	10
L.map_18	103.5	0.499	103.64	0.4941	12	103.75	0.5161	103.64	0.4957	10
L.map_19	102.75	0.5073	102.93	0.5037	12	103.75	0.5002	102.5	0.5018	10
L.map_20	103.25	0.491	103	0.4951	10	104	0.5042	103,5	0.4993	10
L.map_21	102.5	0.5078	102.86	0.4985	10	103.75	0.5154	103.64	0.5013	14
L.map_22	102.75	0.4966	103.71	0.4997	10	103.75	0.5066	103	0.4973	12
L.map_23	102.25	0.5012	103.5	0.5015	12	103.25	0.5044	103.36	0.5022	12
L.map_24	102.5	0.5068	104.36	0.4986	12	104.25	0.511	104.21	0.5006	10
L.map_25	103.25	0.5012	103.29	0.4992	12	104.24	0.5071	104.71	0.5027	10

Table 3. Performance comparisons for original and improved s-boxes based on a sine map.

Name.	Performance Criteria for Original S-box					Performance Criteria for Improved S-box				
	Average Nonlinearity	SAC	BIC-Non.	BIC-SAC	XOR	Average Nonlinearity	SAC	BIC-Non.	BIC-SAC	XOR
S.map_1	101.75	0.5122	104.07	0.4985	14	103.5	0.4924	103.64	0.4911	12
S.map_2	103	0.5046	102.86	0.4964	12	104.5	0.5027	103.64	0.4977	10
S.map_3	102.25	0.4988	102.5	0.498	12	104.25	0.4978	103.93	0.5017	12
S.map_4	103	0.5063	103.79	0.5029	12	104.5	0.5034	103.43	0.5013	12
S.map_5	103.25	0.4973	103.57	0.4978	12	104.5	0.51	103.93	0.5006	12
S.map_6	102.5	0.51	104	0.4967	12	103	0.511	103.64	0.4921	10
S.map_7	103.5	0.501	102.79	0.4991	12	103.75	0.5093	103.5	0.504	10
S.map_8	102.5	0.5002	104.07	0.5005	12	105	0.5083	103.79	0.5029	10
S.map_9	103.75	0.5002	103.57	0.495	12	104	0.5103	103	0.4988	12
S.map_10	101.5	0.4973	103.57	0.4981	10	103.25	0.4934	104	0.499	12
S.map_11	103.75	0.4934	104.29	0.4999	12	104.5	0.5083	104.5	0.4952	12
S.map_12	102	0.5054	103	0.4963	12	103	0.5103	103.43	0.4963	12
S.map_13	102.5	0.4993	104.29	0.4967	14	103.75	0.4971	103	0.501	12
S.map_14	101.5	0.5007	103.64	0.501	10	102.5	0.5056	104.14	0.5003	10
S.map_15	102	0.499	103.71	0.5003	12	102.75	0.498	103.71	0.4957	12
S.map_16	103	0.5002	104.07	0.5026	12	130.25	0.5166	103	0.4925	12
S.map_17	101.75	0.4973	102.86	0.4995	12	102	0.4961	103.43	0.4998	10
S.map_18	103	0.5022	103.07	0.4972	10	103.5	0.4988	103.14	0.499	10
S.map_19	102.75	0.4978	103.43	0.4998	10	104.75	0.4976	103.07	0.5005	12
S.map_20	103.25	0.4973	103.21	0.4959	12	104.75	0.501	103.86	0.5013	12
S.map_21	102.25	0.4934	103.21	0.4978	10	104.5	0.4998	104.14	0.5029	12
S.map_22	103.25	0.5017	102.86	0.4987	12	105	0.5029	105.07	0.5017	10
S.map_23	103	0.5012	103.57	0.5014	12	103.75	0.5107	104.36	0.4997	12
S.map_24	103	0.5078	103.71	0.4994	10	103.75	0.5059	104	0.5007	10
S.map_25	102.75	0.5044	103.14	0.5009	10	103.25	0.5005	103.29	0.5021	12

Table 4. Performance comparisons for original and improved s-boxes based on a circle map.

Name.	Performance Criteria for Original S-box					Performance Criteria for Improved S-box				
	Average Nonlinearity	SAC	BIC-Non.	BIC-SAC	XOR	Average Nonlinearity	SAC	BIC-Non.	BIC-SAC	XOR
C.map_1	102.25	0.5098	102.93	0.495	12	104.25	0.5015	102.79	0.5031	10
C.map_2	103.5	0.5027	103.29	0.501	10	105.5	0.5042	103.64	0.5055	10
C.map_3	102.75	0.5029	104.14	0.4902	12	105.75	0.5005	103.07	0.495	10
C.map_4	103.5	0.4915	103.43	0.4973	10	104.25	0.5005	102.86	0.4974	10
C.map_5	102	0.5115	103.57	0.4965	12	102.5	0.5007	104.07	0.501	12
C.map_6	103.25	0.5105	102.93	0.5022	12	104.5	0.4917	103.29	0.4988	10
C.map_7	102.25	0.498	102.93	0.5024	12	104.25	0.502	103.64	0.4983	12
C.map_8	100.75	0.4998	104.07	0.4961	12	105	0.4954	103.71	0.5026	12
C.map_9	102	0.498	103.07	0.5009	12	102.75	0.5066	103.86	0.4977	10
C.map_10	103.25	0.4978	103.36	0.5017	12	103.5	0.5037	103	0.4986	14
C.map_11	103	0.4946	103	0.5034	14	104.25	0.5007	103.5	0.5004	12
C.map_12	103.5	0.4944	103.29	0.5019	12	105.25	0.4976	103.43	0.4946	10
C.map_13	103.5	0.4932	102.71	0.502	10	104	0.5	103.36	0.4957	12
C.map_14	103.25	0.5061	103.21	0.4982	10	105.5	0.5039	103.29	0.4995	10
C.map_15	102	0.4951	104.63	0.5052	10	104.25	0.5029	103.93	0.5015	10
C.map_16	102.25	0.4968	104	0.4957	10	104	0.5037	104.86	0.5044	10
C.map_17	102.75	0.4939	104.07	0.5012	10	104.5	0.4924	103.79	0.4971	12
C.map_18	102.75	0.5083	102	0.4979	10	103	0.5015	104.21	0.4983	10
C.map_19	103	0.51	103	0.5017	10	104.25	0.4978	103.71	0.4976	10
C.map_20	101.5	0.5078	103	0.5011	12	103.25	0.5034	102.5	0.5025	14
C.map_21	101.75	0.501	102.43	0.4977	10	102.25	0.4993	104	0.4992	10
C.map_22	102	0.5027	103	0.4925	10	102.25	0.5112	103.71	0.5014	10
C.map_23	103	0.4976	103.14	0.5002	10	103.75	0.4937	102.57	0.4995	14
C.map_24	102.75	0.4917	104.57	0.4983	10	103.75	0.4956	104.79	0.5004	12
C.map_25	101.75	0.5171	102.86	0.5014	12	104.75	0.5073	102.79	0.4966	10

Table 5. Performance comparisons for original and improved s-boxes based on a Lorenz system.

Name.	Performance Criteria for Original S-box					Performance Criteria for Improved S-box				
	Average Nonlinearity	SAC	BIC-Non.	BIC-SAC	XOR	Average Nonlinearity	SAC	BIC-Non.	BIC-SAC	XOR
Lorenz_1	101.5	0.4902	103.64	0.4988	10	103.75	0.4973	103.21	0.5041	12
Lorenz _2	103.25	0.5044	103.29	0.5063	12	105	0.5037	102.79	0.4989	12
Lorenz _3	101.75	0.5063	103.36	0.4911	12	103	0.4998	103	0.4985	12
Lorenz _4	102.75	0.5042	103.5	0.5005	12	104.25	0.5027	103.21	0.5013	10
Lorenz _5	103.75	0.5095	104.86	0.4928	12	104.25	0.5024	103	0.5039	12
Lorenz _6	103.5	0.4944	103.79	0.5015	12	105.5	0.4937	104	0.4991	10
Lorenz _7	102.5	0.5027	103.21	0.4959	12	106.25	0.4929	104.07	0.499	14
Lorenz _8	102.25	0.4978	103.14	0.5002	14	103.25	0.4912	103.21	0.5029	12
Lorenz _9	102.25	0.4954	104.21	0.5	12	105.5	0.499	103.07	0.4957	12
Lorenz _10	103.25	0.5029	103.36	0.4959	12	103.75	0.5068	103.43	0.5028	12
Lorenz _11	101.5	0.5002	104.07	0.5018	10	103.75	0.4961	103.07	0.5036	12
Lorenz _12	101.25	0.5085	103.71	0.4981	10	103	0.5015	103.29	5033	12
Lorenz _13	101	0.5029	103.64	0.4989	10	105	0.5039	103.21	0.4993	12
Lorenz _14	102.75	0.4934	103.93	0.4938	12	104.5	0.4988	103.5	0.4992	12
Lorenz _15	103.25	0.4995	103	0.4996	10	103.5	0.4985	103.29	0.5045	10
Lorenz _16	103	0.4961	103.07	0.4987	14	103.25	0.5071	104.07	0.5008	12
Lorenz _17	103.75	0.5093	103.57	0.5015	12	104.25	0.4917	103.86	0.4958	14
Lorenz _18	102.75	0.5068	103.07	0.4994	10	103.25	0.4939	103.29	0.499	12
Lorenz _19	101.25	0.4998	102.79	0.5029	12	103	0.491	104.79	0.5029	10
Lorenz _20	103.25	0.5098	102.79	0.5015	10	104.25	0.499	103.57	0.4973	10
Lorenz _21	103.5	0.4973	103.21	0.4959	12	103.75	0.4971	102.71	0.4985	12
Lorenz _22	103.25	0.5046	103.21	0.4964	12	104.24	0.4988	103.29	0.4986	12
Lorenz _23	102.75	0.5017	103	0.504	12	103.75	0.5007	103.86	0.4991	12
Lorenz _24	103.5	0.5005	103.14	0.5007	10	103.75	0.4978	102.43	0.4964	12
Lorenz _25	102.75	0.5017	104.14	0.5003	12	105.5	0.4993	103.71	0.4993	12

Table 6. Performance comparisons for original and improved s-boxes based on the Labyrinth Rene Thomas system.

Name.	Performance Criteria for Original S-box					Performance Criteria for Improved S-box				
	Average Nonlinearity	SAC	BIC-Non.	BIC-SAC	XOR	Average Nonlinearity	SAC	BIC-Non.	BIC-SAC	XOR
Thomas_1	101.75	0.5046	103.86	0.5018	10	103.5	0.4966	103.86	0.4997	10
Thomas _2	103.25	0.4993	103.29	0.4995	10	104.5	0.4932	103.07	0.4971	12
Thomas _3	102.5	0.5039	104.43	0.4937	12	104	0.5002	103.5	0.5022	12
Thomas _4	103.5	0.5132	104.07	0.4962	12	104	0.5032	102.93	0.4957	12
Thomas _5	102.5	0.5037	103.64	0.4982	12	104	0.5022	103.86	0.5033	14
Thomas _6	103.25	0.51	103.29	0.499	12	104.25	0.5015	103.36	0.4952	10
Thomas _7	103.25	0.4944	103.36	0.4967	12	104.25	0.5034	104.14	0.5047	10
Thomas _8	103	0.5054	102.93	0.502	12	104.75	0.5137	103.57	0.502	12
Thomas _9	103.25	0.4893	103.43	0.4962	12	105.25	0.5088	103.64	0.5017	12
Thomas _10	102	0.4963	104.07	0.4939	12	105.5	0.5095	103.71	0.4992	10
Thomas _11	103	0.5071	102.79	0.4975	10	104	0.502	103.29	0.496	10
Thomas _12	102	0.4976	104.71	0.4963	12	103	0.5149	103.43	0.5031	12
Thomas _13	102.25	0.5037	103.14	0.4941	10	103.5	0.5083	103.5	0.4999	10
Thomas _14	102.75	0.5	103.36	0.5001	10	103	0.4971	103.14	0.5008	12
Thomas _15	103.25	0.5117	102.29	0.4978	10	104	0.5063	104.07	0.4951	12
Thomas _16	103	0.5017	102.64	0.502	10	104	0.5105	103.86	0.5037	12
Thomas _17	101	0.4961	103.07	0.501	12	104.25	0.4897	103.86	0.498	10
Thomas _18	102.5	0.5056	103.86	0.4994	10	103.5	0.5078	103.57	0.5047	10
Thomas _19	103.5	0.4995	103.14	0.5017	10	103.75	0.4924	103.21	0.4967	14
Thomas _20	103	0.5078	103.5	0.4971	10	104.5	0.5012	104.07	0.5006	12
Thomas _21	103.25	0.5095	104	0.4996	12	104	0.5049	103	0.4983	10
Thomas _22	103	0.5027	104.14	0.5009	10	103.75	0.4998	104.21	0.5017	10
Thomas _23	102.5	0.5088	104	0.5021	12	104	0.5056	104.57	0.4983	10
Thomas _24	102.5	0.5051	104.14	0.4969	12	104.5	0.4998	103.5	0.498	10
Thomas _25	103.25	0.4983	102.86	0.5059	12	103.5	0.4951	103.79	0.5001	10

Table 7. Performance comparisons for original and improved s-boxes based on a Chua circuit.

Name.	Performance Criteria for Original S-box					Performance Criteria for Improved S-box				
	Average Nonlinearity	SAC	BIC-Non.	BIC-SAC	XOR	Average Nonlinearity	SAC	BIC-Non.	BIC-SAC	XOR
Chua_1	103.75	0.4922	103.64	0.4988	14	104.25	0.5051	103.64	0.4999	10
Chua _2	103.75	0.4995	103.29	0.494	12	104.75	0.5078	104.21	0.4943	12
Chua _3	102.25	0.4939	103.79	0.506	12	105.5	0.5063	102.86	0.5001	12
Chua _4	103.25	0.5032	104.57	0.5054	10	105	0.51	103.21	0.5046	10
Chua _5	103.5	0.4954	103	0.5028	12	103.75	0.4956	103.5	0.4948	10
Chua _6	103.5	0.5034	103.29	0.502	12	104.25	0.5027	104	0.4973	10
Chua _7	103	0.5027	103.57	0.5024	12	103.75	0.5051	103.21	0.4995	12
Chua _8	102.5	0.5029	104.29	0.5015	10	104	0.5068	103.21	0.4994	10
Chua _9	102.75	0.5059	103.29	0.5011	10	105.25	0.5034	103.5	0.5009	12
Chua _10	103	0.4956	103.43	0.4958	12	104.5	0.5027	103.57	0.4986	12
Chua _11	102.75	0.5022	103.36	0.4971	12	104.25	0.4968	103.79	0.498	12
Chua _12	103.75	0.5039	103.43	0.4999	10	104.75	0.4976	104.07	0.5018	12
Chua _13	101.75	0.498	104.07	0.4981	12	104.75	0.4985	103.57	0.4993	12
Chua _14	102	0.5	103.64	0.4994	12	103.5	0.5024	104.29	0.502	12
Chua _15	102.5	0.5049	104	0.4994	10	103.5	0.5029	103.36	0.5037	12
Chua _16	103	0.4939	103.29	0.4993	14	104.25	0.5081	102.71	0.5006	10
Chua _17	103	0.5044	103.86	0.502	12	105.5	0.4998	103.57	0.4979	12
Chua _18	103	0.4922	103.64	0.5012	12	103.5	0.4983	102.29	0.4979	12
Chua _19	102.75	0.5034	104.57	0.4998	12	104.25	0.5	103.57	0.4931	10
Chua _20	103.25	0.5056	103.79	0.4992	12	103.5	0.4922	102.5	0.4976	12
Chua _21	102.25	0.5007	103.14	0.5065	12	103.5	0.4956	103.07	0.4956	14
Chua _22	103.25	0.4917	103.36	0.4985	10	105	0.5088	103.5	0.4937	12
Chua _23	101.25	0.4995	103.36	0.493	12	103.75	0.4915	103.29	0.4992	12
Chua _24	103	0.5103	102.79	0.4929	14	104.5	0.5007	103.64	0.5042	10
Chua _25	102.75	0.499	103.57	0.4985	12	103.25	0.5034	103	0.5013	12

The statistical properties of the chaotic data used in the s-box generation process are not included in this section. In [35], it is shown that the performance criteria of the s-box structures to be generated using the data which do not show chaotic behavior may be better than the s-box structures generated from chaotic data. In addition, in the code given in Table 1, the initial condition of the logistic map used as the chaotic system was chosen randomly. In other words, the proposed method provides performance improvement, regardless of the statistical properties of the entropy source. This is another strength of the proposed method.

5. Conclusions

Chaotic systems will provide various opportunities for cryptology sciences. Among these, a successful design approach is chaos-based s-box designs. However, the fact that chaos-based s-boxes are worse in terms of performance criteria than designs based on mathematical transformations is a serious problem. This problem is addressed in the study. The question of whether performance improvements of chaos-based designs can be achieved using various postprocessing methods was explored. In the study, the zigzag transformation method, which has a very simple structure, was used. It was observed that the proposed method provides performance improvements in chaos-based s-box structures that have performance characteristics that can be evaluated below average. Since the performance criteria of the chaos-based s-box structures are very close to each other, comparisons were made using the nonlinearity measurement, which is a criterion that can reflect the difference in the best way. In a literature review for the s-box, it was observed that the average value for the nonlinearity value is 103. Therefore, care was taken to ensure that the average nonlinearity value of all the s-box values used in the analysis was below 103. In line with these conditions, 150 different s-box structures were generated. The generated s-box structures were obtained from six different chaotic systems selected from two different chaotic system classes. The reason for using different chaotic systems was to show that the proposed method can be successful for all chaotic systems. All these s-box structures are explicitly presented for the examination of other researchers on a web page [39].

If a general evaluation is made, the advantages of the proposed method are listed below.

- It has been shown that s-box performance criteria can be improved using a postprocessing algorithm.
- The proposed postprocessing algorithm for performance improvements has a simple and elegant structure.
- Speed, computational complexity, and user friendliness are strong features of the proposed method.
- Considering these advantages, it can be said that the proposed postprocessing algorithm is a more convenient method for performance improvement compared to the optimization algorithms described in the literature to date.
- The proposed method can give successful results, regardless of the chaotic system type and class.
- Only the s-box generator should not be considered as the output of the study. It has been shown that new designs can be developed that can be used as a counter measurement to prevent side channel attacks.

Despite these advantages, the proposed postprocessing idea should be based on a more robust foundation in future studies. Some possible avenues for future studies are listed below.

- Many different postprocessing algorithms can be developed to achieve performance improvements. An example is the displacement of s-box rows or columns.
- In this study, postprocessing was applied to only one s-box generator. The success of the proposed method on different s-box generators should be evaluated.
- The postprocessing technique gives successful results for the nonlinearity criteria of 103 and below. However, the question of how performance improvements can be achieved for designs with better nonlinearity measurements should be investigated.
- The fact that the performance improvement is independent of the chaotic system type and class reveals that the proposed method can produce successful outputs from different entropy sources. Performance improvements will be investigated for s-box structures that will be designed in the future using different entropy sources.
- The practical applicability of chaos-based s-box structures in the field of information security should be investigated.
- Applications of the obtained outputs in different fields can be investigated, such as W-MSR-type resilient algorithms, to cope with attacks in complex networks [40,41].

Author Contributions: F.A. and F.Ö. Wrote and edited the manuscript. All authors have read and agreed to the published version of the manuscript.

Acknowledgments: The authors gratefully thank to the Referee for the constructive comments and recommendations which definitely help to improve the readability and quality of the study.

Conflicts of Interest: The authors declare no conflict of interest.

References

1. Li, C.; Zhang, Y.; Yong, E. When an attacker meets a cipher-image in 2018: A year in review. *J. Inf. Sec. Appl.* **2019**, *48*, 1–9. [CrossRef]
2. Özkaynak, F. Brief Review on Application of Nonlinear Dynamics in Image Encryption. *Nonlinear Dyn.* **2018**, *92*, 305–313. [CrossRef]
3. Cho, J.; Kim, T.; Kim, S.; Im, M.; Kim, T.; Shin, Y. Real-Time Detection for Cache Side Channel Attack using Performance Counter Monitor. *Appl. Sci.* **2020**, *10*, 984. [CrossRef]
4. Açıkkapı, M.S.; Özkaynak, F.; Özer, A.B. Side-channel Analysis of Chaos-based Substitution Box Structures. *IEEE Access* **2019**, 79030–79043. [CrossRef]
5. Nyberg, K. Differentially uniform mappings for cryptography. In *Workshop on the Theory and Application of of Cryptographic Techniques*; Springer: Berlin/Heidelberg, Germany, 1994; Volume 765, pp. 55–64.
6. Daemen, J.; Rijmen, V. AES proposal: Rijndael. In Proceedings of the 1st Advanced Encryption Conference, Ventura, CA, USA, 20–22 August 1998; pp. 1–45.

7. Özkaynak, F. Construction of Robust Substitution Boxes Based on Chaotic Systems. *Neural Comp. Appl.* **2019**, *31*, 3317–3326. [CrossRef]

8. Strogatz, S. *Nonlinear Dynamics and Chaos: With Applications to Physics, Biology, Chemistry, and Engineering (Studies in Nonlinearity)*; Westview Press: Boulder, CO, USA, 2001.

9. Kocarev, L.; Lian, S. *Chaos Based Cryptography Theory Algorithms and Applications*; Springer: Berlin/Heidelberg, Germany, 2011.

10. Zhu, C.; Wang, G.; Sun, K. Cryptanalysis and Improvement on an Image Encryption Algorithm Design Using a Novel Chaos Based S-Box. *Symmetry* **2018**, *10*, 399. [CrossRef]

11. Zhang, X.; Wang, X. Multiple-Image Encryption Algorithm Based on the 3D Permutation Model and Chaotic System. *Symmetry* **2018**, *10*, 660. [CrossRef]

12. Ding, L.; Liu, C.; Zhang, Y.; Ding, Q. A New Lightweight Stream Cipher Based on Chaos. *Symmetry* **2019**, *11*, 853. [CrossRef]

13. Demir, K.; Ergün, S. An Analysis of Deterministic Chaos as an Entropy Source for Random Number Generators. *Entropy* **2018**, *20*, 957. [CrossRef]

14. Özkaynak, F. An Analysis and Generation Toolbox for Chaotic Substitution Boxes: A Case Study Based on Chaotic Labyrinth Rene Thomas System. *Iran. J. Sci. Tech. Trans. Elect. Eng.* **2020**, *44*, 89–98. [CrossRef]

15. Cusick, T.; Stanica, P. *Cryptographic Boolean Functions and Applications*; Elsevier: Amsterdam, The Netherlands, 2009.

16. Wu, C.; Feng, D. *Boolean Functions and Their Applications in Cryptography*; Springer: Berlin/Heidelberg, Germany, 2016.

17. Ahmad, M. Random search based efficient chaotic substitution box design for image encryption. *Int. J. Rough Sets Data Anal.* **2018**, *5*, 131–147. [CrossRef]

18. Hussain, I.; Anees, A.; Al-Maadeed, T.A.; Mustafa, M.T. Construction of S-Box Based on Chaotic Map and Algebraic Structures. *Symmetry* **2019**, *11*, 351. [CrossRef]

19. Zahid, A.H.; Arshad, M.J. An Innovative Design of Substitution-Boxes Using Cubic Polynomial Mapping. *Symmetry* **2019**, *11*, 437. [CrossRef]

20. Zhu, S.; Wang, G.; Zhu, C. A Secure and Fast Image Encryption Scheme Based on Double Chaotic S-Boxes. *Entropy* **2019**, *21*, 790. [CrossRef]

21. Liu, H.; Zhao, B.; Huang, L. Quantum Image Encryption Scheme Using Arnold Transform and S-box Scrambling. *Entropy* **2019**, *21*, 343. [CrossRef]

22. Lai, Q.; Akgul, A.; Li, C.; Xu, G.; Çavuşoğlu, Ü. A New Chaotic System with Multiple Attractors: Dynamic Analysis, Circuit Realization and S-Box Design. *Entropy* **2018**, *20*, 12. [CrossRef]

23. Lu, Q.; Zhu, C.; Wang, G. A Novel S-Box Design Algorithm Based on a New Compound Chaotic System. *Entropy* **2019**, *21*, 1004. [CrossRef]

24. Liu, L.; Zhang, Y.; Wang, X. A Novel Method for Constructing the S-Box Based on Spatiotemporal Chaotic Dynamics. *Appl. Sci.* **2018**, *8*, 2650. [CrossRef]

25. Wang, X.; Akgul, A.; Cavusoglu, U.; Pham, V.-T.; Vo Hoang, D.; Nguyen, X.Q. A Chaotic System with Infinite Equilibria and Its S-Box Constructing Application. *Appl. Sci.* **2018**, *8*, 2132. [CrossRef]

26. Wang, X.; Çavuşoğlu, Ü.; Kacar, S.; Akgul, A.; Pham, V.-T.; Jafari, S.; Alsaadi, F.E.; Nguyen, X.Q. S-Box Based Image Encryption Application Using a Chaotic System without Equilibrium. *Appl. Sci.* **2019**, *9*, 781. [CrossRef]

27. Al Solami, E.; Ahmad, M.; Volos, C.; Doja, M.N.; Beg, M.M.S. A New Hyperchaotic System-Based Design for Efficient Bijective Substitution-Boxes. *Entropy* **2018**, *20*, 525. [CrossRef]

28. Islam, F.; Liu, G. Designing S-box based on 4D-4 wing hyperchaotic system. *3D Res.* **2017**, *8*, 9. [CrossRef]

29. Özkaynak, F.; Yavuz, S. Designing chaotic S-boxes based on time-delay chaotic system. *Nonlinear Dyn.* **2013**, *74*, 551–557. [CrossRef]

30. Khan, M.; Shah, T.; Batool, S.I. Construction of S-box based on chaotic Boolean functions and its application in image encryption. *Neural Comp. Appl.* **2016**, *27*, 677–685. [CrossRef]

31. Özkaynak, F.; Çelik, V.; Özer, A.B. A New S-Box Construction Method Based on the Fractional Order Chaotic Chen System. *Signal Image Video Proc.* **2017**, *11*, 659–664. [CrossRef]

32. Zahid, A.H.; Arshad, M.J.; Ahmad, M. A Novel Construction of Efficient Substitution-Boxes Using Cubic Fractional Transformation. *Entropy* **2019**, *21*, 245. [CrossRef]

33. Tanyıldızı, E.; Özkaynak, F. A New Chaotic S-Box Generation Method Using Parameter Optimization of One Dimensional Chaotic Maps. *IEEE Access* **2019**, 117829–117838. [CrossRef]
34. Anees, A.; Hussain, I. A Novel Method to Identify Initial Values of Chaotic Maps in Cybersecurity. *Symmetry* **2019**, *11*, 140. [CrossRef]
35. Özkaynak, F. On the Effect of Chaotic System in Performance Characteristics of Chaos Based S-box Designs. *Phys. A Stat. Mech. Appl.* **2020**, 124072. [CrossRef]
36. Stoyanova, B.; Ivanova, T. CHAOSA: Chaotic map based random number generator on Arduino platform. *AIP Conf. Proc.* **2019**, *2172*, 090001. [CrossRef]
37. Zhu, S.; Zhu, C.; Wang, W. A New Image Encryption Algorithm Based on Chaos and Secure Hash SHA-256. *Entropy* **2018**, *20*, 716. [CrossRef]
38. Yang, C.-H.; Chien, Y.-S. FPGA Implementation and Design of a Hybrid Chaos-AES Color Image Encryption Algorithm. *Symmetry* **2020**, *12*, 189. [CrossRef]
39. Available online: http://www.kriptarium.com/symmetry.html (accessed on 3 November 2019).
40. Shang, Y. Hybrid consensus for averager-copier-voter networks with non-rational agents. *Chaos Solitons Fractals* **2018**, *110*, 244–251. [CrossRef]
41. Shang, Y. Consensus of hybrid multi-agent systems with malicious nodes. *IEEE Trans. Circuits Syst. II Express Briefs* **2019**. [CrossRef]

Concept Paper

Dynamic Symmetry in Dozy-Chaos Mechanics

Vladimir V. Egorov [†]

Russian Academy of Sciences, FSRC "Crystallography and Photonics", Photochemistry Center, 7a Novatorov Street, Moscow 119421, Russia; egorovphotonics@gmail.com
† This article is dedicated to my sister Elena.

Received: 27 September 2020; Accepted: 10 November 2020; Published: 11 November 2020

Abstract: All kinds of dynamic symmetries in dozy-chaos (quantum-classical) mechanics (Egorov, V.V. Challenges 2020, 11, 16; Egorov, V.V. Heliyon Physics 2019, 5, e02579), which takes into account the chaotic dynamics of the joint electron-nuclear motion in the transient state of molecular "quantum" transitions, are discussed. The reason for the emergence of chaotic dynamics is associated with a certain new property of electrons, consisting in the provocation of chaos (dozy chaos) in a transient state, which appears in them as a result of the binding of atoms by electrons into molecules and condensed matter and which provides the possibility of reorganizing a very heavy nuclear subsystem as a result of transitions of light electrons. Formally, dozy chaos is introduced into the theory of molecular "quantum" transitions to eliminate the significant singularity in the transition rates, which is present in the theory when it goes beyond the Born–Oppenheimer adiabatic approximation and the Franck–Condon principle. Dozy chaos is introduced by replacing the infinitesimal imaginary addition in the energy denominator of the full Green's function of the electron-nuclear system with a finite value, which is called the dozy-chaos energy γ. The result for the transition-rate constant does not change when the sign of γ is changed. Other dynamic symmetries appearing in theory are associated with the emergence of dynamic organization in electronic-vibrational transitions, in particular with the emergence of an electron-nuclear-reorganization resonance (the so-called Egorov resonance) and its antisymmetric (chaotic) "twin", with direct and reverse transitions, as well as with different values of the electron–phonon interaction in the initial and final states of the system. All these dynamic symmetries are investigated using the simplest example of quantum-classical mechanics, namely, the example of quantum-classical mechanics of elementary electron-charge transfers in condensed media.

Keywords: quantum mechanics; molecular quantum transitions; singularity; dozy chaos; dozy-chaos mechanics; charge transfer; condensed matter; direct and reverse processes; optical band shapes; Egorov resonance

---

1. Introduction

A new physical theory—dozy-chaos mechanics or quantum-classical mechanics [1–4]—is designed to describe elementary physico-chemical processes, taking into account the chaotic dynamics of their transient state. The simplest version of quantum-classical mechanics is the quantum-classical mechanics of elementary electron transfers in condensed media [5,6]. This theory arose about twenty years ago [5,6] and proved its efficiency in explaining the optical spectra of polymethine dyes and their aggregates [3–11] and other physico-chemical phenomena [2,12–14]. The very first attempts to create it [15–18], which later turned out to be its particular cases [2,4,5,7–9], were undertaken more than thirty years ago. Quantum-classical mechanics can be considered as a kind of "generalization" of quantum mechanics, in which a new property of the electron is revealed [1,2,19]. This new property arises for an electron when it forms chemical bonds between atoms and consists in the appearance as a result of this ability to provoke chaos in the vibrational motion of nuclei in the process of molecular quantum transitions. The theoretical discovery of this unique ability of the electron made it possible to find out the reason for the reorganization of the structure of the nuclear subsystem of the molecule and molecular systems as a result of electronic transitions in them. In other words, the discovery of the ability of an electron to create chaos in the motion of nuclei in a transient molecular state made it possible to explain how a light electron manages to shift the equilibrium positions of vibrations of very heavy nuclei, which occurs as a result of the redistribution of the electron charge during molecular "quantum" transitions. This chaos is called dozy chaos [7,8,20], since it occurs only in a transient molecular state and is absent in the initial and final adiabatic molecular states. As a result of the appearance of dozy chaos, the energy spectrum of electrons and nuclei in the transient state becomes continuous, which indicates the classical nature of the motion of electrons and nuclei in this state, while the initial and final states are quantum states that differ sharply from each other in the electronic and nuclear structure. For this reason, dozy-chaos mechanics can also be called quantum-classical mechanics [1–3,19], and the electron itself, which creates chaos in the transient state, can be called a quantum-classical electron [19]. Consequently, the molecular "quantum" transition can be called the quantum-classical molecular transition.

Formally, dozy chaos arises, in theory, as a result of replacing the infinitesimal imaginary addition $i\gamma$ ($\gamma > 0$) in the energy denominator of the spectral representation of the full Green's function of an electron-nuclear system with a finite value [5–8,20]. This procedure of changing the quantity γ is forced and is associated with the elimination of an essential singularity that exists in the rates of molecular transitions if their dynamics are considered beyond the Born–Oppenheimer adiabatic approximation and the Franck–Condon principle [21–26]. The quantity γ can be considered as the width of the electron-nuclear energy levels in the transient molecular state, which ensures the exchange of energy and motion between electrons and nuclei in the transient state. However, as the comparison of theoretical results with experimental data on the optical spectra of polymethine dyes and their aggregates shows, the value of γ turns out to be much larger than the value of the vibrational quantum $\hbar\omega$ of nuclei: $\gamma >> \hbar\omega$ [1–9,19,20]. This circumstance points to the fact that the exchange of energy and motion between electrons and nuclei is so intense that it leads to chaos in their joint motion in a transient state. This chaos is the dozy chaos that we discussed above, and the quantity γ is called dozy-chaos energy [7,8,20].

Note that the well-known imaginary, damping gamma terms in the standard theory of radiation–matter interactions [27,28] are related to removing resonance singularities in perturbation theory. In quantum-classical mechanics, we are talking about the elimination of an essential singularity in the rates of electron-nuclear(-vibrational) transitions, which arises when taking into account the full-fledged electron-nuclear motion in the transient state, that is, when considering the electron-nuclear motion beyond the Born–Oppenheimer adiabatic approximation and the Franck–Condon principle. This motion is singular due to the incommensurability of the masses of light electrons and heavy nuclei and regardless of whether it is resonant or non-resonant. This is the fundamental novelty of our problem and our approach to its solution, where it becomes necessary to damp the singular dynamics in molecular systems, in comparison with the standard theory of radiation–matter interactions, where it becomes necessary to damp only resonances in atomic systems. Moreover, our imaginary gamma term already exists in the energy denominator of the total electron-nuclear(-vibrational) Green's function, by definition, as an infinitely small quantity. To eliminate the singularity in the rates of molecular transitions, which, as indicated above, exists within the framework of quantum mechanics, this gamma-term is simply assumed not to be infinitely small but finite, and thus becomes the dozy-chaos energy γ. For details of the discussion of this issue, see [2–4,7,8].

Dozy chaos is a mix of chaotic motions of the electronic charge, nuclear reorganization, and the electromagnetic field (dozy-chaos radiation) via which electrons and nuclei interact in the transient state. Apparently, the main mechanism for the occurrence of dozy chaos is associated with the interaction of an electron with optical phonons (see more details in Section 3 in [1]).

The emergence of chaos in dynamical systems is usually associated with the presence of any nonlinear interactions in them. In quantum-classical mechanics [2], the electron–phonon interaction in the original Hamiltonian is assumed to be linear (see term $\sum_\kappa V_\kappa(\mathbf{r})q_\kappa$ in Equation (1), Section 2) and has the same form as in the standard theory of many-phonon transitions [29], on the basis of which it was built. The condition $\gamma >> \hbar\omega$ arising in the complete Green's function of the system (see above) leads to its modification and, therefore, takes the whole theory beyond the scope of quantum mechanics. Therefore, it presents a challenge to solve the inverse problem, namely, using the modified Green's function or/and the general result for the rate constant of quantum-classical transitions (see Section 3), to find the form of the original non-Hermitian Hamiltonian [1] that corresponds to such a modified Green function or/and our overall result for the rate constant. In this non-Hermitian Hamiltonian obtained from the solution of the inverse problem, the electron–phonon interaction can turn out to be nonlinear [1]. Thus, the successful solution of the inverse problem will make it possible to clarify, in more detail, the nature of dozy chaos. On the other hand, it is also a challenge to register dozy chaos in an experiment, for example, using X-ray free-electron lasers [2,20].

The quantum-classical electron that provokes dozy chaos can be considered as some organizing physical principle in nature [19], and quantum-classical mechanics itself, and in this particular case, the quantum-classical mechanics of elementary electron transfers in condensed media, can be considered as the physical theory in which this organizing principle was discovered in science.

In any fundamental physical theory, as a rule, some kind of symmetry laws arises. Dozy-chaos mechanics, or in other words, quantum-classical mechanics, is no exception. The purpose of this concept review of the dozy-chaos mechanics of elementary charged particle (electron or proton) transfers in condensed media is to draw attention to a certain set of symmetries that arise in theory and are associated with various features and modes of charge-transfer dynamics. We call this set of symmetries dynamic symmetry in dozy-chaos mechanics.

2. On Dozy-Chaos Mechanics of Elementary Electron Transfers

The Hamiltonian for describing the elementary electron transfers in condensed media has the form [1–9]:

$$H = -\frac{\hbar^2}{2m}\Delta_\mathbf{r} + V_1(\mathbf{r}) + V_2(\mathbf{r}-\mathbf{L}) + \sum_\kappa V_\kappa(\mathbf{r})q_\kappa + \frac{1}{2}\sum_\kappa \hbar\omega_\kappa\left(q_\kappa^2 - \frac{\partial^2}{\partial q_\kappa^2}\right) \tag{1}$$

where 1 and 2 are the indices of the electron donor and acceptor, respectively; m is the effective mass of the electron; $\mathbf{r}$ is the electron's radius vector; q_κ are the real normal phonon coordinates; ω_κ are the eigenfrequencies of normal vibrations; κ is the phonon index; $\sum_\kappa V_\kappa(\mathbf{r})q_\kappa$ is the electron–phonon coupling term. In comparison with the Hamiltonian in the standard theory of many-phonon transitions (see [29]), in the theory of elementary electron transfers, the Hamiltonian is complicated merely by an extra electron potential well $V_2(\mathbf{r}-\mathbf{L})$ set apart from the original well $V_1(\mathbf{r})$ by the distance $L \equiv |\mathbf{L}|$ [5,6]. The nuclear reorganization energy E associated with the reorganization of the structure of the nuclear subsystem of the molecular system during electronic transitions in it (see Section 1), in this case, during elementary electron transfers in condensed matter, is defined as follows [2–4]

$$E = \frac{1}{2}\sum_\kappa \hbar\omega_\kappa\widetilde{q}_\kappa^2 \tag{2}$$

where $\widetilde{q}_\kappa$ are the shifts of the normal phonon coordinates q_κ, which correspond to the shifts in the equilibrium positions of the nuclei, caused by the presence of an electron in the medium on the donor 1 or on the acceptor 2.

The solution to the problem is sought by Green's function method:

$$G_H\left(\mathbf{r},\mathbf{r}';\,q,q';\,E_H\right) = \sum_s \frac{\Psi_s(\mathbf{r},q)\,\Psi_s{}^*\left(\mathbf{r}',q'\right)}{E_H - E_s - i\gamma} \tag{3}$$

where $\Psi_s(\mathbf{r}, q)$ are the eigenfunctions of the total Hamiltonian H of the system—in our case, the Hamiltonian (1); $(\mathbf{r}, q)$ is the set of all electronic and nuclear (phonon) coordinates; E_s are the eigenvalues of H and E_H is the exact value of the total energy of the system; $i\gamma$ ($\gamma > 0$) is the standard, infinitesimally small imaginary additive—the energy denominator vanishes when $\gamma = 0$; the aforementioned singularity in the rates of "quantum" transitions is eliminated by replacing γ in the energy denominator of Green's function (3) with a finite quantity [2–9]. The general formula for the rate constant of electron photo-transfers is obtained using the technique first described by Egorov [15,16], which generalizes the generating polynomial technique of Krivoglaz and Pekar [30,31] in the theory of many-phonon processes [29]; see the review article [2] for details.

3. General Formula for the Rate Constant of Electron Photo-Transfers

The general result for the rate constant (optical absorption) K is expressed in terms of Green's function of the elementary electron-charge transfers and two generating functions (see [2,5]):

$$K \propto \sum_{\omega_1=-\infty}^{\infty} \sum_{\omega_1'=-\infty}^{\infty} G^{\mathrm{E}}(\omega_1, L) G^{\mathrm{E}} * \left(\omega_1', L\right)$$

$$\times \frac{1}{(2\pi i)^3} \oint \frac{dx}{x^{\omega_1+1}} \oint \frac{dy}{y^{\omega_1'+1}} \oint \frac{dz}{z^{\omega_{12}+1}} Q(\overline{n}_1; x, y, z) S(\overline{n}_1; x, y, z) \tag{4}$$

where the contours encircle the points $x = 0$, $y = 0$, and $z = 0$, correspondingly. Green's function of the elementary electron-charge transfers $G^{\mathrm{E}}(\omega_1, L)$ and the generating functions $Q(\overline{n}_1; x, y, z)$ and $S(\overline{n}_1; x, y, z)$ can be found in [2,5], where $\overline{n}_1 \equiv \overline{n}_{\kappa 1, l 1}$ (Planck's distribution function) is as follows

$$\overline{n}_{\kappa 1, l 1} = \left[\exp\left(\hbar\omega_{\kappa, l}/k_{\mathrm{B}}T\right) - 1\right]^{-1} \tag{5}$$

The energy $\hbar\Omega$ of the absorbed photon and the heat energy $\hbar\omega_{12} > 0$ are related by the law of conservation of energy:

$$\hbar\Omega = J_1 - J_2 + \hbar\omega_{12} \tag{6}$$

J_1 is the electron binding energy on the donor 1 and J_2 is the electron-binding energy on the acceptor 2. The heat energy $\hbar\omega_{12} < 0$ corresponds to the inverse processes relative to optical absorption, i.e., to luminescence [3,29] (see Section 10). The wavelength λ, indicated on the x-axis in the figures below, corresponds to the frequency Ω in Equation (6) by the standard formula $\lambda = 2\pi c/\Omega n_{\mathrm{refr}}$ (c and n_{refr} are the speed of light in vacuum and the refractive index, respectively). The conservation law (Equation (6)) corresponds to the entire shape of the optical band as a whole: by varying the heat energy $\hbar\omega_{12}$, we vary the frequency of light Ω and determine one or another part of the absorption band [2–9,29].

4. The Analytical Result for Optical Absorption Band Shapes and Its Invariance with Respect to the Change in the Sign of Dozy-Chaos Energy γ

From the general result of dozy-chaos mechanics of elementary electron transfers, Equation (4), the expression for the light absorption factor K (the optical extinction coefficient ε [2–9,29] is proportional to K), has been obtained. The obtained expression for K in the framework of the Einstein model of nuclear vibrations in the framework of the Einstein model of nuclear vibrations ($\omega_\kappa = \text{constant} \equiv \omega$), although it is rather complex, is fully expressed in elementary functions and has the following form [2,5,6]:

$$K = K_0 \exp W \tag{7}$$

$$W = \tfrac{1}{2} \ln\left(\frac{\omega\tau \,\sinh \beta_T}{4\pi \,\cosh t}\right) - \frac{2}{\omega\tau}\left(\coth \beta_T - \frac{\cosh t}{\sinh \beta_T}\right)$$

$$+ (\beta_T - t)\frac{1}{\omega\tau\Theta} - \frac{\sinh \beta_T}{4\omega\tau\Theta^2\cosh t} \tag{8}$$

$$1 << \frac{1}{\omega\tau\Theta} \leq \frac{2\cosh t}{\omega\tau \sinh \beta_T} \tag{9}$$

where $\beta_T \equiv \hbar\omega/2k_B T$,

$$t = \frac{\omega\tau_e}{\theta}\left[\frac{AC+BD}{A^2+B^2} + \frac{2\Theta(\Theta-1)}{(\Theta-1)^2+(\Theta/\theta_0)^2} + \frac{\theta_0^2}{\theta_0^2+1}\right] \tag{10}$$

$$|\theta_0| >> \frac{E}{2J_1} \tag{11}$$

$$\theta \equiv \frac{\tau_e}{\tau} = \frac{L\,E}{\hbar\sqrt{2J_1/m}}, \quad \Theta \equiv \frac{\tau'}{\tau} = \frac{E}{\hbar\omega_{12}}, \quad \theta_0 \equiv \frac{\tau_0}{\tau} = \frac{E}{\gamma} \tag{12}$$

$$\tau_e = \frac{\tau}{\sqrt{2J_1/m}}, \quad \tau = \frac{\hbar}{E}, \quad \tau' = \frac{1}{\omega_{12}}, \quad \tau_0 = \frac{\hbar}{\gamma} \tag{13}$$

Here, we use the notation

$$A = \cos\left(\frac{\theta}{\theta_0}\right) + \Lambda + \left(\frac{1}{\theta_0}\right)^2 N \tag{14}$$

$$B = \sin\left(\frac{\theta}{\theta_0}\right) + \frac{1}{\theta_0}M \tag{15}$$

$$C = \theta\left[\cos\left(\frac{\theta}{\theta_0}\right) - \frac{1-\xi^2}{2\theta_0}\sin\left(\frac{\theta}{\theta_0}\right)\right] + M \tag{16}$$

$$D = \theta\left[\sin\left(\frac{\theta}{\theta_0}\right) + \frac{1-\xi^2}{2\theta_0}\cos\left(\frac{\theta}{\theta_0}\right)\right] - \frac{2}{\theta_0}N \tag{17}$$

$$\text{and } \xi \equiv \left(1 - \frac{E}{J_1}\right)^{1/2} \quad (J_1 > E \text{ by definition}) \tag{18}$$

and where we finally have

$$\Lambda = -(\Theta - 1)^2 E + \left[\frac{(\Theta-1)\theta}{\rho} + \Theta(\Theta-2)\right]E^{\frac{1-\rho}{1-\xi}} \tag{19}$$

$$M = 2\Theta(\Theta-1)E - \left[\frac{(2\Theta-1)\theta}{\rho} + 2\Theta(\Theta-1)\right]E^{\frac{1-\rho}{1-\xi}} \tag{20}$$

$$N = \Theta\left[\Theta E - \left(\frac{\theta}{\rho} + \Theta\right)E^{\frac{1-\rho}{1-\xi}}\right] \tag{21}$$

$$E \equiv \exp\left(\frac{2\theta}{1+\xi}\right), \quad \rho \equiv \sqrt{\xi^2 + \frac{1-\xi^2}{\Theta}} \tag{22}$$

The factor K_0 becomes

$$K_0 = K_0^e K_0^p \tag{23}$$

where

$$K_0^e = \frac{2\tau^3 J_1}{m}\frac{\left(A^2+B^2\right)\rho^3\Theta^4\xi}{\theta^2\left[(\Theta-1)^2+\left(\frac{\Theta}{\theta_0}\right)^2\right]^2\left[1+\left(\frac{1}{\theta_0}\right)^2\right]}\cdot\eta \tag{24}$$

$$\text{and } \eta \equiv \exp\left(-\frac{4\theta}{1-\xi^2}\right) \tag{25}$$

and

$$K_0^p = \frac{1}{\omega\tau}\left[1 + \frac{\sinh(\beta_T - 2t)}{\sinh\beta_T}\right]^2 + \frac{\cosh(\beta_T - 2t)}{\sinh\beta_T} \tag{26}$$

Inequalities (9) and (11) are not any significant restrictions on the parameters of the system and associated with items of routine approximations made in the calculations (see [2]). The time scales τ_e, τ, and τ_0, given by Equation (13), control the dynamics of elementary electron-transfer processes. They are discussed in Section 9. The time scale τ' (see Equation (13)) together with the law of conservation of energy (6) and the other parameters of a donor–acceptor system control the dynamics of producing the shape of optical bands [2,4].

Let us consider further the issues related to the change in the sign of the dozy-chaos energy γ. On the one hand, in standard quantum mechanics, where the value of γ is infinitesimal and where this value is introduced formally in order to avoid zero in the energy denominator of the spectral representation of Green's function (see Equation (3)), the sign of γ can be either positive or negative. On the other hand, in quantum-classical mechanics, although the value of γ becomes a finite value, the choice of its sign turns out to be insignificant here too. It is easy to show, for example, that our result for the light absorption factor K, given by Equations (6)–(26), is an even function of γ. For this, it is sufficient to consider those equations that include the dimensionless quantity $\theta_0 = \frac{E}{\gamma}$ (see Equation (12)), in which the reorganization energy E is positive by definition (see Equation (2)). So, it is easy to see that the quantity $t = t(\Theta, \theta_0)$ (see Equation (10)) is an even function of θ_0: in the nontrivial term $AC + BD$, the cofactors A and C are even functions of θ_0, and the cofactors B and D are odd functions of θ_0. Further, the factor K_0^e (see Equation (24)) is obviously an even function of θ_0.

The invariance with respect to the change in the sign of the dozy-chaos energy γ is consistent with the physical case that both the virtual acts of transformation of electron movements and energies into nuclear reorganization movements and energies and the reverse acts occur in the transient dozy-chaos state [4,7–9]. For definiteness, we set $\gamma > 0$ here, there, and everywhere.

5. Potential Box with a Movable Wall. Optical Absorption Band Shapes as Dependent on the Dozy-Chaos Energy γ: From Symmetry to Asymmetry

The reason for the appearance of a singularity in the rates of molecular "quantum" transitions can be seen already from the example of a one-dimensional potential box with a movable wall [1]. The movable wall corresponds to the reorganization of the nuclear subsystem of the molecular system. As indicated above (Section 1), within the framework of quantum mechanics, due to the incommensurability of the masses of electrons and nuclei, the dynamics of nuclear reorganization is singular. Accordingly, if the movable wall of the potential box moves without friction, then this corresponds to an infinitely fast expansion of the potential box during the transition of an electron from the ground state to the first excited state, which leads to a singular "collapse" of their energy levels.

The singularity can be eliminated by assuming that the wall moves with friction [1]. In the exact theory, this assumption corresponds to the introduction of transient chaos into the dynamics of reorganization of the electron-nuclear motion, that is, the introduction of dozy chaos.

In Figure 1, optical absorption band shapes (for $k_B T > \hbar\omega_\kappa/2$), as dependent on the dozy-chaos energy γ, are computed from Equations (6)–(26). At high energies γ, the band shape is close to symmetric and is Gaussian-like (see Section 6). With a decrease in the value of γ, in the red region of the spectrum, a peak appears against the background of a Gaussian-like band, which, with decreasing γ, shifts more and more to the red region of the spectrum and becomes more and more pronounced. Thus, with a decrease in the value of the dozy-chaos energy γ, the band shape transforms from symmetric to asymmetric.

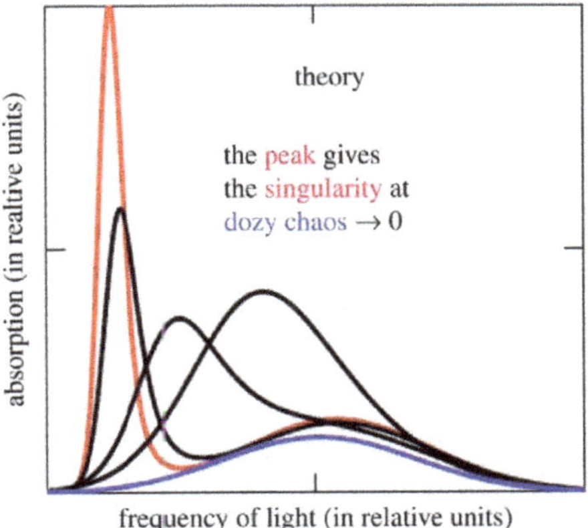

Figure 1. Singularity in the rate of molecular quantum transitions: the optical absorption band shape dependent on the dozy chaos available to a given quantum transition; the band shape with the strongly pronounced peak (J-band) corresponds to the least dozy chaos [9]. The dozy-chaos-dependent optical absorption band is displaced to the red spectral region and narrowed. The position, the intensity, and the width of the optical absorption band are determined by the ratio between the dozy-chaos energy γ and the reorganization energy E (see Section 4). The smaller the value of γ is, the higher the degree of organization of the molecular "quantum" transition and the higher the intensity and lower the width of the optical band. The position of the wing maximum is determined by the energy E, whereas the position of the peak is determined by the energy γ [9].

6. Passage to the Limit to the Standard Theory of Many-Phonon Transitions and the Symmetry of the Standard Result. The Reason for the Asymmetry of the Optical Absorption Band Shape in Dozy-Chaos Mechanics

The limit passage from expressions (6)–(26) for the optical absorption K to the standard result in the theory of many-phonon transitions [29] can be realized by letting the dozy-chaos energy γ tend to infinity ($\theta_0 = E/\gamma \to 0$ according to Equation (12)) in Equation (10) for t (see Figure 3 in [2]) and to zero ($\theta_0 \to \infty$) in Equation (24) for K_0^e (see Equation (162) in [2]). An equation of the standard type for the optical absorption K (for $k_B T > \hbar\omega_\kappa / 2$) is thus obtained [2,5]:

$$K = \frac{a^2 \hbar}{\sqrt{4\pi\lambda_r k_B T}}\exp\left(-\frac{2L}{a}\right)\exp\left[-\frac{(\hbar\omega_{12} - \lambda_r)^2}{4\lambda_r k_B T}\right] \tag{27}$$

where $a \equiv \hbar/\sqrt{2mJ_1}$ and $\lambda_r \equiv 2E$. A formula of this type was obtained by Markus in his electron-transfer model [32–37] and is often called the Marcus formula, and the energy λ_r is called the reorganization energy of Marcus. Similar and more general formulas were previously obtained in the theory of many-phonon transitions (see [29,38]) for optical transitions by Huang and Rhys [39] and Pekar [40–42] (see also Lax [43] and Krivoglaz and Pekar [30]), and for nonradiative transitions, by Huang and Rhys [39] and Krivoglaz [31].

The result in the standard theory of many-phonon transitions, given by Equation (27) and corresponding to high (i.e., room) temperatures, is a symmetric Gaussian function for the shape of the optical absorption band. It completely neglects the dynamics of the transient molecular state. This result corresponds to the high values of the dozy-chaos energy γ in dozy-chaos mechanics (see Figure 1). Physically, large values of γ in dozy-chaos mechanics correspond to a pronounced chaos in the transient state and, hence, a weak organization of the quantum-classical molecular transition (see Section 1). With a decrease in the dozy-chaos energy γ, the transient state becomes less chaotic and the organization of the quantum-classical transition increases, which is manifested in the appearance of a narrow optical absorption peak in the red spectral region and a strong asymmetry of the absorption band shape (see Figure 1).

We also note that the half-width of the Gaussian function for the shape of the optical absorption band (Equation (27))

$$w_{1/2} = 2\sqrt{2\ln 2}\sqrt{2\lambda_r k_B T} \tag{28}$$

is determined both by the individual properties of the "donor-acceptor + medium" system, which are expressed in the reorganization energy λ_r, and by the properties of an ensemble of these systems, which are expressed in temperature T. In other words, even within the framework of the well-known standard theory of many-phonon transitions [29,38], the effects of homogeneous and inhomogeneous broadening in the optical band cannot be separated. The introduction into the theory of a new "homogeneous effect" in the form of the dozy-chaos energy γ in Green's function of the system (Equation (3)) further confuses homogeneous and inhomogeneous effects in the shape of an optical band, greatly complicating the analytical result for it (cf. Equation (27) and Equations (6)–(26)). A discussion of the physical meaning of each of the terms included in this complex analytical result (Equations (6)–(26)) can be found in [2,4]. Our complex result gives a greater variety of optical band shapes (see, e.g., Figure 1) compared to the two band shapes, Lorentzian and Gaussian, which are the result of homogeneous and inhomogeneous effects known from the standard quantum theory of spectral line broadening. These two differences can only be understood in an open quantum system framework where the quantum system is coupled to an external classical bath. In contrast to the standard quantum theory, where the dynamics of quantum transitions are not considered, in quantum-classical mechanics, this bath, which is already quantum here, enters the entire closed quantum "donor-acceptor + medium" system (see the last, phonon term in the Hamiltonian (1)) and becomes classical only in a dynamic (chaotic) transient state (see Green's function (3) with $\gamma >> \hbar\omega$).

7. The Egorov Resonance

One of the main results of quantum-classical mechanics is a dynamic electron-nuclear-reorganization resonance (the so-called transferon resonance) [5,6] (see also [2]) or, according to [10], the Egorov resonance [1,10]

$$(2\tau_e)^{-1} = \tau^{-1} \tag{29}$$

where τ_e is the characteristic time of motion of the electron in the donor–acceptor system and τ is the characteristic time of motion of the reorganization of nuclear vibrations in the environment. These times are given by the following equations

$$\tau_e = \frac{L}{\sqrt{2J_1/m}} \tag{30}$$

where L is the distance between the donor and the acceptor of an electron (see Section 2; L is equal to the length of the polymethine chain—the main optical chromophore of polymethine dyes, (see Section 7 in [1]) [3–9]; J_1 is the binding energy of the electron on the donor 1 (see Section 3; electronic energy of the ground state of the dye) [3–9], and

$$\tau = \frac{\hbar}{E} \tag{31}$$

where E is the energy of reorganization of the nuclear vibrations in the medium (see Section 2, Equation (2)). Equations (30) and (31) are a part of Equation (13) (see Section 4).

8. Implementation of the Egorov Resonance in the Quasi-Symmetric Serious of Optical Band Shapes of a Representative Polymethine Dye

Experimentally, the dynamic electron-nuclear-reorganization resonance (the Egorov resonance, see Section 7) manifests itself, for example, in polymethine dyes [3–9], namely, in the resonance nature of the dependence of the shape of the optical absorption band on the length of the polymethine chain L (see Figure 2). The optical band with $n = 3$ corresponds to the Egorov resonance or is close to it.

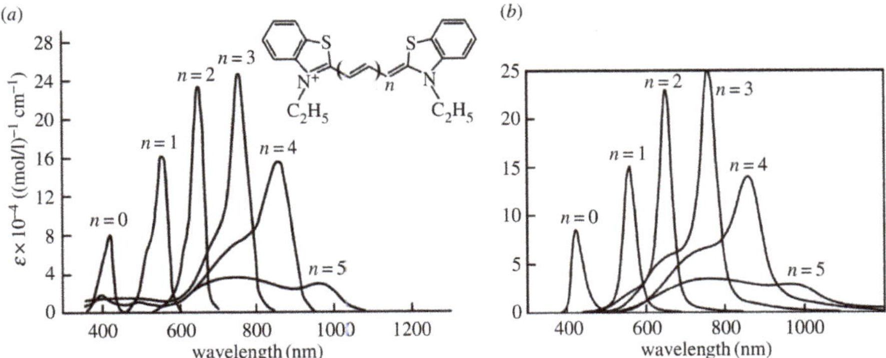

Figure 2. Experimental [44,45] (**a**) and theoretical [6] (**b**) monomers' optical absorption spectra, dependent on the length of the polymethine chain $L = 2(n + 2)d$, where d are certain roughly equal bond lengths in the chain (thiapolymethinecyanine in methanol at room temperature; ε is the extinction coefficient) [4,9]. The optical absorption band with $n = 3$ corresponds to the dynamic electron-nuclear-reorganization resonance (the Egorov resonance, see Section 7) or is close to it. (Original citation)—Reproduced by permission of The Royal Society of Chemistry. For the short chains ($n = 0, 1, 2, 3$), the tunnel effects, associated with the quantity η in Equation (25), can be neglected ($\eta = 1$). For the long chains ($n = 4, 5$), the tunnel effects are small but they must be taken into account ($\eta < 1$). The absorption bands are computed by Equations (6)-(26) with $\eta \leq 1$ instead of the Gamow tunnel factor (Equation (25)) when fitting them to the experimental data of Brooker and co-workers [44] (**a**) in terms of wavelength λ_{max}, extinction ε_{max}, and half-width $w_{1/2}$ with a high degree of accuracy. The following parameters of the "dye + environment" system are used [6]: $m = m_e$, where m_e is the electron mass; $d = 0.14$ nm; $\omega = 5 \times 10^{13}$ s^{-1}; $n_{refr} = 1.33$; for $n = 0, 1, 2, 3, 4, 5$, one has $J_1 = (5.63, 5.40, 4.25, 3.90, 3.74, 3.40)$ eV, $J_1 - J_2 = (1.71, 1.31, 1.11, 0.90, 0.74, 0.40)$ eV, $E = (0.245, 0.248, 0.256, 0.275, 0.297, 0.496)$ eV, and $\gamma = (0.402, 0.205, 0.139, 0.120, 0.129, 0.131)$ eV, respectively; for $n = 0, 1, 2, 3$, factor $\eta = 1$, and for $n = 4, 5$, factor $\eta = 0.55, 0.1$, respectively; $T = 298$ K.

To fit the theoretical result for the optical bands, which is given by Equations (6)–(26) to the corresponding experimental data (Figure 2a), we need estimated numerical values for the ground-state energies of the dye monomers, J_{1M}, and also for the energy gaps between their ground and excited states, $J_{1M} - J_{2M}$. These estimates follow from literature data [5,46–49]: $J_{1M} \cong 5$ eV and $J_{1M} - J_{2M} \cong 1$ eV. In addition, we need estimated numerical values for the reorganization energy of the nuclear environment of dye monomers, E_M. The estimate of E_M is found from the Egorov resonance (see Equations (29)–(31)) from the length of the optical chromophore [5,6]: $\frac{\sqrt{2J_{1M}/m}}{2L_M} = \frac{E_M}{\hbar}$, where L_M is the length of the optical chromophore of the dye monomers ($L_M = 10d$, $d = 0.14$ nm; see the caption to Figure 2).

Under resonance conditions (Equation (29)), the motion of the reorganization of the nuclei of the medium significantly contributes to the electronic transition in the optical π-electron chromophore—the polymethine chain with $n = 3$ as compared to the electronic transition in the optical π-electron chromophores of polymethine dyes with $n \neq 3$. As can be seen from the numerical data in the caption to Figure 2, the series of the dozy-chaos energies γ has a minimum at $n = 3$. Therefore, the appearance of the resonant band corresponding to $n = 3$ can also be interpreted as the transfer of chaos (dozy chaos) from the peak of a band into its wing(s) by a chaotic motion of the quantum-classical π-electronic state of the polymethine chain embedded in the medium as a result of the transition from "non-resonant" chains with $n \neq 3$ to the "resonant" chain with $n = 3$.

Thus, the presence of symmetry in the shape of an optical band at high (room) temperatures is associated with a primitive, Franck–Condon picture of the dynamics of molecular "quantum" transitions. The loss of this symmetry and the appearance of a peak against the background of a wide band wing are related, as already noted in Section 6, to the effect of self-organization of transition dynamics in dozy-chaos mechanics, which is expressed, in particular, in the "pumping" of dozy chaos from one part of the optical band to another part (from the peak region to the wing). Therefore, the series for the shape of the optical bands of a representative polymethine dye, thiapolymethinecyanine, depending on the length of its polymethine chain, has a quasi-symmetric character with respect to the Egorov resonance (see Figure 2), which corresponds to the most organized quantum-classical transition.

9. The Egorov Resonance and Its "Antisymmetric Twin"

The five principal parameters of the problem, viz. electron mass m, electron-donor binding energy $J_1 \equiv J$, distance between the donor and the acceptor L, environmental reorganization energy E, and dozy-chaos energy γ, may be combined into three quantities:

$$\tau_e = \frac{L}{\sqrt{2J/m}}, \ \tau = \frac{\hbar}{E}, \ \text{and} \ \tau_0 = \frac{\hbar}{\gamma} \tag{32}$$

having a time dimension (cf. Equations (13)) and representing two physically meaningful resonances [6]:

$$(2\tau_e)^{-1} = \tau^{-1} \ \text{and} \ (2\tau_e)^{-1} = \tau_0^{-1} \tag{33}$$

The former resonance is between the extended electron motion and the ordered constituent of the environmental nuclear reorganization motion, i.e., it is the Egorov resonance (cf. Equations (29)–(31)). The latter is between the electron motion and, conversely, the chaotic constituent of nuclear reorganization. Since the dozy-chaos energy γ can be considered, in a sense, as the imaginary part of a complex reorganization energy in which the reorganization energy E is its real part [9], then this second resonance can be considered as some antisymmetric twin with respect to the Egorov resonance. Both of these resonances can be regarded as the simplest dynamic invariants for the transient state. The dynamic resonance-invariants are alternatives to the Born–Oppenheimer adiabatic invariants (potential energy surfaces). In other words, these two resonances are the simplest manifestation of the relationship between electron and nuclear movements in the transient state.

Details of the transient-state-dynamics interpretation based on the Heisenberg uncertainty relation can be found in [6–8]. In particular, according to this interpretation, in the simplest cases, elementary electron transfers can be considered as a motion of a free electron–phonon quasiparticle, the so-called transferon, corresponding to the Egorov resonance, or, alternatively, as a motion of a free electron–phonon antiquasiparticle, the so-called dissipon, corresponding to the antisymmetric twin of the Egorov resonance.

10. Symmetry between Optical Absorption and Luminescence in the Standard Theory and Its Violation in Dozy-Chaos Mechanics as a Consequence of the Dynamic Organization of Quantum-Classical Transitions

10.1. Luminescence and Absorption Spectra. Their Mirror Symmetry

According to the standard theory of many-phonon processes [29], which ignores the dynamics of the transient state, the transition from absorption spectra to luminescence spectra is carried out by changing the sign before the heat energy $\hbar\omega_{12}$. Then, the luminescence and absorption spectra appear to be mirror-symmetric with respect to the "pure electronic" transition line $\hbar\Omega = J_1 - J_2$. If we apply this standard rule to the shapes of optical absorption bands obtained according to quantum-classical mechanics from Equations (6)–(26) and shown in Figure 1, then it can be seen that with decreasing chaos (with decreasing the dozy-chaos energy γ), which corresponds to improving the dynamic self-organization of the "quantum" transition, the luminescence and absorption bands narrow and shift towards each other and towards the "pure electronic" transition line (see Figure 1 in [3]). In other words, with the improvement of the dynamic self-organization of quantum-classical transitions, the corresponding luminescence and absorption bands are narrowed and their Stokes shift is reduced. A detailed discussion of the physics related to the nature of the change in the position and shape of the optical bands with decreasing quantity of γ can be found in [4,9].

10.2. Optical Spectra, Nature of the Small Stokes Shift, and Dynamic Asymmetry of Luminescence and Absorption

A striking example of the considered molecular "quantum" transitions, with the dynamics of their transient states taken into account, are the "quantum" transitions in the basic optical chromophore of J-aggregates of polymethine dyes embedded in a solvent—in the system "J-aggregate + environment" [4–8,10,11].

Figure 3 compares the results of the experiment [50] with the result of fitting them to the theoretical result (6)–(26) for optical absorption and the same theoretical result, in which the sign in the heat energy $\hbar\omega_{12}$ is changed to negative (Section 10.1), for luminescence (fluorescence). In the experiment, a very small Stokes shift was obtained for the J-band [50]. Therefore, we are forced to assume that the energy gap between the ground and excited electron states for fluorescence is greater than this gap for optical absorption [3]. This fact means that at the very initial stage of spontaneous emission, the binding energy of the electron in the excited state of the molecule, before the electron creates a photon, decreases markedly. This effect can be associated with the spontaneous loosening of the excited electronic state immediately before the act of production of a photon by an electron during spontaneous emission [3]. (For polymethine dyes and J-aggregates, the universal effect of spontaneous dynamic loosening is abnormally strong due to the very long π-electron systems in which the quantum-classical transitions under consideration occur [3].) Apparently, nothing of the kind occurs with optical absorption. In other words, with respect to the loosening effect, the processes of optical absorption and luminescence are asymmetric.

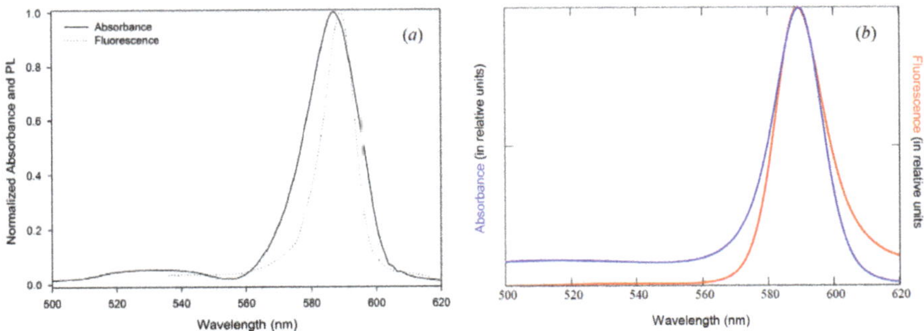

Figure 3. Experimental [50] (**a**) and theoretical [3] (**b**) absorption and fluorescence spectra of J-aggregates. In the analytical result for the shape of the optical bands (Equations (6)–(26)), the transition from absorption spectrum to fluorescence spectrum is carried out by changing the sign before the heat energy $\hbar\omega_{12}$. See details in the Egorov, Vladimir (2018), Mendeley Data, V2, https://doi.org/10.17632/h4g2yctmvg.2.

10.3. Luminescence and Absorption Spectra. Their Mirror Asymmetry

It can be seen from Figure 3 that when the sign changes only in the heat energy $\hbar\omega_{12}$, the theoretical absorption and luminescence spectra turn out to be symmetric with respect to each other (see Section 10.1), while the experiment shows their mirror asymmetry. According to quantum-classical mechanics [3], when passing from optical absorption to luminescence, the sign should be changed not only in the heat energy $\hbar\omega_{12}$ but also in the quantity L (the distance between the donor and acceptor in the elementary electron-charge transfers). The change in the sign of *L* corresponds physically to the reverse motion in space of the electron charge in the luminescence process relative to the absorption process. After that, the luminescence and absorption spectra cease to be mirror-symmetric with respect to the "pure electronic" transition line $\hbar\Omega = J_1 - J_2$ and, as we can see from Figure 4, the theory reproduces well the asymmetry of the absorption and luminescence spectra, which is observed in the experiment. This mirror asymmetry of the spectra is a consequence of taking the chaotic dynamics of the transient state of quantum-classical transitions into account, and it manifests itself under conditions of fairly weak dozy chaos, that is, under conditions of a sufficiently high degree of self-organization of quantum-classical transitions.

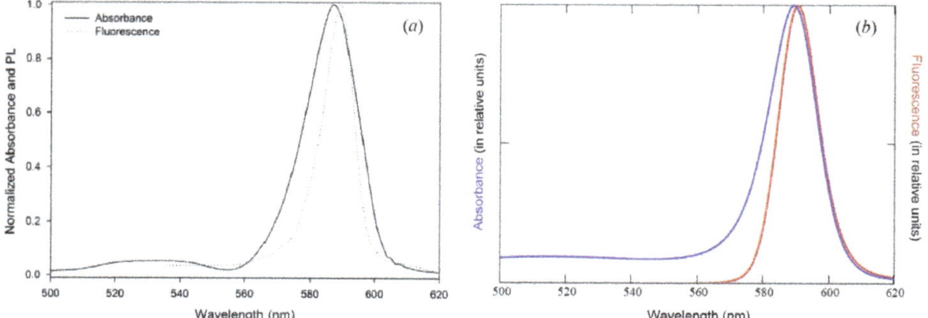

Figure 4. (**a**) The same as in Figure 3a. (**b**) Theoretical absorption and fluorescence spectra [3], fitted to the experimental data [50] (see (**a**)) in the J-aggregates. In the analytical result for the shape of the optical bands (Equations (6)–(26)), the transition from absorption spectrum to fluorescence spectrum is carried out by changing the sign before the heat energy $\hbar\omega_{12}$ and before the length of the optical chromophore (electron-charge-transfer distance) *L* as well. See details in the Egorov, Vladimir (2018), Mendeley Data, V2, https://doi.org/10.17632/h4g2yctmvg.2.

11. A Simplified Version of Dozy-Chaos Mechanics—Nonradiative Transitions

All of the results of the optical spectra match, generally, weak dozy chaos ($\gamma << E$). Strong dozy chaos ($\gamma \geq E$) leads to the elucidation of important patterns in the reactions of proton transfers [12,51] and comparatively fresh temperature-dependent effects on electron transfers in Langmuir–Blodgett films [13,52]. In the case of strong dozy chaos, the dynamics of quantum-classical transitions become weakly dependent on dozy chaos, and the electronic component of the complete electron-nuclear amplitude of transitions can be fitted by the Gamow tunnel exponential, dependent on the transient phonon environment. This elementary method permit us to evade the consideration of the imaginary additive $i\gamma$ in the spectral representation of the complete Green's function and to word the physical nature of the transient state, not in the concept of dozy chaos but in the concept of a large number of tunnel and over-barrier energy states providing the "quantum" transition of an elementary charged particle. This method was worked out [16] long before the development of quantum-classical (dozy-chaos) mechanics [2–9], and now we can say that the concept of a large number of tunnel and over-barrier states is a simplified version of the concept of dozy chaos.

The general result for the rate constant in the simplified version of dozy-chaos mechanics *K* is expressed in terms of the the Gamow tunnel exponential, dependent on the transient phonon-environment-energy $\hbar\omega_1$ and one generating function [15,16]:

$$K \propto \sum_{\omega_1=-\infty}^{\infty} \sum_{\omega_1'=-\infty}^{\infty} G_0(\omega_1, L) G_0 * (\omega_1', L) \times \frac{1}{(2\pi i)^3} \oint \frac{dx}{x^{\omega_1+1}} \oint \frac{dy}{y^{\omega_1'+1}} \oint \frac{dz}{z^{\omega_{12}+1}} S(\bar{n}_1; x, y, z) \qquad (34)$$

where the contours encircle the points $x = 0$, $y = 0$, and $z = 0$, correspondingly (cf. Equation (4)). The Gamow tunnel exponential is

$$G_0 = G_0(\alpha, L) = \exp(-\alpha L) \tag{35}$$

where the function $\alpha = \alpha(\omega_1)$ is given by the following formula:

$$\alpha \equiv \alpha(\omega_1) = [2m(J + \hbar\omega_1)]^{1/2}/\hbar \tag{36}$$

(here, there, and everywhere, $J \equiv J_1$). The generating function is as follows:

$$S(\bar{n}_1; x, y, z) = \exp\left\{-\sum_\kappa \widetilde{q}_\kappa^{\,\leftarrow}(2\bar{n}_{\kappa 1} + 1) + \tfrac{1}{2}\sum_\kappa \widetilde{q}_\kappa^{\,2}[(\bar{n}_{\kappa 1} + 1)(x^{\omega_\kappa} y^{\omega_\kappa} + 1)z^{\omega_\kappa}\right.$$
$$\left. + \bar{n}_{\kappa 1}(x^{-\omega_\kappa} y^{-\omega_\kappa} + 1)z^{-\omega_k}]\right\} \tag{37}$$

The result (34) applies to both optical and nonradiative processes. In the case of optical processes, the heat energy $\hbar\omega_{12}$ is determined from the law of conservation of energy (6) ($\hbar\omega_{12} > 0$—absorption and $\hbar\omega_{12} < 0$—luminescence), where the frequency $\Omega \equiv 0$ in the cases of nonradiative endothermic and exothermic processes:

$$\hbar\omega_{12} = J_2 - J_1 < 0 \tag{38}$$

and

$$\hbar\omega_{12} = J_1 - J_2 \equiv -\hbar\omega_{21} > 0 \tag{39}$$

From the general result for the rate constant in the simplified version of dozy-chaos mechanics, Equations (34)–(39), in the framework of the Einstein model of nuclear vibrations ($\omega_\kappa = \text{constant} \equiv \omega$), the simple expression for the rate constant K has been obtained [16]:

$$K \propto \exp\left\{-\frac{2L}{a} - \frac{2E}{\hbar\omega}\left[\coth\frac{\hbar\omega}{2k_BT} - \frac{\cosh t}{\sinh(\hbar\omega/2k_BT)}\right]\right.$$
$$\left. + \left(\frac{\hbar\omega}{2k_BT} - t\right)\frac{\omega_{12}}{\omega} - \frac{\hbar\omega\,\sinh(\hbar\omega/2k_BT)}{4E\cosh t}\left(\frac{\omega_{12}}{\omega}\right)^2\right\} \tag{40}$$

(cf. Equations (7) and (8)), where $\exp\left(-\frac{2L}{a}\right)$ is the Gamow exponential (cf. Equation (27)) and

$$t = \frac{\omega L}{\sqrt{2J/m}} \tag{41}$$

(cf. Equations (10) and (13)). If, as in the case of the complete theory for optical processes (Sections 2–4), we assume that the expression for the rate constant of the reverse process K_{rev} is obtained by changing the sign in the heat energy $\hbar\omega_{12}$ and in the donor–acceptor distance L (see Sections 10.1 and 10.3) in corresponding expression for the rate constant of the direct process K in the considered simplified version of dozy-chaos mechanics, then, applying this position to Equation (40), we obtain

$$\frac{K_{\text{rev}}}{K} = \exp\left(-\frac{\hbar\omega_{12}}{k_BT}\right)\exp\left(\frac{4L}{a}\right) \equiv K_{\text{eq}}^{-1}\exp\left(\frac{4L}{a}\right) \tag{42}$$

where $K_{\text{eq}}\exp\left(-\frac{4L}{a}\right)$ is the equilibrium constant of charged-particle-transfer reactions in the simplified version of dozy-chaos mechanics. In the limit $L \to 0$, we obtain from Equation (42) the well-known detailed balance relationship in statistical physics and in the standard theory of many-phonon transitions [29,31].

12. The Simplified Version of Dozy-Chaos Mechanics: Proton-Transfer Reactions. On Symmetry in the Brönsted Relationship

Grounded on the simplified version of dozy-chaos mechanics [16] (Section 11), in 1990, a theoretical description of the basic experimental patterns in the Brönsted relationship [51] for the reactions of proton transfer (acid-base catalysis) was given [12]. The Brönsted relationship was found by Brönsted and Pedersen in 1924 (see [51]). The theory in [16] is immediately appropriate to the explanation of electron transfers. To explain the reactions of transfers of heavy charged particles (proton transfers), the result of thermic fluctuations of the potential barrier transparence must be considered because of fluctuations in the barrier width. In contrast to the elementary proton transfer, the electron-transfer process is insensitive to small fluctuations in the barrier width due to the large size of the electronic wave function in the initial and final states. The analytical formulas for the proton-transfer rate constants are obtained. In acid catalysis, the empirical Brönsted relationship is

$$\lg K^{(\mathrm{acid})} = \alpha \lg K_{\mathrm{eq}}^{\mathrm{emp}} + a \tag{43}$$

where $K^{(\mathrm{acid})}$ is the rate constant, $K_{\mathrm{eq}}^{\mathrm{emp}}$ is the empirical equilibrium constant, and α and a are constants. In base catalysis, the empirical Brönsted relationship is

$$\lg K^{(\mathrm{base})} = \beta \lg K_{\mathrm{eq}}^{\mathrm{emp}} + b \tag{44}$$

The theoretically-obtained Brönsted coefficients α and β (the Einstein model of nuclear vibrations $\omega_\kappa = \mathrm{constant} \equiv \omega$) for direct (acid catalysis) and inverse (base catalysis) reactions [12]

$$\alpha = \frac{1}{2} + \frac{L\,k_{\mathrm{B}}T}{\hbar(2J/m)^{1/2}} - \left[1 - \frac{E\sinh t}{2J\sinh(\hbar\omega/2k_{\mathrm{B}}T)}\right] \frac{2m(k_{\mathrm{B}}T)^2}{\hbar^2\gamma_{\mathrm{b}}} \tag{45}$$

and

$$\beta = \frac{1}{2} - \frac{L\,k_{\mathrm{B}}T}{\hbar(2J/m)^{1/2}} + \left[1 - \frac{E\sinh t}{2J\sinh(\hbar\omega/2k_{\mathrm{B}}T)}\right] \frac{2m(k_{\mathrm{B}}T)^2}{\hbar^2\gamma_{\mathrm{b}}} \tag{46}$$

(t is given by Equation (41) and γ_{b} is barrier rigidity) are symmetric relative to $\frac{1}{2}$ and meet the generally known empirical equality [53,54]

$$\alpha + \beta = 1 \tag{47}$$

(which directly follows from the Brönsted relationships (43) and (44)).

13. The Simplified Version of Dozy-Chaos Mechanics: Symmetrization of the Amplitude and Rate Constant of the Transition for the Case of Different Electron–Phonon Interactions on the Donor and Acceptor

Until now, both in the case of the complete theory for optical processes (Sections 2–4) and in the case of its simplified version for nonradiative processes (Section 11), we have considered the case of the same electron–phonon interaction when a light charged particle, in particular an electron, is localized on the donor or on the acceptor. In other words, it was assumed that the reorganization energy $E \equiv E_1 = E_2$ (Equation (2)). For example, "quantum" transitions and the corresponding shapes of optical bands in polymethine dyes are well described by the case of the same value of the electron–phonon interaction on the donor and on the acceptor, because charge alternation occurs in the polymethine chain upon optical excitation [1,3–6]. In this section, we will briefly consider the case of different electron–phonon interactions when an electron is localized on a donor or acceptor. This corresponds to different magnitude shifts of the normal phonon coordinates $\widetilde{q}_{\kappa 1}$ and $\widetilde{q}_{\kappa 2}$ (in the case of the same interaction, $\widetilde{q}_\kappa \equiv \widetilde{q}_{\kappa 1} = -\widetilde{q}_{\kappa 2}$ [2]) and an obvious redefinition of the reorganization energy E (Equation (2)):

$$E_{1,2} = \frac{1}{2}\sum_\kappa \hbar\omega_\kappa \widetilde{q}^2_{\kappa 1,2} \tag{48}$$

where $E_1 \neq E_2$. For example, in the case of nonradiative processes, the change in sign in the heat energy $\hbar\omega_{12}$ (Equations (38) and (39)) and in the donor–acceptor distance $L \equiv |\mathbf{L} \equiv \mathbf{L}_{12}|$ (Equation (1)) is associated with the permutation of indices 1 and 2 in the reverse order. The assumption $E_1 \neq E_2$ leads to asymmetry with respect to the permutation of indices 1 and 2 in the expression for the rate constant of transitions and the loss of connection between forward and reverse processes, expressed in Equation (42). To restore this connection, it is necessary to symmetrize the expression for the amplitude and rate constant of electron transfers with respect to different values of the electron–phonon interaction at the donor and at the acceptor, which leads to the case of reorganization energies $E_1 \neq E_2$.

The symmetrization method proposed in [14,17,18] consists of the fact that, in addition to the transition amplitude [2–4]

$$A_{12} = \langle \Psi_2(\mathbf{r} - \mathbf{L}, q)|V|\Psi_1(\mathbf{r}, q)\rangle \tag{49}$$

which, in view of taking the wave function Ψ_2 in the Born–Oppenheimer adiabatic approximation $\Psi_2 = \Psi_2^{BO}$ and taking into account the entire dynamics of the transition only in the wave function $\Psi_1 = G\,\widetilde{V}\,\Psi_1^{BO}$ (G is Green's function of the Hamiltonian $H - \widetilde{V}$, $\widetilde{V} \equiv \sum_\kappa V_\kappa(\mathbf{r})(q_\kappa - \widetilde{q}_\kappa)$ [2–4]), can be called the amplitude of the transition on the acceptor A_{12}^a, we introduce into the theory also the amplitude

$$A_{12}^d = A_{21}^a \tag{50}$$

in which, on the contrary, the wave function Ψ_1 is taken in the adiabatic approximation $\Psi_1 = \Psi_1^{BO}$, and the entire dynamics of the transition are taken into account only in the wave function $\Psi_2 = G\widetilde{V}\,\Psi_2^{BO}$. This new amplitude A_{12}^d can be called the amplitude of the transition on the donor. Then, the half-sum of these two amplitudes is taken as the total transition amplitude:

$$A_{12} = \frac{A_{12}^d + A_{12}^a}{2} = A_{21} \tag{51}$$

Since the symmetrization is carried out only with respect to the electron–phonon interaction, in Equation (51), the permutation of indices 1 and 2 in the quantity $\mathbf{L}_{12}$ is not performed and the sign of $L \equiv |\mathbf{L}| \equiv |\mathbf{L}_{12}|$ does not change.

Using Equation (51), for the case of different electron–phonon interactions on the donor and acceptor in the framework of the Einstein model of nuclear vibrations ($\omega_\kappa = \text{constant} \equiv \omega$), the simple analytical expression for the rate constant has been obtained [17,18]:

$$K \propto \frac{1}{2}\left[\left(\frac{E_1 e^{-t}+E_2 e^t}{E_1 e^t+E_2 e^{-t}}\right)^{\frac{\omega_{12}}{2\omega}} + \left(\frac{E_1 e^{-t}+E_2 e^t}{E_1 e^t+E_2 e^{-t}}\right)^{-\frac{\omega_{12}}{2\omega}}\right]$$
$$\times \exp\left\{-\frac{2L}{a} - \frac{E_1+E_2}{\hbar\omega}\coth\frac{\hbar\omega}{2k_BT} + \frac{\sqrt{(E_1 e^{-t}+E_2 e^t)(E_1 e^t+E_2 e^{-t})}}{\hbar\omega\,\sinh(\hbar\omega/2k_BT)}\right.$$
$$\left. +\left(\frac{\hbar\omega}{2k_BT}-t\right)\frac{\omega_{12}}{\omega} - \frac{\hbar\omega\,\sinh(\hbar\omega/2k_BT)}{2\sqrt{(E_1 e^{-t}+E_2 e^t)(E_1 e^t+E_2 e^{-t})}}\left(\frac{\omega_{12}}{\omega}\right)^2\right\} \tag{52}$$

where $e^{\pm t} \equiv \exp(\pm t)$. Substituting $E_2 = E_1 \equiv E$ into Equation (52), we obtain Equation (40) for the rate constant in the case of the same electron–phonon interaction on the donor and acceptor. It is easy to see that Equation (52) satisfies the relationship of detailed balance in the simplified version of dozy-chaos mechanics (Equation (42)).

14. Conclusions

In this final section, we will list all kinds of symmetries in dozy-chaos mechanics of elementary electron transfers considered in the article and discuss their physical meaning.

First of all, one should note the symmetry associated with the invariance of the expression for the rate constant of elementary electron transfers with respect to sign reversal in the dozy-chaos energy γ (Section 4). This invariance is consistent with the physical case that both the virtual acts of transformation of electron movements and energies into nuclear reorganization movements and energies and the reverse acts occur in the transient dozy-chaos state [4,7–9].

The result in the standard theory of many-phonon transitions [29], corresponding to high (that is, room) temperatures, is a symmetric Gaussian function for the shape of the optical absorption band. It completely neglects the dynamics of the transient molecular state. This result corresponds to the high values of the dozy-chaos energy γ in dozy-chaos mechanics (see Figure 1). Physically, the high values of γ in dozy-chaos mechanics correspond to the weak organization of the quantum-classical molecular transition (Section 1). With a decrease in the dozy-chaos energy γ, the organization of the quantum-classical transition increases, which is manifested in the appearance of a narrow optical absorption peak in the red region of the spectrum and strong asymmetry of the absorption band (Section 5, Figure 1). In other words, the presence of symmetry in the shape of an optical band at high (room) temperatures is associated with a primitive, Franck–Condon picture of molecular "quantum" transitions. The loss of this symmetry is associated with taking into account the effect of self-organization of the dynamics of transitions in dozy-chaos mechanics, which is expressed, in particular, in the "pumping" of dozy chaos from one part of the optical band (narrow peak) to another part (wide wing).

A series for the shape of optical absorption bands in polymethine dyes, depending on the length of the polymethine chain, has a quasi-symmetric and resonant character, where a certain "average" chain length corresponds to the resonance (Section 8, Figure 2). In theory, this resonance—the "center of symmetry" of the series—is the Egorov resonance (Section 7).

An important illustration of the dynamics of the transient state for the Egorov resonance (Equations (29)–(31)) is a qualitative picture of the dynamics based on the use of the Heisenberg uncertainty relation [6–8] (Section 9). In this picture, a quasiparticle called transferon corresponds to the Egorov resonance. This quasiparticle has an antisymmetric twin—an antiquasiparticle called dissipon (Equation (33)). The transferon is depicted by a narrow optical band and the dissipon by a broad one. Strictly speaking, the dissipon is a certain broad resonance rather than (narrow) resonance proper.

Dozy-chaos mechanics, where the transition from absorption spectra to luminescence spectra is carried out by changing only the sign in the heat energy $\hbar\omega_{12}$, as in the standard theory of many-phonon transitions [29], gives a mirror-symmetric picture of the shapes of absorption and luminescence bands (Section 10.1). However, the need to take into account the dynamics of the "quantum" transition in the theory leads to the need to change the sign in the donor–acceptor distance L as well. This, in turn, leads to the appearance of mirror asymmetry in the pattern of absorption and luminescence band shapes (Section 10.3): transitions with light emission give narrower bands in comparison with absorption bands. Physically, this means that, as a result of taking into account the chaotic dynamics of "quantum" transitions in dozy-chaos mechanics, transitions with emission of photons show themselves to be more organized in comparison with transitions with absorption of photons.

Nonradiative transitions are considered within the framework of a simplified version of dozy-chaos mechanics, in which the electronic component of the complete electron-nuclear amplitude of transitions is fitted by the Gamow tunnel exponential, dependent on the transient phonon environment (Section 11). As in dozy-chaos mechanics for optical processes in its full formulation, this simplified version of dozy-chaos mechanics is considered for the case of the same electron–phonon interactions on the donor and acceptor. Direct and reverse processes turn out to be related not by the standard detailed balance relationship known from statistical physics but by a new, more complex, detailed balance relationship, which, in addition to the standard equilibrium constant, includes an exponential factor with the donor–acceptor distance in the exponent (Equation (42)).

Within the framework of the simplified version of dozy-chaos mechanics and the Einstein model of nuclear vibrations, the previously obtained [12] expressions for the Brönsted coefficients α and β for proton-transfer reactions (Section 12), which satisfy the well-known symmetric relation (Equation (47)), are given.

A simplified version of dozy-chaos mechanics is also considered for the case of electron–phonon interactions on the donor and acceptor of different magnitudes (Section 13), where a special procedure for the symmetrization of the total amplitude of the quantum-classical transition (Equation (51)) and the corresponding rate constant is performed. The analytical result obtained earlier [17,18] for the rate constant of nonradiative transitions (Equation (52)), which satisfies the new detailed balance relationship (Equation (42)), is presented.

In conclusion, we note that it is of interest to generalize dozy-chaos mechanics for optical processes in its full formulation (Sections 2–4) for the case of different electron–phonon interactions on the donor and acceptor, as well as to construct a theory of nonradiative dozy-chaos processes in its full version. An important point in the formulation of the problem in the theory of nonradiative dozy-chaos processes is the determination of the perturbation operator in the amplitude of the transition which causes the nonradiative transition. In the standard theory of many-phonon transitions [29], the well-known operator of nonadiabaticity [29,31] is taken as such an operator (see [2]). It is also of interest to generalize dozy-chaos mechanics to the case of nonlinear optics [1,10,19].

Funding: This work was supported by the Ministry of Science and Higher Education within the State assignment Federal Scientific Research Center "Crystallography and Photonics" Russian Academy of Sciences.

Conflicts of Interest: The author declares no conflict of interest.

Data Availability Statement: The data on which this article is based are available as an online resource with digital object identifier (doi) 10.5061/dryac.t0r3p and at the Egorov, Vladimir (2018), Mendeley Data, V2, https://doi.org/10.17632/h4g2yctmvg.2 [3].

References

1. Egorov, V.V. Dozy-Chaos Mechanics for a Broad Audience. *Challenges* **2020**, *11*, 16. [CrossRef]
2. Egorov, V.V. Quantum-classical mechanics as an alternative to quantum mechanics in molecular and chemical physics. *Heliyon Phys.* **2019**, *5*, e02579. [CrossRef]
3. Egorov, V.V. Quantum-classical mechanics: Luminescence spectra in polymethine dyes and J-aggregates. Nature of the small Stokes shift. *Res. Phys.* **2019**, *13*, 102252. [CrossRef]
4. Egorov, V.V. Nature of the optical band shapes in polymethine dyes and H-aggregates: Dozy chaos and excitons. Comparison with dimers, H*- and J-aggregates. *R. Soc. Open Sci.* **2017**, *4*, 160550. [CrossRef]
5. Egorov, V.V. On electrodynamics of extended multiphonon transitions and nature of the J-band. *Chem. Phys.* **2001**, *269*, 251–283. [CrossRef]
6. Egorov, V.V. Nature of the optical transition in polymethine dyes and J-aggregates. *J. Chem. Phys.* **2002**, *116*, 3090–3103. [CrossRef]
7. Egorov, V.V.; Alfimov, M.V. Theory of the J-band: From the Frenkel exciton to charge transfer. *Phys. Uspekhi* **2007**, *50*, 985–1029. [CrossRef]
8. Egorov, V.V. Theory of the J-band: From the Frenkel exciton to charge transfer. *Phys. Procedia* **2009**, *2*, 223–326. [CrossRef]
9. Egorov, V.V. Optical lineshapes for dimers of polymethine dyes: Dozy-chaos theory of quantum transitions and Frenkel exciton effect. *RSC Adv.* **2013**, *3*, 4598–4609. [CrossRef]
10. Petrenko, A.; Stein, M. Toward a molecular reorganization energy-based analysis of third-order Nonlinear optical properties of polymethine dyes and J-aggregates. *J. Phys. Chem. A* **2019**, *123*, 9321–9327. [CrossRef]
11. Petrenko, A.; Stein, M. Molecular Reorganization energy as a key determinant of J-band formation in J-aggregates of polymethine dyes. *J. Phys. Chem. A* **2015**, *119*, 6773–6780. [CrossRef] [PubMed]
12. Egorov, V.V. Effects of fluctuations in the transparency of the barrier in proton transfer reactions. *Russ. J. Phys. Chem.* **1990**, *64*, 1245–1254.

13. Egorov, V.V. On electron transfer in Langmuir-Blodgett films. *Thin Solid Films* **1996**, *284–285*, 932–935, Erratum. *Thin Solid Films* **1997**, *299*, 190. [CrossRef]
14. Egorov, V.V. Electron transfer in condensed media: Failure of the Born-Oppenheimer and Franck-Condon approximations, collective phenomena and detailed balance relationship. *J. Mol. Struct. THEOCHEM* **1997**, *398–399*, 121–127. [CrossRef]
15. Egorov, V.V. The Method of Generating Polynomials in the Theory of Electron Transfer. In Proceedings of the Second All-Union Conference on Quantum Chemistry of Solid State, Jurmala, Latvia, 8–11 October 1985; Latvia University: Riga, Latvia, 1985; p. 141. (In Russian).
16. Egorov, V.V. Theory of tunnel transfer. *Khimicheskaya Fiz.* **1988**, *7*, 1466–1482.
17. Egorov, V.V. Theory of Acid-Base Catalysis. In Proceedings of the Scientific Research: Physico-Chemical Processes in Energy Converters, Moscow Institute of Physics and Technology: Dolgoprudnyi, Moscow, Russia, 25–26 November 1988; MPhTI: Moscow, Russia, 1989; pp. 4–16. (In Russian).
18. Egorov, V.V. Towards theory of elementary charge transfers in acid-base catalysis. *Russ. J. Phys. Chem.* **1994**, *68*, 221–228.
19. Egorov, V.V. Quantum-classical electron as an organizing principle in nature. *Int. J. Sci. Tech. Soc.* **2020**, *8*, 93–103. [CrossRef]
20. Egorov, V.V. Where and why quantum mechanics ceases to work in molecular and chemical physics. In Proceedings of the European XFEL Theory Seminar, Schenefeld, Hamburg, Germany, 6 March 2018; Available online: https://indico.desy.de/indico/event/20069/ (accessed on 21 February 2018).
21. Born, M.; Oppenheimer, J.R. Quantum theory of the molecules. *Ann. Phys.* **1927**, *84*, 457–484. [CrossRef]
22. Franck, J.; Dymond, E.G. Elementary processes of photochemical reactions. *Trans. Faraday Soc.* **1925**, *21*, 536–542. [CrossRef]
23. Condon, E.U. A theory of intensity distribution in band systems. *Phys. Rev.* **1926**, *28*, 1182–1201. [CrossRef]
24. Condon, E.U. Nuclear motions associated with electron transitions in diatomic molecules. *Phys. Rev.* **1928**, *32*, 858–872. [CrossRef]
25. Condon, E.U. The Franck-Condon principle and related topics. *Am. J. Phys.* **1947**, *15*, 365–374. [CrossRef]
26. Mustroph, H. Potential-Energy Surfaces, the Born-Oppenheimer Approximations, and the Franck-Condon Principle: Back to the Roots. *Chem. Phys. Chem.* **2016**, *17*, 2616–2629. [CrossRef] [PubMed]
27. Davydov, A.S. *Quantum Mechanics*; Pergamon Press: Oxford, UK, 1976.
28. Berestetskii, V.B.; Lifshitz, E.M.; Pitaevskii, L.P. *Quantum Electrodynamics*, 2nd ed.; Elsevier: Amsterdam, The Netherlands, 1982.
29. Perlin, Y.E. Modern methods in the theory of many-phonon processes. *Sov. Phys. Uspekhi* **1964**, *6*, 542–565. [CrossRef]
30. Krivoglaz, M.A.; Pekar, S.I. The shape of the spectra of the impurity light absorption and luminescence in dielectrics. *Tr. Inst. Fiz. Akad. Nauk UKR. SSR* **1953**, *4*, 37–70. (In Russian)
31. Krivoglaz, M.A. The theory of thermal transitions. *Zh. Eksp. Teor. Fiz.* **1953**, *25*, 191–207. (In Russian)
32. Marcus, R.A. On the theory of oxidation-reduction reactions involving electron transfer. I. *J. Chem. Phys.* **1956**, *24*, 966–978. [CrossRef]
33. Marcus, R.A. Electrostatic free energy and other properties of states having nonequilibrium polarization. *J. Chem. Phys.* **1956**, *24*, 979–989. [CrossRef]
34. Marcus, R.A. On the theory of oxidation-reduction reactions involving electron transfer. II. Applications to data on the rates of isotopic exchange reactions. *J. Chem. Phys.* **1957**, *26*, 867–871. [CrossRef]
35. Marcus, R.A. On the theory of oxidation-reduction reactions involving electron transfer. III. Applications to data on the rates of organic redox reactions. *J. Chem. Phys.* **1957**, *26*, 872–877. [CrossRef]
36. Marcus, R.A.; Sutin, N. Electron transfers in chemistry and biology. *Biochim. Biophys. Acta* **1985**, *811*, 265–322. [CrossRef]
37. Marcus, R.A. Electron transfer reactions in chemistry. Theory and experiment. *Rev. Mod. Phys.* **1993**, *65*, 599–610. [CrossRef]
38. Frank-Kamenetskii, M.D.; Lukashin, A.V. Electron-vibrational interactions in polyatomic molecules. *Sov. Phys. Uspekhi* **1975**, *18*, 391–409. [CrossRef]
39. Huang, K.; Rhys, A. Theory of light absorption and non-radiative transitions in F-centres. *Proc. R. Soc. A* **1950**, *204*, 406–423.
40. Pekar, S.I. Theory of F-centers. *Zh. Eksp. Teor. Fiz.* **1950**, *20*, 510–522. (In Russian)

41. Pekar, S.I. To the theory of luminescence and light absorption by impurities in dielectrics. *Zh. Eksp. Teor. Fiz.* **1952**, *22*, 641–657. (In Russian)

42. Pekar, S.I. On the effect of lattice deformations by electrons on optical and electrical properties of crystals. *Uspekhi Fiz. Nauk* **1953**, *50*, 197–252. (In Russian) [CrossRef]

43. Lax, M. The Franck-Condon principle and its application to crystals. *J. Chem. Phys.* **1952**, *20*, 1752–1760. [CrossRef]

44. Brooker, L.G.S.; Sprague, R.H.; Smith. C.P.; Lewis, G.L. Color and constitution. I. Halochromism of anhydronium bases related to the cyanine dyes. *J. Am. Chem. Soc.* **1940**, *62*, 1116–1125. [CrossRef]

45. James, T.H. (Ed.) *The Theory of the Photographic Process*; Macmillan: New York, NY, USA, 1977.

46. Shapiro, B.I. Aggregates of cyanine dyes: Photographic problems. *Uspekhi Khimii* **1994**, *63*, 243–268. [CrossRef]

47. Gurevich, Y.Y.; Pleskov, Y.V. Electrochemistry of semiconductors: New problems and prospects. *Uspekhi Khimii* **1983**, *52*, 563–595. (In Russian) [CrossRef]

48. Shapiro, B.I. Chemical theory of the spectral sensitization of silver halides. *Uspekhi Nauchn. Fotogr.* **1986**, *24*, 69–108. (In Russian)

49. Lenhard, J.R.; Hein, B.R. Effects of J-aggregation on the redox levels of a cyanine dye. *J. Phys. Chem.* **1996**, *100*, 17287–17296. [CrossRef]

50. Aviv, H.; Tischler, Y.R. Synthesis and characterization of a J-aggregating TDBC derivative in solution and in Langmuir-Blodgett films. *J. Lumin.* **2015**, *158*, 376–383. [CrossRef]

51. Brönsted, J.N. Acid and basic catalysis. *Chem. Rev.* **1928**, *5*, 231–338. [CrossRef]

52. Naito, K.; Miura, A. Photogenerated charge storage in hetero-Langmuir-Blodgett films. *J. Am. Chem. Soc.* **1993**, *115*, 5185–5192. [CrossRef]

53. Bell, R.P. *The Proton in Chemistry*; Cornell University Press: Ithaca, New York, NY, USA, 1959.

54. Shapiro, I.O. The Brönsted relationship in proton-transfer reactions. In *Proceedings of Physical Chemistry: Modern Problems*; Kolotirkin, Y.M., Ed.; Khimija: Moscow, Russia, 1987; pp. 128–164. (In Russian)

Publisher's Note: MDPI stays neutral with regard to jurisdictional claims in published maps and institutional affiliations.

MDPI
St. Alban-Anlage 66
4052 Basel
Switzerland
Tel. +41 61 683 77 34
Fax +41 61 302 89 18
www.mdpi.com

Symmetry Editorial Office
E-mail: symmetry@mdpi.com
www.mdpi.com/journal/symmetry